DOLOMITES OF THE MONTEREY FORMATION AND OTHER ORGANIC-RICH UNITS

Edited by

ROBERT E. GARRISON
University of California, Santa Cruz

MIRIAM KASTNER
Scripps Institute of Oceanography
University of California, San Diego

DONALD H. ZENGER
Pomona College, Claremont

April 18-21, 1984

Published by
The Pacific Section
Society of Economic Paleontologists and Mineralogists
Los Angeles, California
U.S.A.

COVER PHOTOGRAPH: Transmission electron micrograph of Monterey Dolomite from the Santa Maria Basin, *Courtesy of Karl Mertz and Ken Pisciotto.*

This symposium volume can be purchased by writing to:
Treasurer, Pacific Section S.E.P.M.
P.O. Box 10359
Bakersfield, CA 93389

Printed by Comet Reproduction Service
Santa Fe Springs, CA 90670

Pacific Section
Society of Economic Paleontologists and Mineralogists

Pacific Section SEPM Officers - 1984

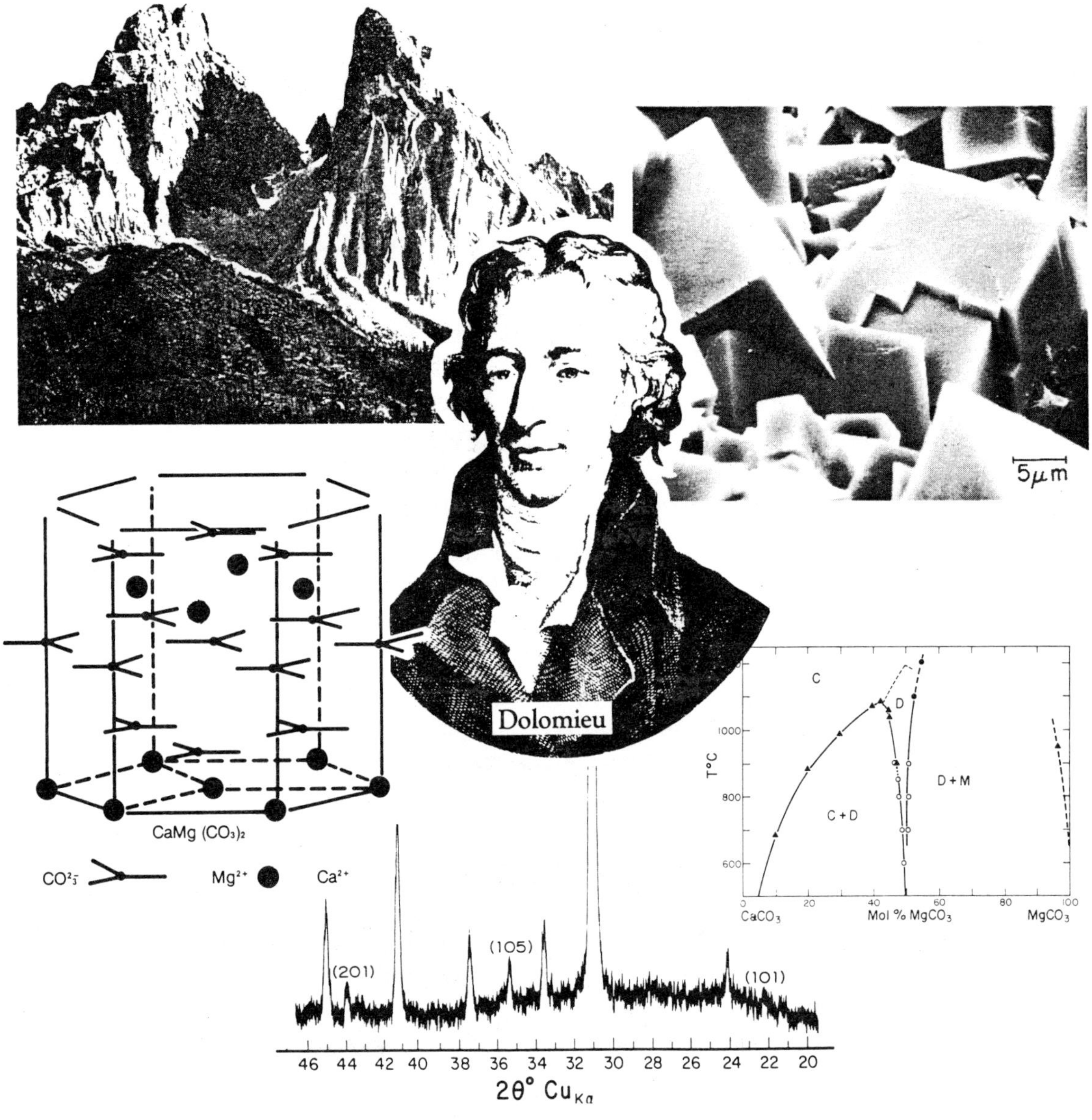

We dedicate this volume on dolomitization in the Monterey Formation to Déodat de Dolomieu (1750-1801), French natural scientist, catastrophist, and revolutionary, who was the first to describe the rock and mineral that were named in his honor. Also keenly interested in volcanism (particularly basaltic rocks) and author of numerous articles and books on geology, Dolomieu was described by one biographer (Taylor, 1971) as being "...esteemed as a judicious inquirer within the framework of existing styles of geological research...". (Figure of Dolomieu from photograph by F. G. Bourrouilh-Le Jan of portrait in Archives de l'Académie des Sciences, Institut de France, courtesy of Messieurs les Secrétaires Perpétuels of that institution.)

Contents

INTRODUCTION

Three recent treatises* serve to emphasize the variety in occurrences of dolomite as well as the range of currently proposed models for dolomitization. Among the varied settings in which dolomite occurs, perhaps none has remained so little discussed or so poorly understood as the association of dolomites and organic-rich sedimentary rocks such as in the Miocene Monterey Formation of California.

Long regarded as interesting curiosities, dolomitic rocks (dolostones or dolomitites) in the Monterey Formation recently have achieved broader recognition as fractured hydrocarbon reservoirs in the Santa Maria Basin and elsewhere. Moreover, work in the past five years has shown that Monterey dolomites: (1) may possess a measurable primary magnetic remanence useful in tectonic and geochronologic studies, and (2) may yield, as insoluble residues, exceedingly well preserved opal-A diatoms which enable precise biostratigraphic assignments. Both of these attributes may result from formation of dolomite during very early diagenesis. Coupled with experimental work showing that dolomitization becomes enhanced in low sulfate or sulfate-free diagenetic environments (as, for example, in zones of sulfate reduction within organic-rich sediments) these findings have stimulated a variety of studies of Monterey dolomites and of dolomites associated with organic-rich sediments and sedimentary rocks elsewhere. In organizing this symposium we have sought to present a progress report of research on this unusual and fascinating variety of dolomite.

The papers in this volume include most of those presented at the symposium, which was held April 19, 1984, at the 59th Annual Meeting of the Pacific Section of the AAPG, SEPM and SEG in San Diego. In addition we were fortunate to receive a few additional papers, some submitted by researchers unable to attend the San Diego meeting. We include also abstracts of two oral presentations at the symposium along with two presentations at a companion symposium on paleomagnetism.

The first three papers report occurrences of dolomites associated with low oxygen sediments in modern ocean basins. G. B. Shimmield and N. B. Price document the occurrence of Holocene dolomite rhombs formed in hemipelagic sediments just beneath the sea floor off Baja California. These authors attribute its formation to a combination of low sedimentation rates and high flux of organic carbon. K. Kelts and J. McKenzie review the occurrence of dolomites associated with low oxygen sediments of two different ages and in two distinct settings. Dolomite layers in Quaternary diatomaceous sediments of the Guayamas Basin slope (Gulf of California) have positive carbon-isotope values and apparently formed in the zone of methanogenesis; dolomites in the upper Miocene Tripoli Formation of Sicily, in contrast, have negative values suggesting genesis in the zone of sulfate reduction. The differences may be explained by differing rates of sedimentation and burial. L. D. Kulm, E. Suess, and T. M. Thornburg describe a variety of dolomites associated with Neogene organic-rich muds and related to rates of subsidence in forearc basins west of Peru. Chemical and isotopic evidence shows that a major mechanism for dolomitization was degradation of organic carbon. Because of continuous deformation in this tectonically active setting, many of these early-formed dolomites show pervasive fracturing.

Discussion of Monterey dolomites begins with a detailed analysis by K. A. Mertz of the sedimentology and diagenesis of the organic-rich lower Monterey Formation in the Santa Lucia Mountains (Salinas Basin). Dolomitization and phosphatization are widespread in these rocks, and Mertz employs field, petrologic and stable isotope data to recreate the environments, temperatures and burial depths of dolomite and apatite genesis. Studying the same suite of rocks, K. Kastner, K. Mertz, D. Hollander and R. Garrison further consider dolomitization and phosphatization in organic-rich sediments. Based on oxygen isotope values, the dolomites formed at approximate temperatures of 30° to 47°C, the apatites at 32° to 79°C. A complex paragenesis is thus indicated, and this may have included initial precipitation of amorphous calcium phosphate which only later, at higher temperatures during burial, crystallized to apatite. Based also on studies of lower Monterey rocks in the Salinas Basin, R. E. Garrison and S. A. Graham describe reworked dolomite concretions in deep-water gravity flow deposits. Isotopic evidence implies these concretions formed at burial depths of 100 meters or more, thus their exhumation and redeposition indicates complex processes of slope erosion and reworking.

In areas further to the south, R. E. Kablanow, R. C. Surdam and D. Prezbindowski discuss the petrology, isotope geochemistry and origin of Monterey dolomites in the Pismo and Huasna Basins. Isotopic analyses provide the basis for interpreting environments and burial depths of dolomitization, with variations in carbon isotopic compositions interpreted as due largely to

---

*Land, L. H., 1983. Dolomitization: Amer. Assoc. Petroleum. Geologists, Education Course Notes Series No. 24, 20 p.

Zenger, D. H., Dunham, J. B., and Ethington, R. L. (Editors), 1980, Concepts and models of dolomitization: SEPM Spec. Public. No. 28, 320 p.

Zenger, D. H., and Mazzullo, S. J. (Editors), 1982, Dolomitization: Benchmark Papers in Geology/65, Hutchinson Ross Publ. Co., Stroudsburg, Pa., 426 p.

in Garrison, Robert E.; Kastner, Miriam and Zenger, Donald H., eds., 1984, **Dolomites of The Monterey Formation and Other Organic-Rich Units:** Pacific Section S.E.P.M., Vol. 41, p. 1-3.

variations in burial rates for different lithofacies. In a companion paper these authors also document the occurrence of secondary solution porosity in Monterey dolomites, adding an important new consideration to the assessment of these rocks as hydrocarbon reservoirs.

Two papers discuss dolomites and dolomitization in the Santa Maria Basin. N. J. Henderson, G. A. Crawford, J. B. Dunham and R. H. Osborne provide a detailed description and analysis of dolomites recovered in cores of the Sisquoc Formation which lies stratigraphically above the Monterey Formation in this basin. Two types of dolomites are present: nodular ferroan dolomite and disseminated non-ferroan dolomite. Isotope values suggest the former grew at relatively shallow burial depths in the zone of methanogenesis whereas the latter formed at somewhat greater depths, and there are suggestions of primary facies control on the form in which dolomite occurs. J. S. Compton and R. Siever describe dolomites in different members of the Monterey Formation in the western Santa Maria Basin. Like Henderson et al., they make a clear linkage between the type of dolomite and the character of the host sediment, and they also discuss field and petrologic criteria for determining the relative times of dolomitization. Both pre- and post-compactional dolomites are present in the rocks they discuss.

The next two papers deal with more general aspects of Monterey dolomites. J. Kushnir and M. Kastner present very detailed chemical and isotopic analyses of a dolomite concretion and a dolomite layer. Isotopic values in the concretion decrease systematically from the center outward whereas in the layer these values vary only slightly without any obvious trend. The concretion thus formed progressively during burial, while dolomitization in the layer was rather rapid within a narrow range of burial depths and temperatures. This study demonstrates that subsequent chemical or isotopic exchange has not occurred in these dolomites; it also shows the necessity for carefully controlled as opposed to random sampling. C. M. Isaacs, focusing on disseminated dolomite in Monterey rocks of the Santa Maria and Santa Barbara areas, shows that this, rather than beds and concretions, is the most abundant form in which the mineral dolomite occurs. Based on carbon isotope values she suggests that organic-geochemical reactions may have been much less involved in the formation of disseminated dolomite than in the genesis of dolomite beds and concretions.

The final two papers deal with diatom preservation and paleomagnetism. C. W. Lagle outlines a procedure for recovering siliceous microfossils from dolomites; this paper should serve to introduce many geologists to the very exciting prospects for biostratigraphic age dating of Monterey rocks, including cherty and porcelaneous parts of the Monterey which have previously evaded precise age assignments. J. S. Hornafius focuses on the origin of remanent magnetization in Monterey dolomites, suggesting it is derived from detrital magnetite. His data show that the primary remanence is locked in, probably by very early dolomitization, within a few meters below the sediment-water interface. A particularly intriguing aspect of his work, and one that well illustrates the great potential for further magnetic studies of Monterey rocks, is his discovery within a dolomite layer of an apparent magnetic reversal.

This volume ends with four abstracts. Three of these concern paleomagnetic studies of Monterey rocks and further emphasize the promise of this method for refining our age assignments and tectonic reconstructions. The fourth abstract summarizes results of magnesium mass balance calculations for the Monterey Formation.

The view which emerges from this symposium contains the following elements and uncertainties. Much or most of the dolomite in organic-rich rocks may be in the form of disseminated dolomite crystals; formation of individual dolomite crystals in organic-rich sediments may begin very early during diagenesis, within a few tens of centimeters below the sea floor, but isotopic evidence suggests most of the disseminated dolomite formed later at relatively large burial depths. More prominent in stratigraphic successions, however, are beds and concretions of dolomitic rock (dolostone or dolomitite) which appear to form within unlithified host sediment by precipitation and replacement, at burial depths of tens to hundreds of meters. Oxygen isotopes provide an estimate of the temperatures of formation from which burial depths may also be estimated; carbon isotopes indicate the nature of the carbon reservoirs, hence allow assignment to diagenetic zones of organic carbon degradation and, more indirectly, permit estimates of rates of burial. Variability of isotopic values, however, may be large within a single dolomite bed or concretion, and several papers in this volume illustrate the necessity for carefully patterned sampling and show how random sampling may cause misleading conclusions at best, totally erroneous ones at worst.

Phosphatic nodules and peloids commonly occur spatially associated with the dolomites. Preliminary isotopic evidence suggests a complex paragenetic history; in some cases the apatite formed before dolomite, in other cases later and in still other cases the two phases may have formed simultaneously. Phosphate may have initially precipitated as amorphous material which subsequently crystallized to apatite at higher temperatures.

Early-formed dolomite concretions and layers may preserve opal-A siliceous microfossils and magnetic signals. They may also experience fracturing beginning shortly after their formation; these fractures may be caused by volume changes during diagenesis and/or by tectonic stresses. Such early-formed dolomite bodies may, in addition, be exhumed and redeposited through complex combinations of slope erosion and redeposition. Among other processes which modify the dolomites is dissolution which may create secondary porosity of economic importance.

Many unresolved issues remain. For example, we still have no clear explanation of why dolomite beds and concretions tend to be restricted to specific stratigraphic horizons within the host sediment. We remain uncertain about the effects of apparently widespread disseminated dolomite crystals on important rock properties such as brittleness and porosity. We have not yet fully integrated organic and inorganic geochemical data in achieving a comprehensive picture of how the dolomite forms. Nor do we yet completely comprehend the relationship between dolomitization and other forms of diagenesis, most notably phosphatization and silica phase changes.

Fundamental questions remain also regarding the sources and mechanisms of magnetization. These kinds of problems challenge us to continue, expand and refine the types of combined field, petrological, geochemical, and paleomagnetic studies reported in this volume.

We express our appreciation to the individuals and organizations who helped us in organizing this symposium and in preparing this volume. Ken Pisciotto, President of the SEPM's Pacific Section during 1983, encouraged us from the outset. Rheinhard Suchsland provided invaluable assistance in the preparation and editing of the volume. J. Garrison, S. Harris, D. Hollinger, R. Marshall and K. Mertz helped with the typing and editing. Grants from the Petroleum Research Fund (PRF #13953-AC2) and the Chevron Oilfield Research Co. helped us to defray some of the editing costs, and we are grateful to those organizations. We are grateful also for the support of our institutions, Pomona College, the Scripps Institution of Oceanography, and the University of California, Santa Cruz. Finally, we wish to acknowledge the hard work and persistence of the speakers and authors who made this enterprise possible.

Robert Garrison
Miriam Kastner
Donald H.
Zenger

RECENT DOLOMITE FORMATION IN HEMIPELAGIC SEDIMENTS OFF BAJA CALIFORNIA, MEXICO

G.B. Shimmield and N.B. Price, Grant Institute of Geology, University of Edinburgh, West Mains Road, Edinburgh, EH9 3JW, Scotland, U.K.

ABSTRACT

Early diagenetic dolomite occurring as euhedral rhombohedral (5-40 µm.) has been identified in shallow 0-40 cm.) cores from hemipelagic sediments off the coast of Baja California, Mexico. The lateral distribution of dolomite appears to be confined only to these sediments, whilst shelf and oceanic sediments show little or no evidence of dolomite precipitation. Its vertical distribution within the cores appears to be confined to the zone of reduction. Dolomite is found to be unstable in oxidising sediments, suggesting that it is not a stable phase in sea water. Mineralogy and composition (56 mole % $CaCO_3$) show it to be dolomite rather than "protodolomite". Coexistence of other diagenetic phases with the dolomite indicates that its formation predates glauconite and manganese carbonate, and postdates some pyrite and barite. We have attributed the precipitation of dolomite to a low sediment accumulation rate and a high flux of $C_{org.}$ In such sediments extensive biomixing occurs allowing abundant Mg^{2+} ions to reach the surface of reduction - the locus of dolomite precipitation. Dolomitization of hemipelagic sediments of the eastern Pacific is discussed.

INTRODUCTION

The problem of the origin of dolomite, and the chemical conditions necessary for precipitation or replacement of limestones, has been debated over the last century (Correns, 1939; Fairbridge, 1957; Milliman, 1974). Many of these reports have considered dolomite genesis to have taken place under unusual environmental mechanisms due to evaporation and ground water control (Curtis et al., 1963; Deffeyes et al., 1964; Illing et al., 1965; Shinn et al., 1965; Von der Borch, 1965; Hsu and Siegenthaler, 1969; Folk and Land, 1975). Interest in the problem has been revived in recent years as extensive dolomitization has been found to occur in several deep-sea cores which are mainly confined to the ocean margins (Kulm et al, 1981; Pisciotto and Mahoney, 1981; Kelts and McKenzie, 1982). In these sediments there has been recognition that the dolomite is associated with organic carbon ($C_{org.}$). Much of the interest in the latter sediments has been concerned with the timing of dolomite growth. The conclusions that have been reached, supported by laboratory work (Baker and Kastner, 1981; Kushnir and Kastner, 1982), indicate that the mineral can form as an early diagenetic phase, but could be restricted by the presence of SO_4^{2-} ions. Dolomite rhombohedral have been noted in surface sediments for decades (Fairbridge, 1957; Taft, 1961; Bonatti, 1966; Milliman, 1974; Behrens and Land, 1972), but due to their lack of abundance, and the possible detrital origin of some of them, there has been little study in respect of their genesis in the marine environment and their relation to more deeply buried dolomites.

In order to gain some insight into the timing of dolomite growth during sediment burial it is imperative that Recent occurrences be carefully examined and attempts made to relate them to other diagenetic phases which are more easily interpreted within the sequence of diagenesis.

Here, we attempt to observe the distribution, composition and mineral structure of dolomite in shallow subsurface cores taken from the eastern Pacific off the coast of Baja California, Mexico.

GENERAL ENVIRONMENTAL SETTING

The Pacific margin of the Baja California peninsula, between latitudes 27°N and 23° N, lies within the high productivity zone bordering the eastern Pacific (Figure 1). A semi-arid climate characterizes the region with an annual rainfall of only 12 cm. (Aschman, 1959) mostly occurring during the summer and autumn. The coastline around Magdalena Bay consists mainly of large sand dunes running parallel to the coast and extending some 20 km. back into the hinterland. The promontories forming Cabo San Lazaro, Cedros Island and the Vizcaino peninsula further to the north, are relict ophiolitic terranes and contribute amphibolitic detritus into the coastal environment. Here placer deposits of phosphorite are being commercially exploited at the present. The low lying terrain, lack of major rivers and development of coastal lagoons limits the amount of terrigenous input to the neighboring shelf.

The hydrography of the region is dominated by two main current systems: the California Current and the warmer North Equatorial water (Reid et al, 1958). The intense seasonal upwelling and associated biological productivity within this area promotes a marked oxygen minimum zone between 200 and 1000 meters depth (D'Anglejan, 1967) with a minimum dissolved O_2 concentration of <0.5 ml/l. This surface productivity is reflected in the organic carbon

in Garrison, Robert E.; Kastner, Miriam and Zenger, Donald H., eds., 1984, **Dolomites of The Monterey Formation and Other Organic-Rich Units:** Pacific Section S.E.P.M., Vol. 41, p. 5-18.

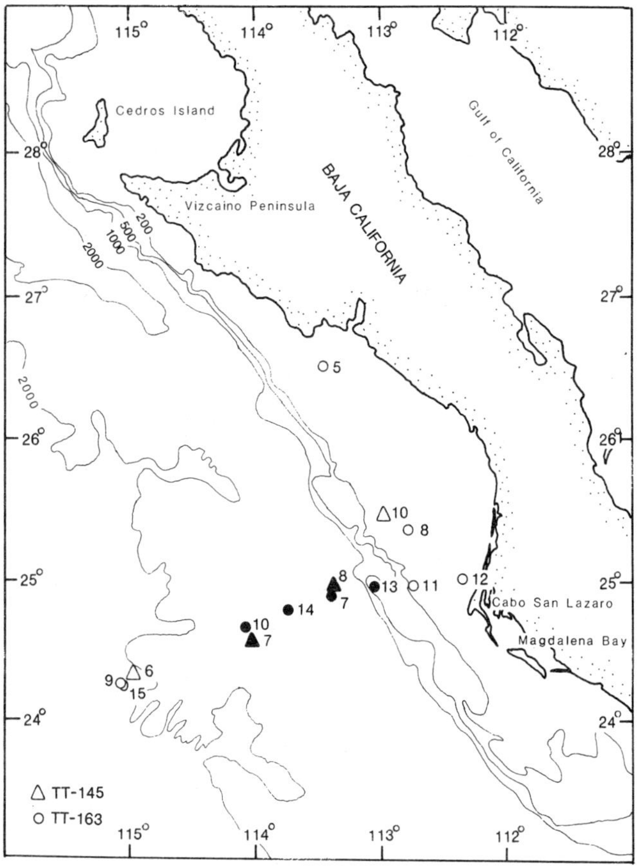

Figure 1 Core location map. Shaded symbols represent the occurrence of dolomite in the sediment. (Bathymetry in fathoms).

content of the sediments which reaches a maximum over the outer shelf. The combination of high C_{org} matter flux and low terrigenous accumulation rate for this continental margin makes it ideal to study the diagenetic processes involved without the obscuring effects of a large detrital component common in many other sediments.

The Baja California shelf and slope is a continuation of the continental borderland developed of southern California (Emery, 1960) resulting in a series of longitudinal intrashelf and slope basins and banks. Published investigations of sediments associated with this complicated topography (D'Anglejan, 1967; Jahnke et al., 1983; Sawlan and Murray, 1983) show that the basins tend to be characterized by greenish muds whilst the longitudinal banks display silts and foraminiferal sands.

METHODS OF STUDY

Sediment samples were collected during the R/V Thomas G. Thompson (University of Washington) cruises TGT-145 and TGT-163 in 1979 and 1980. Fourteen shallow (0-40 cm.) box-cores were sampled from a range of environments along a broad transect at roughly latitude 25°N that encompasses the continental rise, slope and shelf (Fig. 1). These box-cores were subsampled using a 2.5 inch acetate liner and horizontally sectioned into 1 cm. and 2 cm. intervals aboard ship and stored wet at 4° C.

The bulk, dried sediment was investigated for its mineralogy and geochemistry. Subsamples of ground sediment, as glass mounts, were x-ray diffracted using Cu Kα radiation to identify the bulk mineralogy of the sediment. Selected mineral extractions were treated in the same manner to investigate their crystal structure.

Major element content was determined by X-ray fluorescence (XRF) techniques on fused powders using a Cr anode following a technique similar to Norrish and Hutton (1969) and calculating the effects of mass absorption. Salt corrections were made by measuring the water content of the sediment and assuming constant salinity of $35^{0}/00$. Analytical precision (2σ, n=8) for Mn and Mg was 2.57% and 1.49% respectively.

Organic carbon was determined by combustion using a Leco induction furnace on residues tat were acidified, using 50% HCI at 60°C, in Leco ceramic crucibles to remove carbonate carbon. The precision (2σ, n=9) of the method was 4.52%.

Investigation of the dolomite and other authigenic phases involved examination of the total sediment, the silt fraction, and hand-picked grains. These included using a scanning electron microscope (SEM) with an energy dispersive analysing system (EDAX) to observe textural features of the dolomite and its host sediment. Accurate analysis of the dolomite (for CA, Mg, Fe, Sr, Ba, Si, Al, Na, K and S) was achieved using electron microprobe techniques on individual dolomite rhombs in polished grain mounts.

SEDIMENT DISTRIBUTION

Cores taken on the continental shelf show sediments to be grey-green, fine sands and silts. Characteristically, these sediments contain much quartz and amphibole and are generally carbonate-poor, except for local accumulations of foraminiferal debris. They are of moderate $C_{org.}$ content (1-5 - 2.0%). Cores from the shelf break and continental rise comprise carbonate - and biogenic silica-poor muds and silts with no trace of H_2S within the interval sampled. Typically, these sediments have a red-brown surface oxidising layer over underlying grey-green sediment. This colour change has been attributed to oxidation/reduction of Mn and Fe at depth, although recent evidence (Lyle, 1983) has shown that Fe alone may be the cause of this colour change. The depth of the colour change broadly relates to the $C_{org.}$ of the sediment; organic-poor (0.45%) outer continental rise sediments tend to have an oxidising zone in excess of 30 cms. thickness. Landwards, with increasing $C_{org.}$ content, the depth of this boundary decreases to a few centimeters and at the shelf break sediments tend to be reducing throughout.

Within these sediments a range of authigenic minerals has been identified and include dolomite, glauconite, barite, apatite, manganese carbonate, and possibly gypsum. The origin of

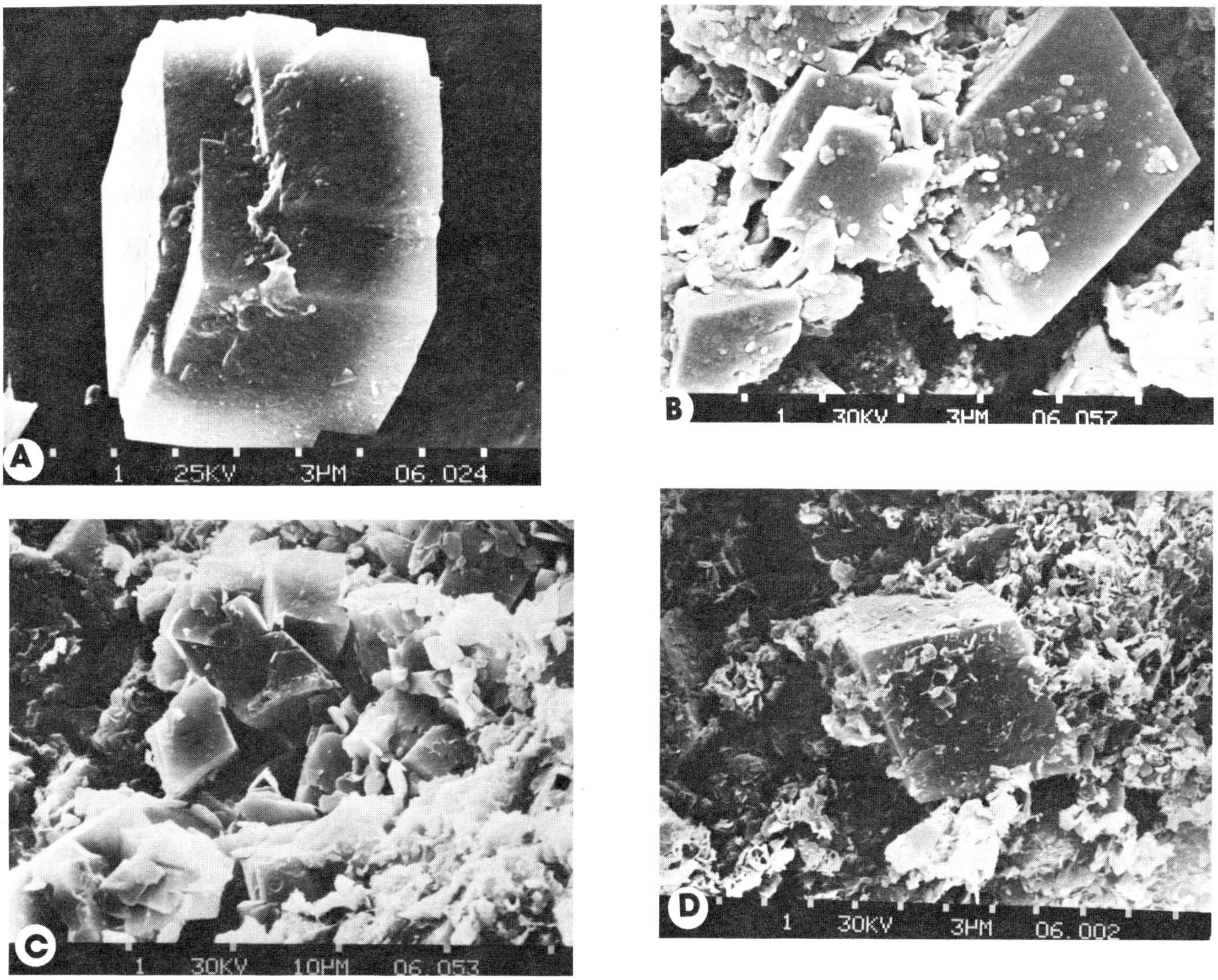

Figure 2 SEM photographs showing dolomite morphology:
(A) Grain mount of rhombohedral dolomite with step dislocations (163-13).
(B) Interpenetrating rhombs (163-13).
(C) Interpenetrating rhombs, morphology suggests *in situ* precipitation (163-13).
(D) Isolated rhomb with small intergrowth at top left of central face. Note adhering Fe-rich smectites (163-7).

gypsum in unknown. It could be a diagenetic phase, an oxidation product, or indeed, an artifact.

PHYSIOGRAPHY AND CHEMISTRY OF THE DOLOMITE

The dolomite occurs as discrete rhombohedra, or with small interpenetrating crystals, (Figure 2: A,B,C,D) growing displacively in the sediment. The rhombs vary in size from one or two microns up to a maximum of 40 µm, and show no evidence of size sorting in the sediment. Occasionally, minute inclusions of pyrite and claymaterical can be seen. Crystal terminations are sharp and well defined which indicate that no abrasion or other transport effects have taken place. Close inspection of certain surfaces of the dolomite grains sometimes show dislocations (Figure 2A) due to step growth along a spiral path (Nielsen, 1964). Such dislocations indicate that dolomite growth could take place at very low levels of supersaturation (<1.01) where two-dimension surface nucleation would be nearly impossible (Burton et al., 1951; Berner, 1981). Visual observation on the dolomite using the SEM and cathode luminescence failed to reveal any evidence of an earlier carbonate precursor phase. Foraminiferal tests are unaltered and do not have dolomite associated with them. Spatial arrangement of the rhombs in the sediment indicate direct precipitation, rather than some physico-chemical alteration from another carbonate. There is, however, a close association between growth of dolomite and other diagenetic phases such as glauconite, manganese carbonate and pyrite. These will be discussed below.

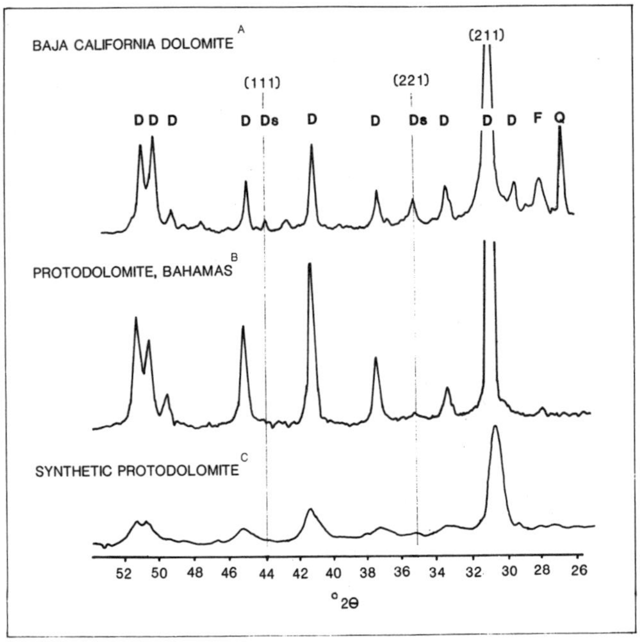

Figure 3 X-ray diffractograms of dolomite (A) and comparison with published spectra of protodolomites (B: Illing et al., 1977; C: Glover and Sippel, 1967) all using Cu Kα radiation. The superstructure reflections (111) and (221) reveal that the Baja California variety is an ordered dolomite. (D-dolomite, Ds-dolomite superstructure reflection, F-feldspar, Q-quartz)

Chemical and mineralogical examination shows that the dolomite from Baja California is not a "protodolomite" as defined by Graf and Goldsmith, (1956) and Gaines, (1977). This is illustrated in Figure 3 where the (211) and (111) superstructure reflections are clearly visible, confirming the degree of ordering compared with synthetic protodolomite or other naturally occurring protodolomites.

The Mg/Ca ratio of the dolomite was estimated using XRD by measuring the shift of the (211) reflection relative to a quartz internal standard (Goldsmith and Graf, 1958) giving a composition of 56 mole % $CaCO_3$. The use of a fluorite internal standard and slow chart scans of $1/8^o$ 2θ/min. showed the a_o and c_o unit cell parameters to be 4.8291 A and 16.1521 A respectively. These have been plotted in Figure 4 and compared to other published dolomites and protodolomites. The Baja California dolomite hexagonal unit cell parameters are consistent with ordered dolomite (Land, 1980) and contrast with the measurements on protodolomites that show expanded parameters departing from the calcite - magnesite trend. It would appear that the excess of calcium over that of stoichiometric dolomite does not result in disordering.

Electron microprobe analyses of dolomite from two cores (163-7 and 163-13, Figure 1) are shown in Table 1. Results for the Ca/Mg ratios, shown as a histogram in Figure 5, confirm the 56 mole % $CaCO_3$ obtained by the XRD method. Both stations have seemingly similar dolomite compositions, and show a smaller range of excess Ca than that found in IPOD core material from the Gulf of California by Kelts and MacKenzie, (1982). While Sr tends to show reasonably consistent values, and somewhat lower than those of Behrens and Land, (1972) (800 ppm) from Baffin Bay, Texas, they do suggest incorporation of the element within the dolomite structure; the greater variability of Ba may be due to impurities such as barite.

Si and Al contents of the dolomites (Figure 6) are, in all probability, due to the presence of clay inclusion as the Si/Al ratio (3.33) is identical to that of the host sediment. Fe contents appear to be in excess, both of the level of S that can be accommodated as pyrite (FeS_2), barite ($BaSO_4$), and Fe-rich smectite. The analyses suggest that a very small amount of Fe (<0.6 wt.%) is incorporated within the dolomite lattice.

DISTRIBUTION IN THE SEDIMENTS

Visual observation of smear slides of surface and subsurface samples show that dolomite distribution, along the transect studied, appears to be limited to those sediments associated with the hemipelagic environment (Figure 1). Distal cores (145-6, 163-9, and 163-15) have only trace amounts in them, whilst the shelf silts are devoid of dolomite. As a first approximation, the quantity of dolomite in the bulk sediment has been estimated from the excess Mg over that associated with clays. Table 2 shows that the upper limit of dolomite in the subsurface sediment of five cores broadly correlates with the level of $C_{org.}$

In order to determine, more reliably, the vertical distribution of dolomite in these sediments the silt-size fraction at 5 cm. intervals was extracted by wet sieving. An estimate of the dolomite content in this fraction was made by line counting 300 grains (Galehouse, 1969). Results of this are shown in Fig. 7 together with $C_{org.}$ and solid phase Mn in the sediment. The vertical distribution of Mn is used to illustrate the division of the sediment into oxidising and reducing regimes. Incorporation of the Mn into a surface sediment usually occurs as MnO_2 and upon burial and bacterial activity is rapidly reduced to Mn^{2+} at some level in the sediment. The redox of this transition is broadly similar to, but precedes, the reduction of Fe^{3+} to Fe^{2+}. Figure 7 shows very clearly the different depths of this redox change for Mn in the four cores. Inspection of the $C_{org.}$ contents show that those sediments with high organic matter have shallow redox changes (eg. 163-7). The vertical distribution of dolomite appears to be related to these redox boundaries and/or $C_{org.}$ within the sediment. It is present in significant amounts only in the reduced sediment, and here is broadly correlatable with the abundance of $C_{org.}$. Within the oxidising zone authigenic dolomite is either absent or present in small quantities. Moreover, inspection of the dolomite within oxidising sediment shows evidence of corrosion, (Figure 8D) whereas dolomite at depth occurs as perfect rhombs (Figure 2A,B,C,D). Sawlan and Murray (1983) present pore water NO_3 data from two cores, TGT-145-7 and TGT-145 - 8 with similar solid phases Mn

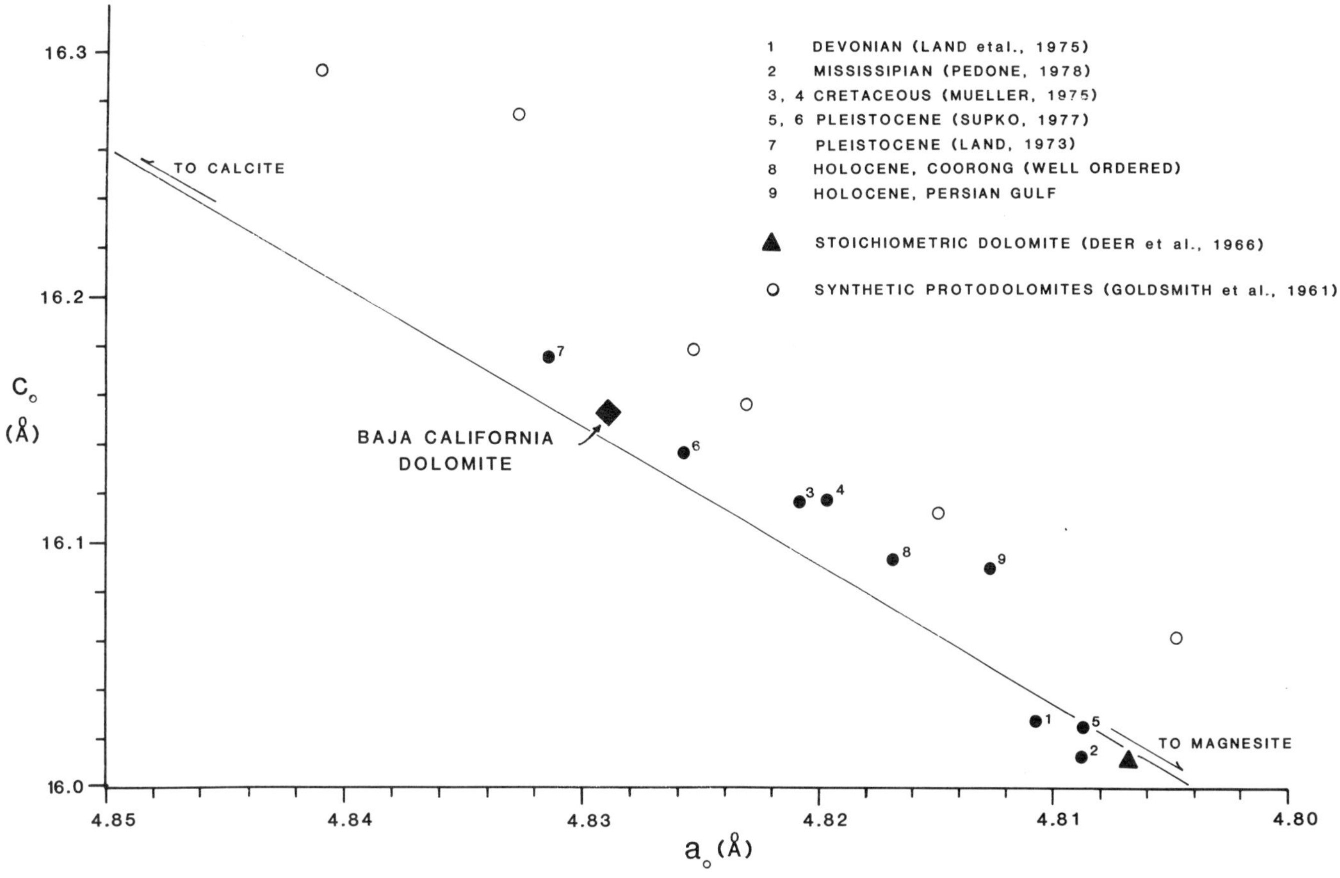

Figure 4 A plot of unit cell parameters a_o and c_o (in A) of Baja California dolomite, and published dolomites of various ages. The sloping line extrapolates to values for pure calcite and magnesite. Expansion of the unit cell due to disordering will result in an increase of the a_o and c_o parameters relative to this line (after Land, 1980).

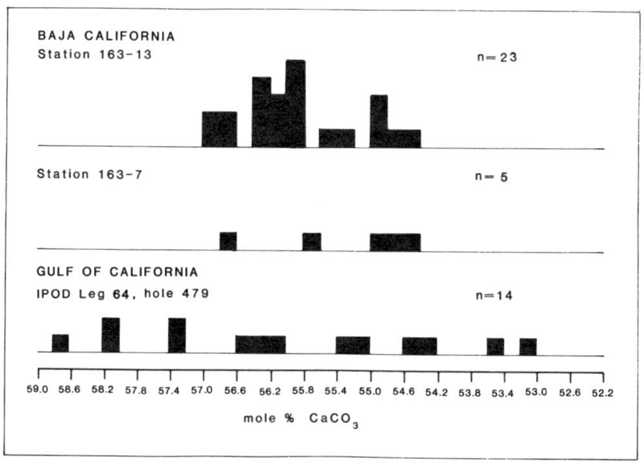

Figure 5 A histogram of the distribution of analyses for mole % $CaCO_3$ in Baja California dolomite, stations 163-13 and 163-7, and the Gulf of California, IPOD Leg 64, hole 479 (Kelts and McKenzie, 1982) performed by electron microprobe. (n=the number of analyses)

profiles. They show that NO_3 is completely consumed at depths greater than 40 cms. and 20 cms. respectively. Only on total depletion of O_2 and NO_3 (Gieskes, 1975) will the process of sulphate reduction predominate indicating that in these cores dolomite occurs in the presence of SO_4^{2-} ions.

Hemipelagic sediments along the studied transect show a maximum development of dolomite of only 3.5 wt.%. The variability of dolomite within these cores is broadly related to the level of $C_{org.}$ in the sediment. At station 163-13 the content of dolomite is unusually high and estimated at 25 wt. % in its deeper parts (Table 2) in this organic-rich sediment. The distribution of sediments within this core shows that the upper 20 cms. represents a lag deposit displaying concentrations of phosphorite, pyrite, barite and glauconite, whereas clays are poorly represented. At depth these same minerals are highly diluted by normal clay accumulation.

ELECTRON MICROPROBE ANALYSIS OF BAJA CALIFORNIA DOLOMITE

Core number and depth interval	Ca	Mg	Fe	Si	Al	Na	K	Ba	S	Sr
13:15-16	21.07	9.93	.47	1.22	.35	.13	.11	n.d.	n.d.	412
13:15-16	22.52	10.92	.39	.35	.09	.06	.04	n.d.	n.d.	431
13:15-16	22.74	10.67	.27	.74	.25	.08	.10	n.d.	n.d.	545
13:15-16	22.06	11.04	.77	.26	.08	.05	.04	n.d.	n.d.	736
13:15-16	21.82	10.06	.59	.43	.08	.05	.02	n.d.	n.d.	n.d.
13:15-16	22.98	11.43	.12	.59	.11	.12	.05	n.d.	n.d.	n.d.
13:15-16	23.00	10.85	.16	.27	.08	.07	.05	n.d.	n.d.	n.d.
13:15-16	22.78	10.92	.22	.27	.09	.05	.03	n.d.	n.d.	n.d.
13:18-19	22.50	10.63	1.19	.50	.15	.06	.08	1161	960	n.d.
13:18-19	23.72	10.93	.36	.25	.08	.06	.06	450	740	n.d.
13:18-19	23.15	11.08	.43	.21	.08	.06	.06	1348	700	n.d.
13:20-21	23.17	11.12	.15	.32	.06	.04	.08	508	150	n.d.
13:20-21	22.64	10.76	1.10	.32	.15	.05	.08	590	350	n.d.
13:20-21	23.02	11.47	.45	.83	.25	.15	.11	450	740	n.d.
13:21-22	22.81	10.81	.23	.87	.26	.08	.13	758	630	376
13:21-22	23.12	11.36	.12	.17	.09	.06	.05	391	380	298
13:21-22	23.08	11.07	.19	.36	.09	.05	.08	616	440	464
13:22-24	22.47	10.72	.25	.55	.17	.07	.10	1405	680	390
13:22-24	23.36	10.85	.69	.36	.13	.07	.08	697	330	317
13:22-24	20.89	10.03	1.09	2.53	.32	.06	.14	688	280	221
13:25-26	22.58	10.75	.90	.62	.23	.07	.13	797	290	265
13:25-26	23.48	11.66	.15	.37	.08	.06	.07	892	630	339
13:25-26	21.00	10.66	.55	.27	.07	.03	.05	862	650	479
7:30-34	23.08	11.65	.06	.21	.09	.06	.10	437	460	339
7:30-34	23.75	11.45	.01	.06	.03	.02	.06	651	310	166
7:30-34	23.47	11.81	.17	.61	.31	.14	.07	625	60	n.d.
7:30-34	23.55	10.92	.22	.14	.06	.03	.09	275	n.d.	n.d.
7:30-34	19.70	9.80	.11	1.94	.83	.15	.09	208	n.d.	n.d.
Detection limit	.015	.08	.033	.014	.009	.014	.01	560	122	127
Precision (±2)	.175	.08	.056	.027	.016	.017	.019	573	277	163

Table 1. Electron microprobe analysis of Baja California dolomite. Major elements as wt.%; analysis of Ba, S and Sr in ppm. Counting statistics are given in the same units. (n.d.=not determined)

Station (TGT-163)	9	10	14	7	13
20cm.					
C_{org}	0.34	1.10	1.03	1.33	3.60
excess Mg	0.01	0.20	0.19	0.29	18.73
Dolomite	0.09	1.76	1.68	2.53	18.73
25cm.					
C_{org}	0.29	1.05	1.05	1.42	4.27
excess Mg	0.04	0.14	0.13	0.29	2.95
Dolomite	0.25	1.22	1.13	2.53	25.70

Table 2. $C_{org.}$ determined by combustion in an induction furnace. Mg determined by x-ray fluoresence and salt-corrected. Excess Mg derived from an average pelagic clay Mg/Al ratio of 0.25. Dolomite content calculated from the excess Mg value assuming a stoichiometry for the dolomite of $Mg_{0.44}Ca_{0.56}Co_3$. All values in weight %.

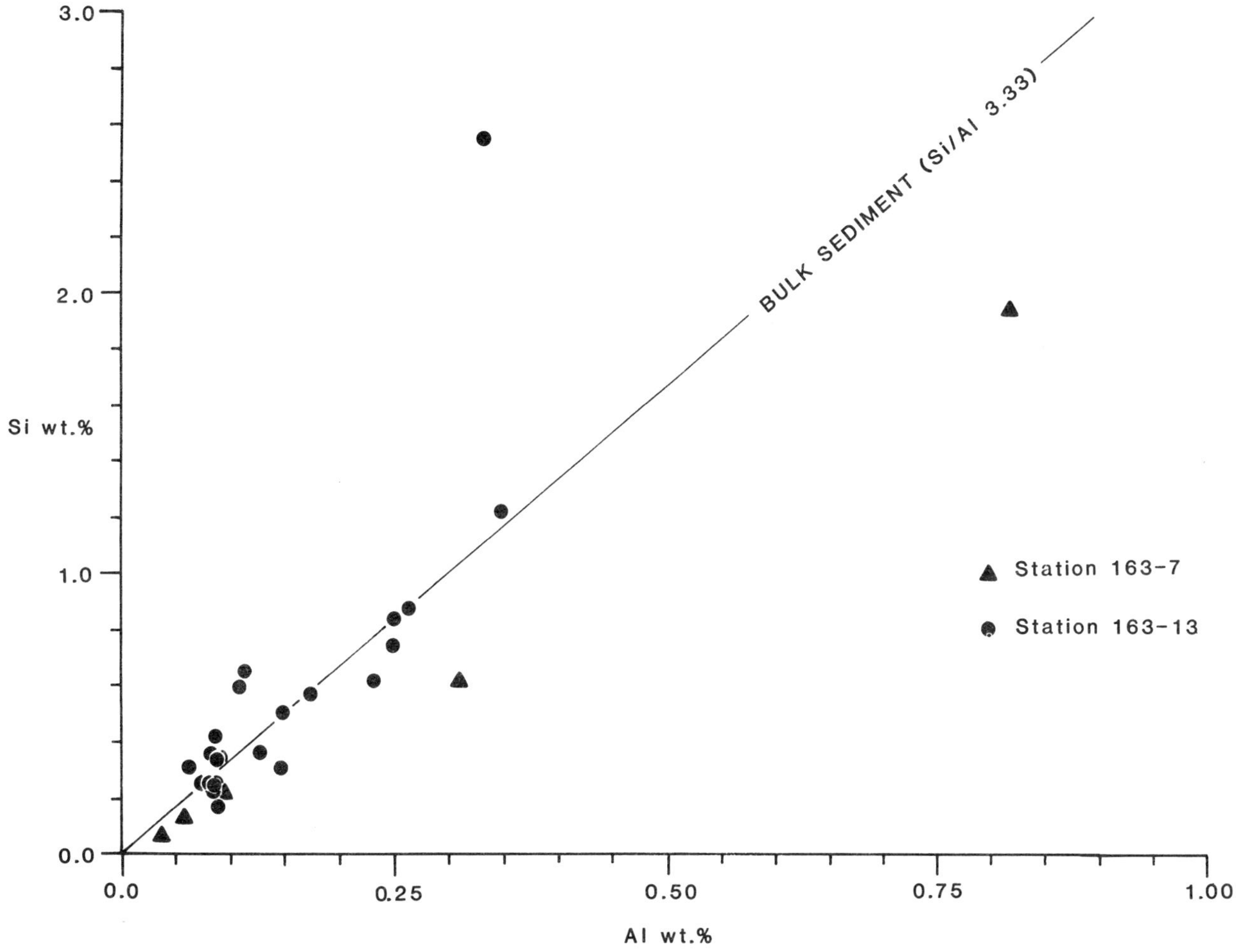

Figure 6 A plot of the Si and Al content of Baja California dolomites as measured from electron microprobe analyses. The Si/Al ratio (3.33) of bulk sediment based on 50 surface (0-10 cm.) analyses (XRF) is depicted by the line.

THE RELATION OF DOLOMITE TO OTHER AUTHIGENIC PHASES

All the above evidence suggests that dolomite is an early diagenetic phase. In order to understand the genesis of this dolomite it is important to relate it to the occurrences of other diagenetic minerals that have been identified by SEM and EDAX, and XRD techniques.

The coexistence of rare pyrite inclusions within the dolomite does suggest some sulphate reduction has occurred before the formation of dolomite. The existence of minor quantities of Ba within the dolomite also implies some barite formation before its crystallization. Recent reports have described the early formation of manganese carbonate (Pedersen and Price, 1982; Sawlan and Murray, 1983) in hemipelagic sediments which appears as a result of Mn reduction and bicarbonate production during burial before SO_4^{2-} exhaustion. In cores 163-10 and 145-7 $MnCO_3$ has been identified by EDAX (Figure 8A,B) towards their base (25-30 cm.) and its formation depth in these sediments is consistent with that calculated on thermodynamic grounds by Sawlan (1982). Examination of the carbonate reveals that dolomite can be associated with this mineral phase (Figure 8C) and that its depth of formation may be prior to that of the $MnCO_3$.

The association of dolomite with glauconite, in 163-13, is of particular interest as this mineral pair provides a valuable clue to the timing of dolomite growth, and its stability in the marine environment. Glauconite occurs both in the lag and the underlying clay in this core. Typically, it forms greenish-grey, very friable, fissured pellets usually 2 to 5 mm. in diameter (Figure 9A). In this respect they are identical to newly formed, "nascent" glauconite that has been identified in other Recent sediments (Odin and Matter, 1981). The structure of the glauconite seen by SEM and XRD analysis supports our contention that it is recently formed. Diffraction patterns show the mineral to be poorly-crystalline and lacking a well defined peak at 10 A; instead a broad hump is seen at

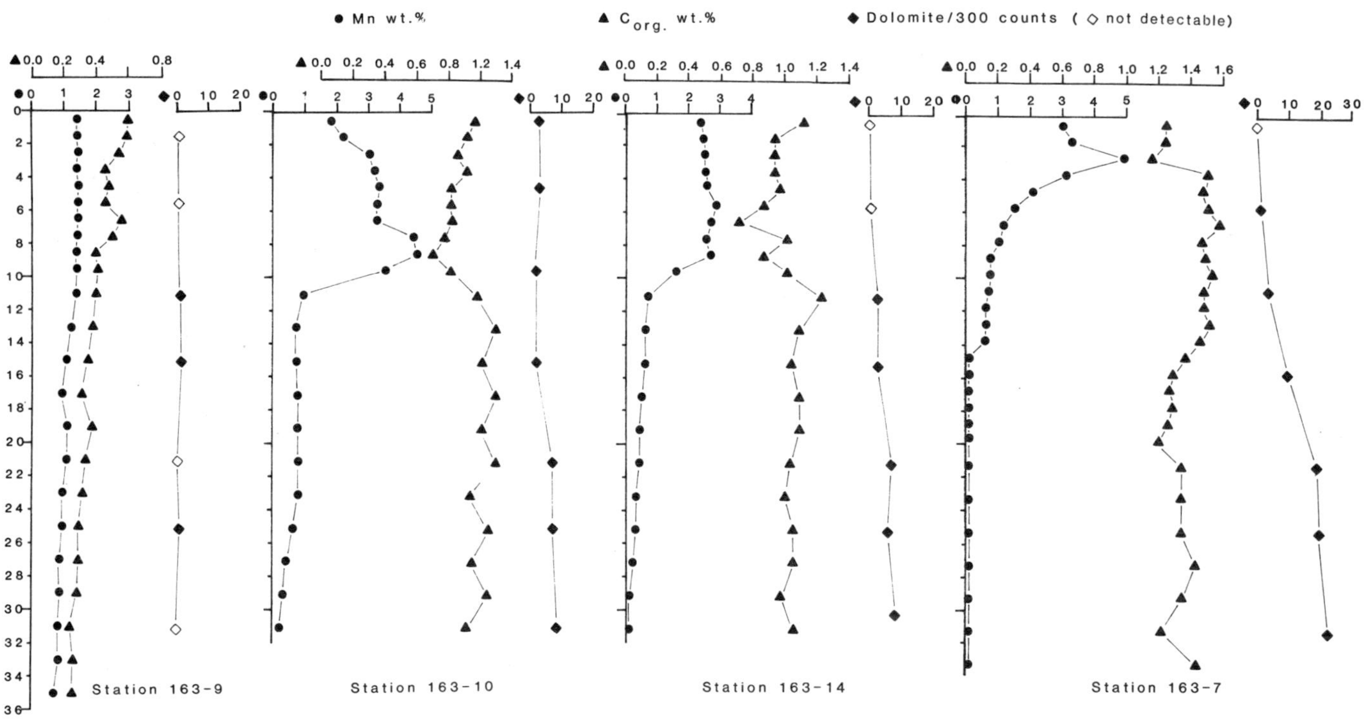

Figure 7 Vertical distribution of the frequency of dolomite estimated from counting 300 grains, solid phase Mn (wt.%) and $C_{org.}$ (wt.%) in four cores off Baja California (163-9, 163-10, 163-14, 163-7).

this wavelength. The mineral appears to be composed of mixed layer smectite phases as noted by Odin and Matter, (1981). SEM observations confirm a box-like morphology typical of early smectite phases (Figure 9B). Dolomite is seen to be intimately associated with this glauconite (Figure 9C,D) which acts as a non-corrosive cement binding the earlier formed dolomite together. Broken grains mounted in the SEM show that up to 70% of the inside of these pellets is composed of perfectly formed dolomite rhombs (Figure 9D). Glauconite extracted from this lag deposit shows dolomite to be unstable at the surface of the pellet. Here, dolomite is only recognised as highly pitted remains or rhombohedral impressions (Figure 10A,B,C,D). Clearly, formation of the dolomite proceeds that of the glauconite. Within the lag deposit only, dolomite at the surface of glauconite pellets is unstable and prone to dissolution.

POSSIBLE ORIGINS OF DOLOMITE AND THE SOURCE OF MAGNESIUM

Reports of trace amounts of dolomite rhombohedral (5-40 um.) in unconsolidated deep-sea sediments have existed over the last century (Correns, 1939; Fairbridge, 1957; Bonatti, 1966). Various conflicting views have been proposed on their origins, including detrital sources, formation due to hydrothermal metamorphism and submarine volcanism, and organic formation. Recent discoveries have focused attention on the association of dolomite with organic-rich sediment (Friedman and Murata, 1979; Pisciotto and Mahoney, 1981; Kelts and McKenzie, 1982). The formation of dolomite off Baja California seems to be consistent with these latter observations. Dolomite in IPOD cores from the Gulf of California (Kelts and McKenzie, 1982) shows that it is associated with organic-rich horizons extending from Late Pleistocene to Late Pliocene. Similar dolomite-bearing hemipelagic sediments extending down to the Miocene have been noted on the Pacific side of Baja California (Pisciotto and Mahoney, 1981). Kelts and McKenzie, (1981). Kelts and McKenzie (1982) have interpreted dolomite formation to be a result of low temperature, early diagenesis with primary precipitation of dolomite occurring in the interstitial spaces of anoxic, diatomaceous oozes. They claim that the process takes place in a subsurface system, closed to sea water exchange and characterized by low sulphate concentration. Experimental data on the replacement of calcite and aragonite by dolomite (Baker and Kastner, 1981) also suggest that dolomite formation is favored by low dissolved sulphate concentrations. Here, it is implied that the common site of dolomite formation is where the dissolved sulphate concentration is reduced by microbial sulphate reduction. Kushnir and Kastner (1982) from adsorption experiments of SO_4^{2-} on carbonates show that the residence time of adsorbed SO_4^{2-} is very short. In fact 50% of the adsorbed SO_4^{2-} is desorbed or exchanged in about ten minutes.

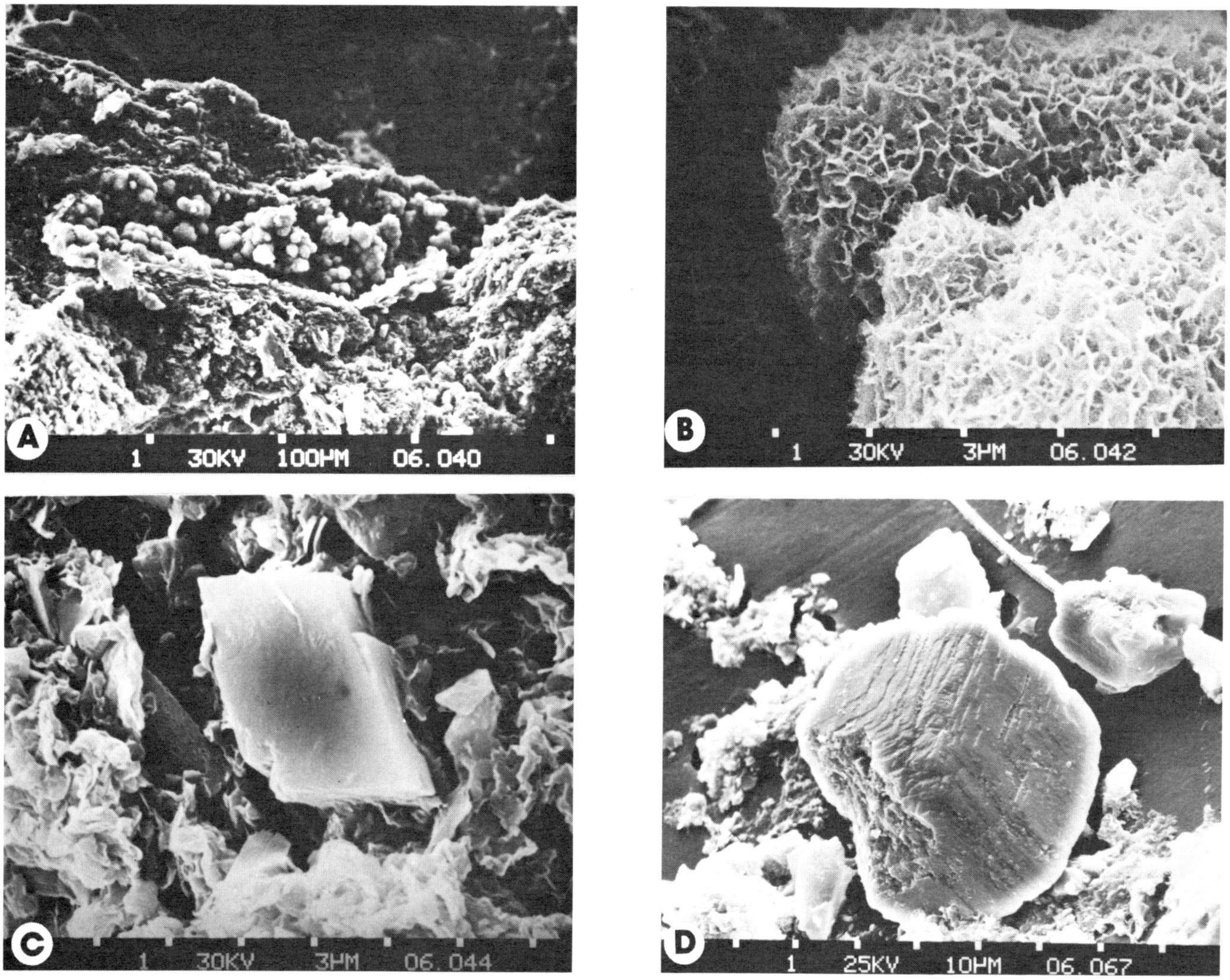

Figure 8 SEM photographs of the occurrence of dolomite in hemipelagic sediments and its association with other diagenetic phases.
(A) Botryoidal manganese carbonate growing in a cavity (163-10, 28-30 cm. depth).
(B) Close-up of the same manganese carbonate.
(C) Dolomite at the edge of the cavity surrounded by manganese carbonate.
(D) Dolomite rhomb showing evidence of dissolution from the oxidised zone of 163-14.

The observations on morphology, location, and mineral associations indicate that the Baja California dolomite formation is not related to detrital input, evaporative reflux (Adams and Rhodes, 1960), evaporative pumping (Hsu and Siegenthaler, 1969), mixed water or schizohaline (Folk and Land, 1975) processes. Hence, our interpretation of dolomite formation must be related to normal hemipelagic accumulation.

The first appearance of significant dolomite in the Baja California sediment, beneath the zone of Mn reduction and often in the presence of NO_3^-, clearly indicates that dolomite nucleation and growth can occur in the presence of SO_4^{2-} ions. Moreover, the formation of dolomite rhombs predating the development of glauconite pellets also suggests growth in the immediate subsurface environment. While this dolomite may usually occur in small quantities, its preponderance in the glauconite suggest development within this zone can be extensive. Glauconite typically forms in a non-suphidic environment common under fluctuating oxic-anoxic conditions (Odin and Matter, 1981; Berner, 1981). One of the major characteristics of glauconite is the excess of Fe^{3+} over Fe^{2+} (Burst, 1958; Odin and Matter, 1981; Ireland et al., 1983). Ideally, glauconite is most likely to occur in reducing microenvironments in an overall oxidising regime (Odin and Matter, 1981). In sediments of high sulphide activity, or post sulphide activity, it is extremely unlikely that sufficient Fe^{3+} ions would be present to account for the amount of glauconite seen.

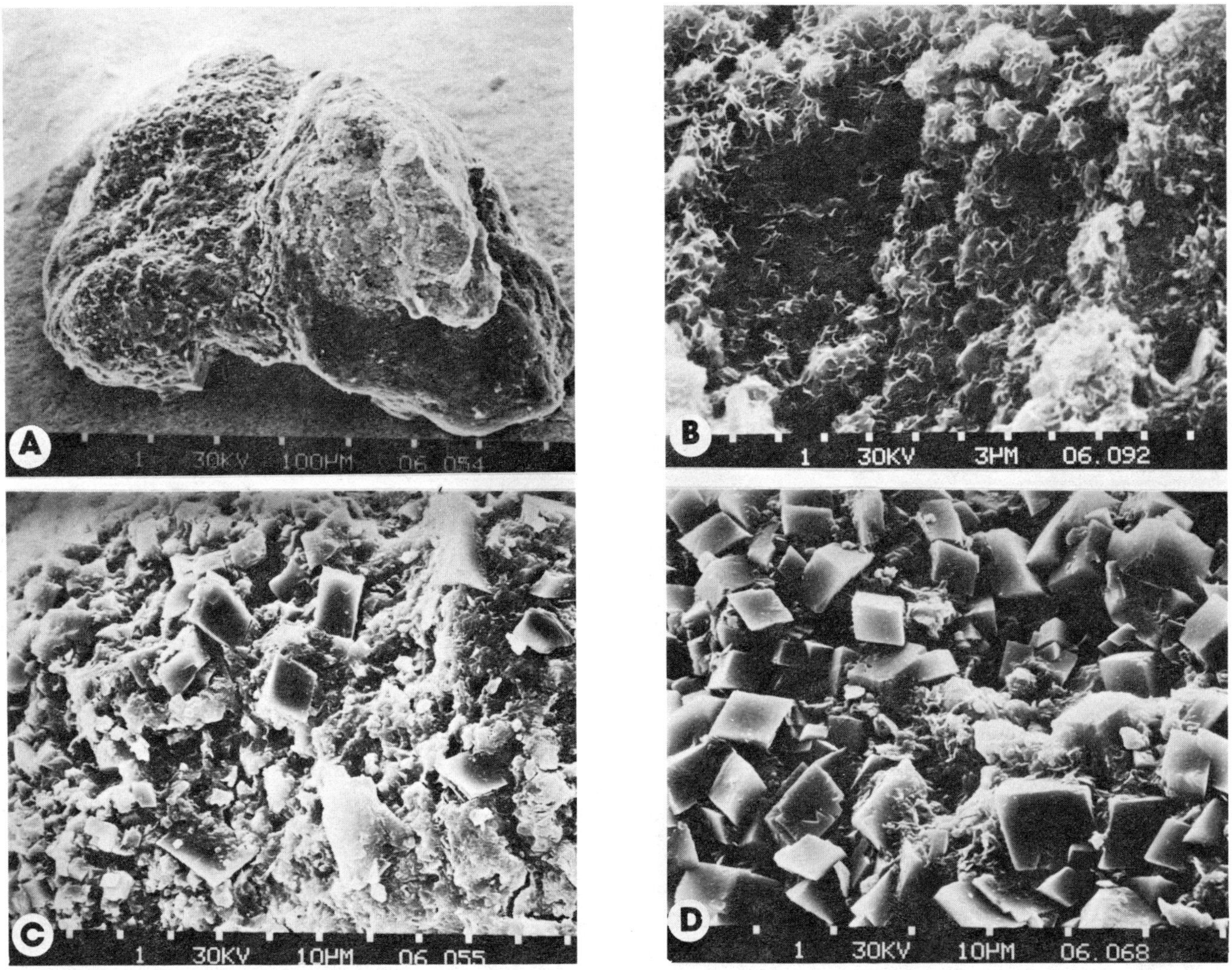

Figure 9 SEM photographs of the dolomite/glauconite association in 163-13.
(A) Grain mount of a glauconite pellet (20 cm. depth).
(B) Broken inside surface of a glauconite pellet showing box-like, smectitic clays (20 cm. depth).
(C) Surface of a glauconite pellet at depth (24 cm.), showing good euheral dolomite with no evidence of corrosion.
(D) A broken glauconite pellet showing 70% dolomite. Some glauconite is visible at the centre of the dolomite cluster (20 cm. depth).

The association of dolomite with early diagenetic manganese carbonate formed from the precipitation of Mn^{2+} under mildly reducing conditions is also consistent with SO_4^{2-} being present in the sediment. The inclusion of pyrite is some dolomite rhombs does suggest sulphate reduction, although it is not known if this is due to local reduction in microenvironments or an overall reduction of the sediment.

Thermodynamic and stability calculations have suggested that dolomite could be the stable phase in sea water (Garrels et al., 1960; Baker and Kastner, 1981). Our observation of dolomite dissolution in the uppermost sediments, particularly 163-13, implies that this is probably not correct. Within a lag deposit pore water conditions relating to alkalinity, Mg^{2+} and Ca^{2+} activity would not be significantly different from that of the overlying sea water.

The relationship between the amount of dolomite and organic matter may be significant both in respect of our understanding of dolomite formation and the source of Mg. It has been suggested that the source of Mg for dolomitization could come from sea water, organic matter or ion exchange in sediments (Milliman, 1974). With deeply formed dolomite the sea water source of Mg is obviously somewhat limited (Kelts and McKenzie, 1982), and here one may resort to an organic or ion exchange source of Mg involving clay minerals for dolomite precipitation or replacement within a sedimentary sequence. Organic-rich, diatomaceous sediments are known to have Mg concentrations in excess of those

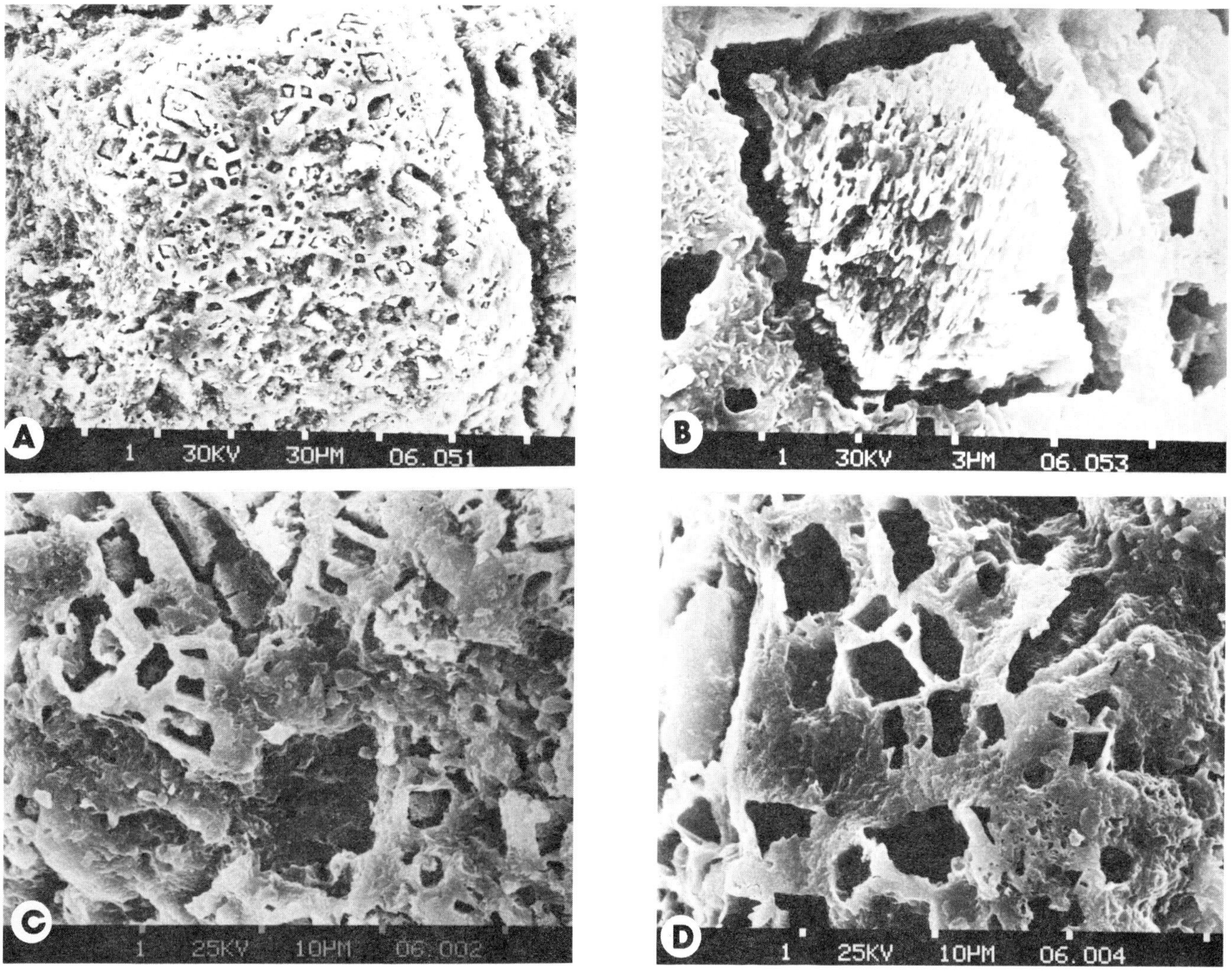

Figure 10 SEM photographs of dolomite dissolution at the surface of 163-13 in glauconite pellets from the lag deposit.
(A) Surface of the pellet with dolomite rhombs beginning to corrode.
(B) Close-up of the dolomite in (A).
(C) Glauconite with rhombohedral pits in which there is some relict dolomite remaining.
(D) Close-up of the rhombohedral pit from which all the dolomite has been completely corroded away.

held in terrigenous silicates, but even in such instances the amount of organically bound Mg tends to be modest and rarely exceeds 1 or 2% of the sediment (Calvert, 1976). With lower C_{org} found in the sediments off Baja California this source of Mg would be somewhat limited.

Preliminary measurements on accumulation of Baja California hemipelagic sediments indicate rates of 0.5 to 1.5 cm./1000 years for those sediments existing in the reducing zone. Moreover, excess ^{230}Th measurements show that the oxidising layer may be entirely bioturbated. Estimates of bioturbation in Pacific sediments suggest the top 10 to 20 cms. may be entirely biomixed every 1000 years. Certainly the rate of biomixing is considerably greater than sediment accumulation (Turekian et al., 1978). The profiles of solid phase Mn in Figure 7 could be interpreted in terms of a surface bioturbated layer and an underlying zone of steady state accumulation. If extensive bioturbation occurs in the Baja California sediment, irrigation of this surface could bring sufficient Mg within the entrained waters for significant amounts of dolomite to form at, or near, this redox boundary. Due to the low sedimentation rate Mg may also reach this zone by diffusing in from the overlying sea water (Gieskes, 1975). The presence of minor amounts of dolomite within the oxidising layer could be caused by precipitation within microenvironments, or alternatively, could have formed at the locus of dolomite precipitation within the reduced sediment and been subsequently transported by benthic activity.

The mechanism of dolomitization within the uppermost reduced sediment, or at the redox boundary, is unknown, but higher alkalinity in the reduced sediment may be one factor leading to its precipitation. Elevated Mg/Ca ratios that have previously been considered important in the formation of supratidal dolomite are not a limiting requirement. This view is also held by Baker and Kastner (1981) and Kelts and McKenzie (1982) in the formation of deep-sea dolomite.

Variation in input of $C_{org.}$ to the sediments will affect the subsurface environment in two ways. Firstly, it is likely to produce variation in the total dissolved C in the reduced sediment which may have some bearing on the dolomite formation. Secondly, and perhaps more importantly, it may induce a range of benthic activity and corresponding entrainment of sea water Mg to the zone of reduction leading to dolomitization. The occurrence of dolomite in many ancient sediments is often associated with bioturbation as seen in the dolomite bands of the Monterey Formation of southern California (Friedman and Murata, 1979).

Laboratory experiments on dolomite kinetics (Gaines, 1980) show it to be extremely slow growing. Within the rapidly accumulating sediments of the shelf, conditions conducive to dolomite growth are changing too rapidly for precipitation to take place. Here, incipient dolomite formed would be rapidly buried before significant crystal growth could be achieved. Within organic-poor oceanic red clays a low rate of sediment mixing from bioturbation again reduces the likelihood of significant quantities of dolomite. It would seem that the slowly accumulating hemipelagic sediments, with high $C_{org.}$ contents due to elevated biological productivity as is seen here, represent an ideal locus for dolomite precipitation. In the eastern Pacific the generally low terrigenous supply coupled wit high biological productivity, suggests that this area of the ocean in highly conducive towards early diagenetic dolomite formation. Reports of some rhombohedral dolomite in late Cenozoic hemipelagic sediments from IPOD cores (Pisciotto and Mahoney, 1981; Kelts and McKenzie, 1982) and in dredge hauls (Kulm et al, 1981) suggests they may have formed in the manner described.

Dolomite formation in deeply buried sediments undoubtedly occurs. Such late formation has been evidenced by the isotopic composition of the mineral and its pervasive growth in certain horizons. We suggest that the earliest formed dolomite, as is seen in the immediate subsurface environment off Baja California, may act as a substrate for further dolomite precipitation, especially during methanogenesis after sulphate has been consumed. Growth of this dolomite is obviously limited by the amount of residual sea water and organic Mg within the sediment and the extent of ion exchange involving Mg from clay minerals, during diagenesis. Such dolomite is not likely to be uniformly precipitated within the sediment, but is going to be confined to those horizons with prior dolomite. In the exceptional case (as in 163-13) of very high $C_{org.}$ and benthic activity, early dolomite formation can be extensive.

SUMMARY AND CONCLUSIONS

Our observations on the distribution of dolomite in subsurface sediments off Baja California show:

1. Dolomite precipitation occurs principally in hemipelagic sediments seen from the shelf break to the continental rise. Sediments on the shelf and in the oceanic setting show little or no dolomite genesis. Within the hemipelagic environment dolomite abundance broadly correlates with the level of $C_{org.}$ in the sediment.
2. Vertical distribution of dolomite in hemipelagic shallow cores show that within the silt fraction dolomite forms in reduced sediment, as defined by the distribution of solid phase Mn. Here, dolomite abundance again appears to correlate with the amount of $C_{org.}$ in the sediment. Trace quantities of dolomite within oxidising sediments show evidence of dissolution.
3. Structurally, the dolomite is well-ordered and therefore dolomite, not protodolomite. Chemically, it contains 56 mole % $CaCO_3$, found by XRD and electron microprobe analysis.
4. The association of other diagenetic minerals and dolomite shows that whereas pyrite and barite can predate dolomite precipitation the growth of glauconite, occurring as "nascent", grey-green, friable, fissured pellets, occurs after dolomite precipitation. Early diagenetic manganese carbonate appears to postdate the formation of dolomite. Some of these associations clearly indicate that dolomite forms as an early diagenetic mineral in the presence of SO_4^{2-} ions.
5. The relationship between dolomite and C_{org} within these sediments has been interpreted in terms of levels of benthic activity allowing sea water Mg to be entrained within the sediment, and its precipitation as dolomite near the zone of reduction.
6. We believe that this early formed dolomite may be a useful substrate for further dolomite growth during burial.
7. The combination of slow total sediment accumulation and high surface productivity, as occurs along the eastern margin of the Pacific ocean, appears to produce an ideal site for early diagenetic dolomite.

ACKNOWLEDGEMENTS

We wish to thank the University of Washington and crew of the R/V *Thomas G. Thompson* for their assistance in obtaining core material for this work. We would like to thank, in particular, Dr. Jim Murray for his encouragement and support in this work.

We would also like to thank Andrew Ross and the staff of the M.R.C. Electron Microscope Unit at Western General Hospital, Edinburgh for the use of their facilities.

This research was sponsored by the Natural Environment Research Council, U.K.

REFERENCES CITED

Adams, J.E. and Rhodes, M.L., 1960, Dolomitization by seepage refluxion: Am. Assoc. Petroleum Geologists Bull., v. 44, p. 1912-1920.

Aschman, H., 1959, The Central Desert of Baja California: Demography and Ecology; Univ. Calif. Pres, Berkeley, Calif., 315 pp.

Baker, P.A. and Kastner, M., 1981, Constraints on the formation of sedimentary dolomite: Science, v. 213, p. 214-216.

Behrens, E.W. and Land, .S., 1972, Subtidal Holocene dolomite, Baffin Bay, Texas: Jour. Sed. Petrology, v. 42, p. 155-161.

Berner, R.A., 1981, A new geochemical classification of sedimentary environments: Jour. Sed. Petrology, v. 51, p. 359-365.

_____, 1980, Early Diagenesis: Princeton Univ. Press, Princeton, 241 pp.

Bonatti, E., 1966, Deep sea authigenic calcite and dolomite: Science, v. 153, p. 534-537.

Burst, J.F., 1958, Mineral hereogeneity in "glauconite" pellets: Am. Mineralogist, v. 43, p. 481-497.

Burton, W.K., Cabrera, N. and Frank, F.C., 1951, The growth of crystals and equilibrium structure of their surfaces: Royal Soc. Lond. Philos. Trans. v. A-243, p. 299-358.

Calvert, S.E., 1976, The mineralogy and geochemistry of nearshore sediments: in Riley, J.R. and Chester, R. (eds.) Chemical Oceanography Vol. 6, Academic Press, London, p. 187-271.

Correns, C.W., 1939, Pelagic sediments of the North Atlantic Ocean: in Trask, P.D. (ed.) Recent Marine Sediments, Am. Assoc. Petroleum Geologists, p. 373-395.

Curtis, C.D., Evans, G., Kinsman, D.J.J., and Shearman, D.J., 1963, Association of dolomite and anhydrite in Recent sediments of the Persian gulf: Nature, v. 197, p. 679-680.

D'Anglejan, B.F., 1967, The origin of marine phosphorites off Baja California: Mar. Geol., v. 5, p. 15-44.

Deer, W.A., Howie, R.A. and Zussman, J., 1966, An Introduction to Rock Forming Minerals: Longman, London, 528 pp.

Deffeyes, K.S., Lucia, F.J., and Weyl, P.K., 1964, dolomitization: observations on the island of Bonaire, Netherlands Antilles: Sciences, v. 143, p. 678-679.

Emery, K.O., 1960, The Sea Off Southern California; Wiley, New York, 366 pp.

Fairbridge, R.W., 1957, The dolomite question: in LeBlanc, R.J. and Breeding, J.G. (eds.), Regional Aspects of Carbonate Deposition, Soc. Econ. Paleontologists and Mineralogists Spec. Pub. 15, p. 125-178.

Folk, R.L. and Land, L.S., 1975, Mg/Ca ratio and salinity: Two controls over crystallization of dolomite: Am. Assoc. Petroleum Geologists Bull., v. 59, p. 60-68.

Friedman, I. and Murata, K.J., 1979, Origin of dolomite in Miocene Monterey Shale and related formations in the Temblor Range, California: Geochim. Cosmochim. Acta, v. 43, p. 1357-1365.

Gaines, A.M., 1980, Dolomite kinetics: Recent experimental studies: in Zenger, D.H., Dunham, J.B. and Ethington, R.L., (eds.) Concepts and Models of Dolomitization: Soc. Econ. Paleontologists and Mineralogists Spec. Pub. 28, p. 81-86.

_____, 1977, Protodolomite redefined: Jour. Sed. Petrology, v. 47, p. 543-546.

Galehouse, J.S., 1969, Counting grain mounts: number percentage vs. number frequency: Jour. Sed. Petrology, v. 39, p. 812-815.

Garrels, R.M., Thompson, M.E. and Siever, R., 1960, Stability of some carbonates at 25°C and one atmosphere total pressure: Am. Jour. Sci., v. 258, p. 402-418.

Gieskes, J.M., 1975, Chemistry of interstitial waters of marine sediments: Annual Review of Earth and Planetary Sciences, v.3, p. 433-453.

Glover, E.D. and Sippel, R.F., 1967, Synthesis of magnesium calcites: Geochim. Cosmochim. Acta, v. 31, p. 603-613.

Goldsmith, J.R. and Graf, R.E., 1958, Relation between lattice constants and composition of the Ca-Mg carbonates: Am. Mineralogist, v. 43, p. 84-101.

______,_______,and Heard, H.C., 1961, Lattice constants of the calcium - magnesium carbonates: Am. Mineralogist, v. 46, p. 453-457.

Graf, D.L., and Goldsmith, J.R., 1956, Some hydrothermal synthesis of dolomite and protodolomite: Jour. Geology, v. 64, p. 173-187.

Hsu, K.J. and Siegenthaler, C., 1969, Preliminary experiments on hydrodynamic movement induced by evaporation and their bearing on the dolomite problem: Sedimentology, v. 12, p. 11-25.

Illing, L.V., Wells, A.J., and Taylor, J.C.M., 1965, Penecontemporary dolomite in the Persian Gulf: in Pray, L.C. and Murray, R.C. (eds.) Dolomitization and Limestone Diagenesis: A Symposium, Soc. Econ. Paleontologists and Mineralogists Spec. Pub. 13, p. 487-499.

Ireland, B.J., Curtis, C.D. and Whiteman, J.A., 1983, Compositional variation within some glauconites and illites and implications for their stabilities and origins: Sedimentology, v. 30, p. 769-787.

Jahnke, R.A., Emerson, S.R., Roe, K.K. and Burnett, W.C., 1983, The present day formation of apatite in Mexican continental margin sediments: Geochim. Cosmochim. Acta, v. 47, p. 259-266.

Kelts, K.R. and McKenzie, J.A., 1982, Diagenetic dolomite formation in Quaternary anoxic diatomaceous muds of Deep Sea Drilling Project Leg 64, Gulf of California: Init. Repts, DSDP, v. 64 pt. 2, p. 553-569.

Kulm, L.D., Thornburg, T.M., Schrader, H., Masias, A., Resig, J.M., and Johnson, L., 1981, Late Cenozoic carbonates on the Peru continental margin: Lithostratigraphy, biostratigraphy, and tectonic history: Geol. Soc. America Memoir 154, p. 469-509.

Kushnir, J. and Kastner, M., 1982, Adsorption of sulfate ions on calcite, aragonite and dolomite: Their role in inhibiting dolomitization: (abs.) EOS, v. 63, p. 999.

Land, L.S., 1973, Holocene meteoric dolomitization of Pleistocene limestones, North Jamaica: Sedimentology, v. 20, p. 411-424.

____, 1980, The isotopic and trace element geochemistry of dolomite: the state of the art: in Zenger, D. H., Dunham, J.B. and Ethington, R.L. (eds.), Concepts and Models of Dolomitization: Soc. Econ. Paleontologists and Mineralogists Spec. Publ. 28, p. 87-110.

____, Salem, M.R.I. and Morrow, W.F., 1975,, Paleohydrology of ancient dolomites: Geochemical evidence: Am. Assoc. Petroleum Geologists Bull., v. 59, p. 1602-1625.

Lyle, M., 1983, The brown-green color transition in marine sediments: A marker of the Fe(III) - Fe(II) redox boundary: Limnol. Oceanography, v. 28, p. 1026-1034.

Milliman, J.D., 1974, Marine Carbonates, Spring Verlag, New York, 375 pp.

Mueller, H.W., 1975, Centrifugal progradation of carbonate banks: A model for deposition and early diagenesis, Ft. Terett Formation, Edwards group, Lower Cretaceous, Central Texas: Ph.D. Diss. Univ. Texas at Austin, 300 pp.

Nielsen, A.E., 1964, Kinetics of Precipitation: MacMillan, new York, 151 pp.

Norrish, K. and Hutton, J.J., 1969, An accurate x-ray spectographic method for the analysis of a wide range of geological samples: Geochim. Cosmochim. Acta, v. 33, p. 431-453.

Odin, G.S. and matter, A., 1981, De glauconiarum origin: Sedimentology, v. 28, p. 611-641.

Pederson, T.F. and Price, N.B., 1982, The geochemistry of manganese carbonate in Panama Basin sediments: Geochim. Cosmochim. Acta, v. 46, p. 59-68.

Pedone, V.A., 1978, Petrography, chemistry and crystallography of baroque dolomite, Kingsport, Tennessee: M.A. Thesis, Univ. Texas at Austin, 142 pp.

Pisciotto, K.A. and Mahoney, J.J., 1981, Isotopic survey of diagenetic carbonates, Deep Sea Drilling Project Leg 63: Init. Repts. DSDP, v. 63, p. 595-609.

Ried, J.L., Roden, G.I., and Wyllie, .G., 1958, Studies of the California Current System: Calif. Coop. Oceanic Fisheries Investigation Progr. Rept., 1 July 1956 - 1 Jan 1958, p. 27-57.

Sawlan, J.J., 1982, Early diagenetic remobilization of some transition metals in hemipelagic sediments: Ph.D. Thesis, Univ. of Washington, 251 pp.

Sawlan, J.J., and Murray, J.W., 1983, Trace metal remobilization in the interstitial waters of red clay and hemipelagic marine sediments: Earth and Planet. Sci. Lett., v. 64, p. 213-230.

Shinn, E., Ginsburg, R.N. and Lloyd, R.M., 1965, Recent supratidal dolomite from Andros Island: in Pary, L.C. and Murray, R.C. (eds.) Dolomitization and Limestone Genesis: Soc. Econ. Paleontologists and Mineralogists Spec. Pub. 13, p. 112-123.

Supko, P.R., 1977, Subsurface dolomite in modern carbonate sediments along the southern coastal of Florida: Science, v. 134, p. 561-562.

Taft, W.H., 1961, Authigenic dolomite in modern carbonate sediments along the southern coast of Florida: Science, v. 134, p. 561-562.

Turekian, K.K., Cochran, J.K. and De Master, D.J., 1978, Bioturbation in deep-sea deposits: Rates and consequences: Oceanus, v. 21, p. 34-41.

Von der Borch, C.C., 1965, The distribution and preliminary geochemistry of modern carbonate sediments of the Coorong area, South Australia: aeochim. Cosmochim. Acta, v. 29, p. 781-799.

A COMPARISON OF ANOXIC DOLOMITE FROM DEEP-SEA SEDIMENTS: QUATERNARY GULF OF CALIFORNIA AND MESSINIAN TRIPOLI FORMATION OF SICILY

K. Kelts and J. McKenzie
Geological Inst. ETH/Zurich CH8092 Switzerland

ABSTRACT

Two examples of anoxic dolomite formation in hemipelagic deep-sea diatomaceous sediments are reviewed. Their carbon-isotope ratios from dolomites fall into two distinct fields; positive values (0 to 14 o/oo) were recorded for the Quaternary dolostones from the Guaymas Basin slope DSDP Site 749, Gulf of California, whereas negative values (0 to -22 o/oo) characterize the dolomitic sediments of the Messinian Tripoli Formation of the Central Sicilian Basin. This isotopic separation is related to differences in sedimentation rates in the two areas; higher sedimentation rates (0.5 to 1.5 mm/yr) in the Guaymas Basin quickly convey sediments into the zone of methanogenesis whereas lower rates (0.03-0.4) in the Central Sicilian Basin allowed sediments to remain longer in a zone of sulfate reduction. The distinct isotopic differences in anoxic dolomites could be used as a rough measure of ancient sedimentation rates in sequences such as the Monterey Formation, and hence, an indirect measure of source rock potential of a deep-marine deposit.

INTRODUCTION

A widespread abundance of dolomite formed in the deep sea has only recently been recognized. Most of these occurrences are associated with dysaerobic to anaerobic sediments. We thus use the term anoxic dolomites in a generic sense to designate those dolomites formed by early diagenesis in reducing sediments. Anoxic dolomites are commonly characterized by equigranular micrite crystals showing cement or replacement textures and having carbon isotope ratios of either extremely positive or negative values. The occurrence and formation of anoxic dolomite are consistent with a general theory of dolomite formation related to environments with low sulphate concentrations (Baker and Kastner, 1981).

The Monterey Formation is one classic association of anoxic dolostones intercalated with siliceous sediments (Murata, et al 1969; Friedman and Murata, 1979; Pisciotto, 1981). By analogy with conditions in the modern Gulf of California (Bramlette, 1946) these siliceous sediments are assumed to derive from areas of high diatom productivity. Our recent studies of Quaternary anoxic dolomites from the Gulf of California (Kelts and McKenzie, 1982) indicate that the dolomite there is an early diagenetic product with carbonate ion derived from the bacterial decomposition of organic matter. There seems to be a direct relationship between high productivity, diatom sedimentation, high organic content, anoxic conditions and the sulphate reduction and dolomite precipitation.

In this paper we examine two occurrences of anoxic dolomites in diatomaceous sequences and identify features common to the Monterey Formation. These are DSDP Site 479 in the Gulf of California and the Messinian Tripoli Formation of Central Sicily.

Although occurrences of anoxic dolomite are known from many areas of hemipelagic sediments along the North American continental margin and Gulf of California region (Russell et al.; 1967, Hein, et al, 1979; Pisciotto and Mahoney, 1981; Kelts and McKenzie, 1982; Wada et al., 1982) we review the findings from Site 479 along the Guaymas Basin slope because they present a natural long-term experiment where numerous in-situ geochemical and sedimentological parameters delimit the boundary conditions of anoxic dolomite formation. The Tripoli Formation presents an interesting case providing an opportunity to reevaluate a setting commonly associated with restricted circulation and penesaline conditions. Our premise is that whilst the carbon isotopes give evidence of the diagenetic environment, they perhaps also can be used as indicators of sedimentation rate or productivity.

GULF OF CALIFORNIA

DSDP Site 479 and Sediments

DSDP Leg 64 Site 479 is a modern example of dolomite formation. This site was selected for its present-day location beneath an upwelling zone of fertile waters accompanied by prolific seasonal blooms of diatoms and dinoflagellates. Site 479 was drilled to 440 meters sub-bottom in 766 m of water on the outer marginal slopes of the Guaymas Basin in the central Gulf of California (Curray, Moore, et al. 1982). The hole thus sits within a layer of low-oxygen water which intersects the slope between 500 and 1300

in Garrison, Robert E.; Kastner, Miriam and Zenger, Donald H., eds., 1984, **Dolomites of The Monterey Formation and Other Organic-Rich Units:** Pacific Section S.E.P.M., Vol. 41, p. 19-28.

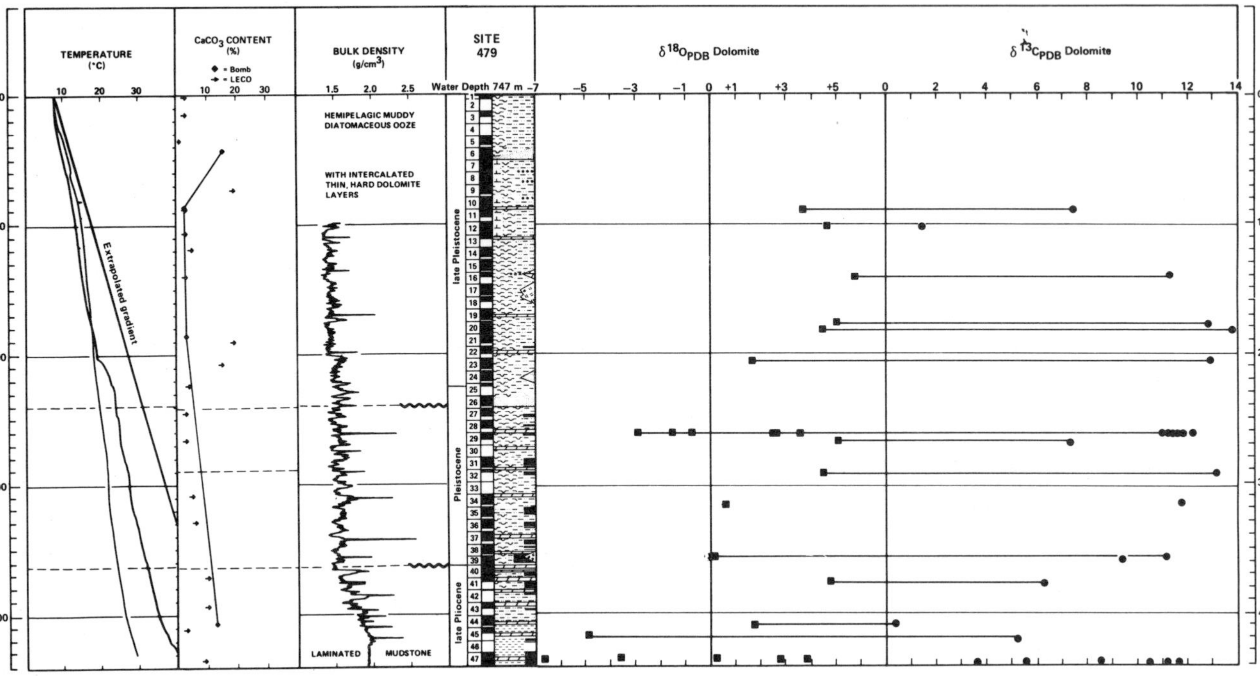

Figure 1. Gulf of California, Guaymas Basin slope DSDP Site 479, a compilation of lithology, host sediment carbonate content, bulk density and temperature logs. The results of isotopic analyses for dolomite samples are shown with tie lines connecting carbon-13 and oxygen-18 values from the same sample (Kelts and McKenzie, 1982).

meters (Calvert, 1966). Its Quaternary history and paleoenvironment are well known as are in-situ geochemical parameters from interstitial waters and gases (Gieskes et al., 1982; Simoneit et al., 1982; Schoell, 1982; Galimov and Simoneit, 1982).

The sediment column comprises mainly olive-drab, rapidly deposited, anoxic hemipelagic, muddy diatomaceous oozes rich in organic carbon (Leclaire and Kelts, 1982). Meter-thick zones of rhythmically laminated and structureless sediment alternate in a manner similar to nearby Site 480 (Schrader et al., 1980), where accumulation rates exceed 500m/my. The section (Figure 1) reflects generally uniform conditions throughout the cored intervals but is divided into 3 units on the basis of two breaks in sedimentation that might mark hiatuses: (1) a coarse sand layer at 240 m and (2) a change in sediment induration which accompanies increased compaction and the disappearance of biogenic silica in laminated mudstones below 364 m. This lower unit has very fine laminations, some with partings of kerogen and sparkling fish scales and others with a few lamellae of a monospecific coccolith ooze (see site chapter, Curray, Moore et al. 1982). The features suggest that this lower unit was similar to the diatomaceous oozes above, but that biogenic opal has been removed by diagenesis.

Sporadic, hard dolomitic fragments and decimeter-thick beds occur within softer oozes to mudstones in all three units. The first recognizable bed was encountered at 88 meters sub-bottom. Others followed in spaced intervals. Using logging results, we could clearly identify at least 19 lithified horizones which are initially spaced 20 meters apart down to 350 m and 5 to 10 meters apart below that depth. This is consistent with increasing compaction. Individual beds do not appear to be thicker than about 30-40 cm; hard dolomitic mudstone layers therefore account for a maximum of only 2-3% of the sedimentary section. At present, we do not know the lateral extent of these horizons, or whether some are concretional. No concentric structures are observed. Thus, we assume by analogy with the Monterey Formation that beds pinch and swell but have kilometer continuity (Bramlette, 1946; Pisciotto, 1981).

Color mottles outline preserved sedimentary structures which show that dolomite may occur as a cement for bioturbated, laminated, or transitional diatomaceous mud facies (Figure 2). The calcium carbonate content of the hemipelagic, diatomaceous ooze, host sediment is generally low (0-10%) with rare minor pulses to 20%. Most of this consists of benthic foraminifers and some nannofossils and planktonic fora-

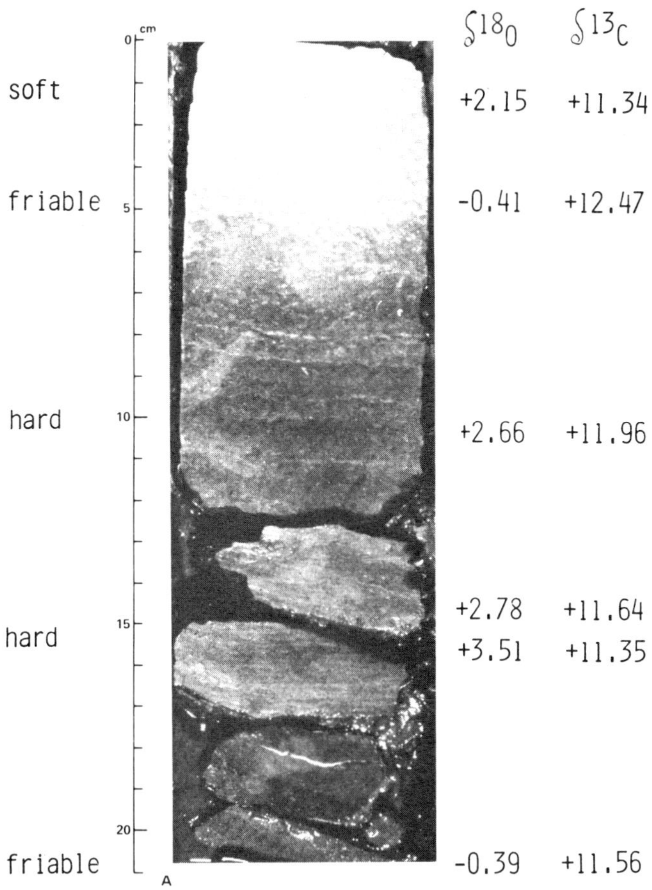

Figure 2. Upper portion of a dolostone layer (Samples 64-479-29-1, 0-21 cm; 263 m subbottom). Isotope results illustrate bimodal gradients away from harder center (approx. 15 cm) (Kelts and McKenzie 1982).

minifers (LeClaire and Kelts, 1982). There were no clear signs of dolomitic mudstones being confined to layers with high precursor carbonate contents, and in some cases, unaltered benthic foraminifers are practically adjacent to dolomitic beds. Monospecific nannofossil lamellae also occur between dolomitic horizons. This pattern of alternating dolostones and host sediment seems to extend into the Miocene sediments presumed to underly the Guaymas margin. Dredge hauls (Moore, 1973) from steep transform scarps contain similar dolostones (Kelts and McKenzie 1982) with Pliocene fauna.

The host sediment shows a relatively uniform terrigenous composition downhole. In addition to the amorphous opal diatom component, sediments are dominantly quartz and feldspar with plagioclase more common than K-feldspars. Downhole trends in opal diagenesis are not evident until below 360 meters. However, clinoptilolite was encountered in traces below 50 meters and indicates at least some active silica diagenesis. Clay minerals (15-30%) include little reactive smectite and show the same composition in host sediments and acid residues of dolostone. It should be noted that scattered dolomite crystals are present as well as a few pellets of higher magnesium calcite (2-12%). Primary features of the sediments, such as sedimentary structures, microfossils, or magnetic vectors are commonly best studied in diagenetic dolomite layers, even and especially where later diagenesis may have altered these in the host sediment.

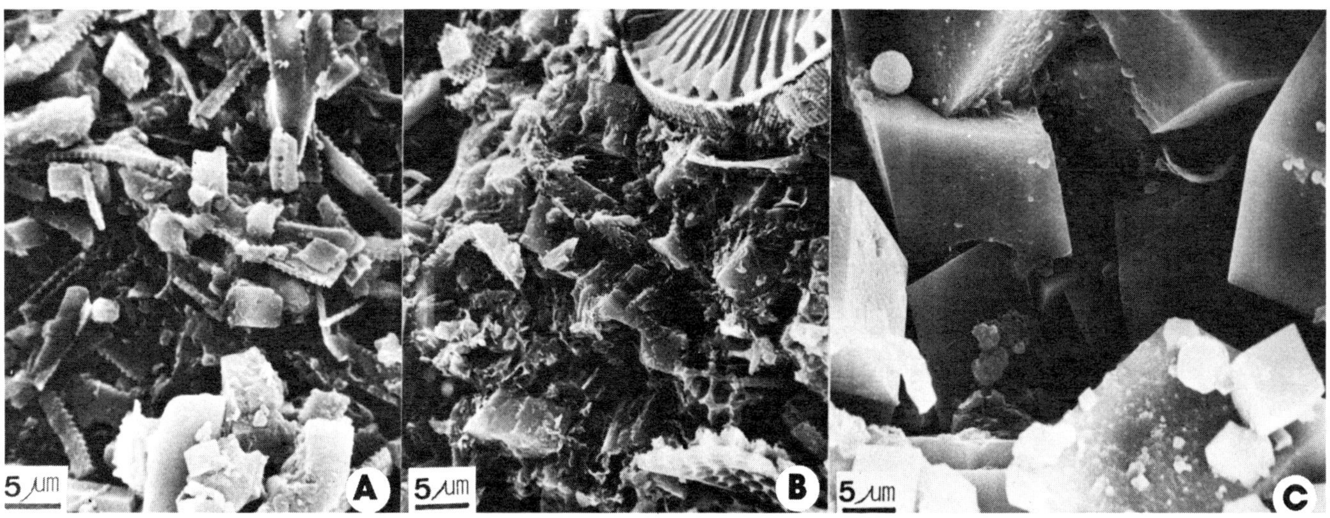

Figure 3. Comparison of SEM microtextures (all at same scale) across the dolostone bed from Figure 2 (64-479-29-1, 0-25 cm; 263 m): (A) 1 cm below top of bed: soft friable diatom mudstone with scattered 1-3 micron crystals. (B) 24.5 cm below top of bed: hard dolomitic mudstone near lower portion of bed with relict diatom frustules cemented by closely spaced small (5 micron) crystals. (C) Bed center: brittle dolostone with euhedral smooth-surfaced intergrown rhombs (5-20 microns).

Gradients

The petrological findings from Site 479 are similar to those from Neogene samples recovered along the outer continental margin of Baja California. The most interesting feature of undisturbed dolomitic mudstone layers, is the present of gradients in texture, mineralogy, induration and isotopes across beds. Centers may be hard dolostone with bulk densities greater than 2.5 g/cm^3 and up to 80% carbonate. Commonly outer margins are soft, yellowish-gray friable mudstone with carbonate contents down to a few percent Some layers have vug-like small pits in the hard centers which might derive from gas bubbles or dissolution.

The X-ray determined compositions from the (211) peak of dolomite crystals ranges from calcium-rich to stoichiometric (42-50% $MgCO_3$) and ordering peaks are present but weak. Where measurable, the outer rind of a sample had consistently higher calcium values as did the odd bits of soft dolomitic mudstone found in the host sediment. Calcite co-existing with dolomite is a low-magnesium variety as would be expected from calcareous microfossils. High-magnesium varieties are very rare, but were encountered near the top of another basinal site (Site 478); Curray, Moore et al., 1982) as yellow, soft pellets. Iron might be expected to enter the lattice of diagenetic dolomite in reducing sediments but microprobe analyses (Kelts and McKenzie, 1982) of the Site 479 layers failed to show any significant trace of iron or manganese. Pyrite however, is an abundant accessory in oozes and dolomite layers, a suggestion that dolomite may have formed after available iron had been locked into sulfides. Total carbonate values of some of the dolostones exceeds the amount of pore space (about 60%) available at shallow depths to 80 meters (Einsele, 1982) suggesting the incorporation of at least some precursor carbonate.

Among the different samples, we observe a variation of crystal sizes and textural patterns related more to progressive degrees of bed induration or accretion than to downhole trends. The layer in Core 64-479-29-1, from 263 m sub-bottom illustrates aspects of these microtextural gradients (Fig. 2 and 3). Both thin sections and SEM analyses confirm that dolomite is a pore filling cement consisting of individual, tiny (approximately 10 microns), idiomorphic crystals. Diatom frustules are encased or filled with a coarser mosaic. There is no evidence for concomitant corrosion of tests. Dolomite may however delicately pseudomorph some calcareous microfossils as verified by microprobe analysis of the wall of a foraminiferal test. These features as well as the smooth walled rhombohedral crystal surfaces are evidence of very slow, uniform growth processes.

Crystals are small, poorly formed (1-2 microns) and scattered in softer samples. Coexisting coccoliths may show ragged edges from dissolution. Grain sizes and abundance increase in more indurated samples. Well-formed smooth surfaced, sharp-edged, equigranular, interpenetrating rhombs (10-30 microns) are common in cemented but punky samples with dissolution of foraminiferal tests possibly contributing to porosity. In an impervious dolostone sample (Core 64-479-47, 440 meters), crystals coalesce as anhedral mosaics. None of the samples show evidence of recrystallization to sugary dolomite.

Stable Isotopes

Most of the dolomite from Site 479 shows extreme enrichment of Carbon-13 with values as great as +13.8 o/oo and an average of +9.8 o/oo for 30 samples (Fig. 4). In two cases it was possible to measure several samples across a contiguous bed: the laminated to bioturbated examples from Core 29 (Fig. 2 and 3) and Core 47 (Fig. 5). These revealed clear gradients (Fig. 5) related to increasing burial and progressive accretion. Following models of Claypool and Kaplan (1974) and Irwin et al. (1977), we interpret the heaviest Carbon-13 values in the centers as evidence of earlier formation in the zone of active methanogenesis as they also show the lowest isotope temperature ranges. The value of +3.5 o/oo at the outer edge is evidence of formation at greater depths, particularly in conjunction with more negative oxygen-18 values pointing to higher isotope temperatures. The oxygen-18 values for the measured samples, which are generally the hard centers, show temperatures of formation equivalent to, or more commonly, less than those for their present burial depth. Deuser (1970) already noted a gradient across a hand speciment of anoxic carbonate from the Atlantic margin but attributed it to spill over of brines from coastal lagoons.

The Guayamas Slope Model

Among the most significant elements of this setting, we note that high productivity of diatoms and dinoflagellates leads to high sedimentation rates with reactive organic matter. The geochemical consequences include (Figure 6): (1) low-oxygen bottom waters, (2) strongly reducing conditions in

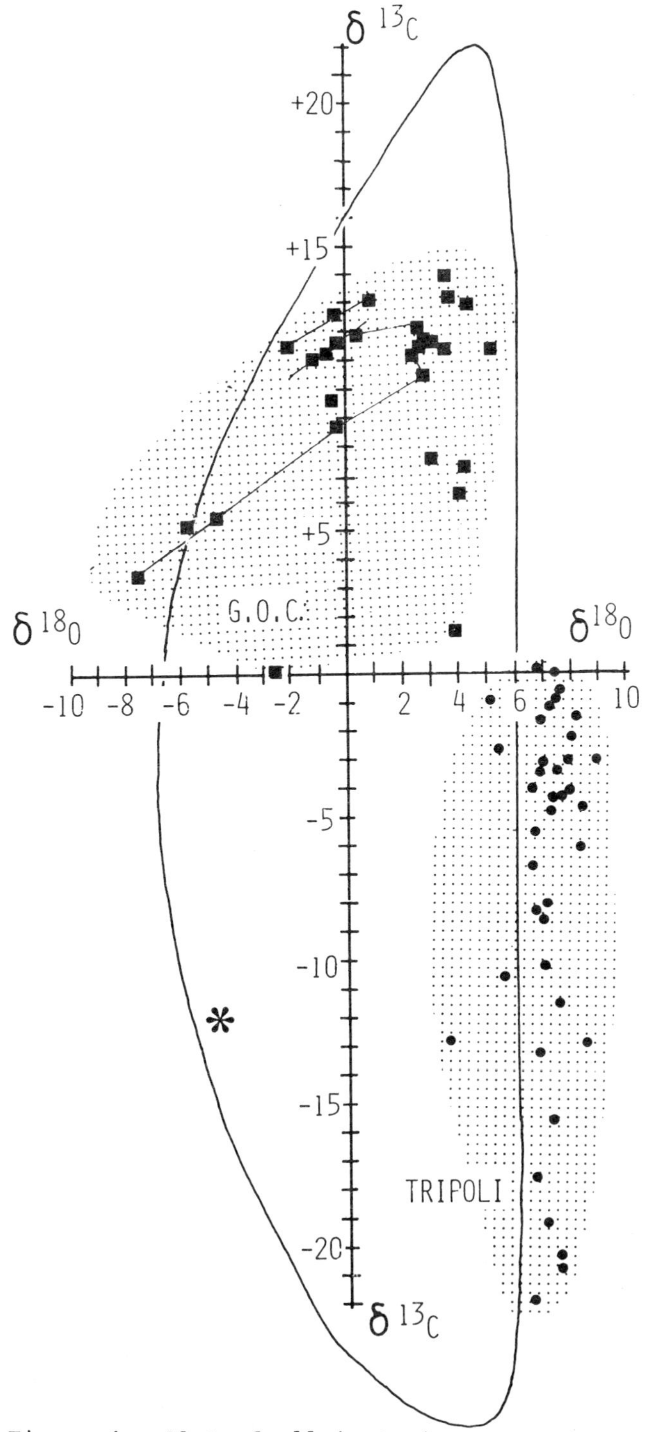

Figure 4. Plot of all isotopic measurements and their areas for dolomite samples from DSDP Hole 479 (Squares: Kelts and McKenzie, 1982) and the Tripoli Formation (Dots). These are compared with the envelope for values from the Monterey Formation dolomites (Pisciotto, 1981; Murata et al., 1969). The asterisk designates a single sample of silt-sized, limpid, euhedral, dolomite cementing glauconite pellet sands from Neogene sediments off shore southern Baja California (Core 64-475-17-5). The Tripoli Fm. values include selected data points from the Sutera section (Bellance et al., 1982/83). Tie lines in the Gulf of California (G.O.C.) field connect samples from single dolomite beds.

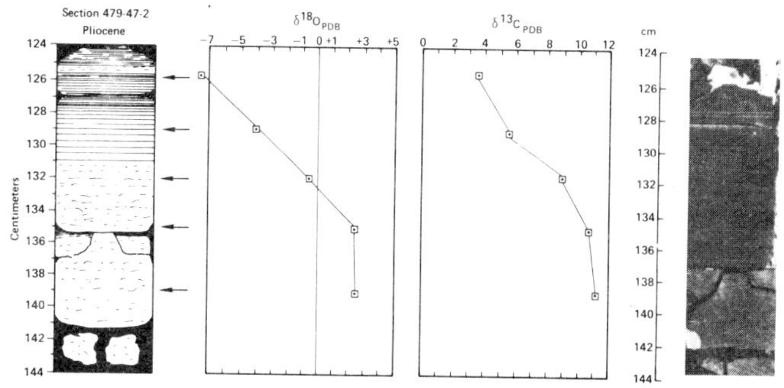

Figure 5. Linear gradients of oxygen-18 and carbon-13 which match textural induration trends across dolostone layer 64-479-47-2, 124-144 cm.

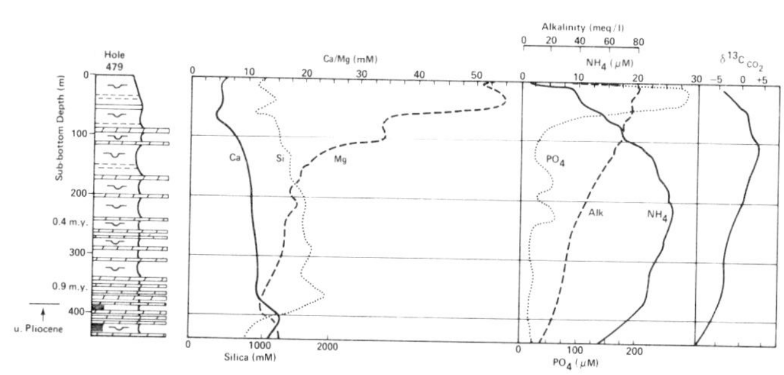

Figure 6. Compilation of lithology with geochemical analysis from Hole 479. Ca, Si, Mg, PO_4, Alk, NH_4 from Gieskes et al., (1982). Carbon-13 from in-situ dissolved carbon dioxide from Schoell (1982).

sediments with low initial carbonate content and high porosity, (3) early formation of pyrite from sulfate reduction, (4) trapped geochemical gradients with high ammonia, phosphate and alkalinities, and (5) prolific production of carbon dioxide and methane in a closed system. The geochemical environment is thus conducive to carbonate mineral precipitation.

Figure 7 summarizes the conceptual model for dolomite formation from DSDP Site 479. The zone of methanogenesis begins only 45-60 cm sub-bottom in the Gulf of California (Goldhaber and Kaplan, 1980). Within a few hundred years, sulfate is completely depleted from the sediment before significant dolomite is formed. As a result of the sulfate reduction, and breakdown of reactive organic matter, alkalinity climbs to over 80 meq/l within a few meters sub-bottom and declines uniformly downsection. This is a probable indicator of concomitant carbonate precipitation. The first conspicuaous dolomite in the sediments was detected around 60 m (50,000 years) although some disseminated rhombs are likely nearer the surface. Such

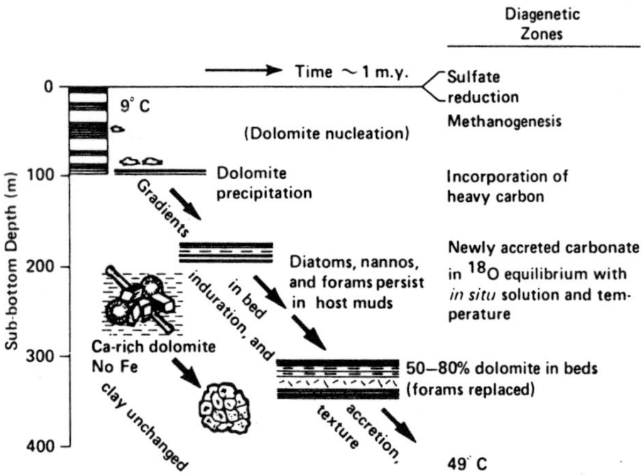

Figure 7. Generalized model for shallow subsurface nucleation and slow accretion of dolomite layers from Site 479, Guaymas Basin slope (Kelts and McKenzie, 1982). Changing burial environments are reflected in gradients in texture, mineralogy and isotopic composition (see text).

rhombs were also identified in basinal sites at a few meters subsurface (Kelts and McKenzie, 1982). The main site of dolomite formation at 50-80 m sub-bottom is also indicated by a rapid drop in magnesium content of pore waters (Gieskes et al., 1982 and Fig. 5), as well as by the correspondence of downhole temperature and carbon dioxide gas compositions and carbon and oxygen isotope signals in dolomite.

Apparently, dolomite continues to form and slowly accrete along layers down to the depth of the lowermost samples recovered. Surprisingly, calcium continues to rise, until a crossover point with magnesium is reached at 360 m sub-bottom. This indicates that elevated Mg:Ca ratios are not a limiting requirement for the growth of dolomite crystals. The absolute amounts of magnesium must limit the amount of dolomite formed to about 2 to 4% of the section. We interpret the mineralogical gradient from more to less stoichimetric compositions as a bed is traversed from center to edge as a result of the declining Mg:Ca ratios in the pore waters rather than a maturation signal. This is because stoichimetric hard centers seem to conformably reflect isotope evidence of shallower environments and not a mixed signal or resetting. From the isotopic evidence of a conveyor-belt type of crystallization with slow progressive burial, we were able to estimate growth rates. Our results (Kelts and McKenzie, 1982) gave 0.1 to 0.6 mm/ty as maximum bed accretion rates. At these low rates, the slow diffusion of magnesium and calcium could replenish some of the pore water reserves used for carbonate precipitation.

The sporadic occurrence of the beds remains a key problem, but may be the result of irregularities in sedimentation, such as levels of slightly higher precursor carbonate. These have been linked to latent climate signals from Pleistocene glacials (Kelts and McKenzie, 1982). The sporadic occurrence of dolostone layers in the Tripoli Formation is also linked to early glacial signals which would also lead to different sedimentation rates (McKenzie et al., 1979/80). Other problems require a mass balance approach including an explanation for calcium ion increases toward the bottom of the Hole 479.

The model (Fig. 6) of a primary precipitation of dolomite during low-temperature, early diagenesis of anoxic, low-carbonate, hemipelagic diatomaceous oozes has numerous implications for the geologic record. Such beds are likely to preserve original structures, textures and components (such as microfossils, or paleomagnetic directions) destroyed by compaction in surrounding mudstones. Evidently, dolomite formation precedes major silica diagenesis. It should be noted that the carbon-13 values of these dolomites cannot be used to study basinal or global shifts in the carbon cycle. Small amounts of anoxic dolomite in chalks might also affect a delicate bulk isotopic record.

THE TRIPOLI FORMATION

The lower Messinian Tripoli Formation of the Central Sicilian Basin comprises cyclic alternations of diatomites and marls that commonly contain micritic dolomite (McKenzie et al., 1979/80; Bennet, 1980). In the Tripoli sediments, dolomite occurs not only diseminated in the marls but also in the diatomites with an increasing percentage towards the top of the formation; it also occurs as discrete beds which are present as rhythmic decimeter-thick dolostone layers with wide lateral continuity. The formation directly underlies the Messinian evaporite complex and the transitional sediments into this complex are invariably dolomitic. All of the Tripoli dolomite is non-stoichimetric with Ca:Ma ratios varying between 52:48 and 57:43. In the marls, micritic dolomite is almost entirely composed of subhedral grains up to 10 microns across with an occasional large euhedral rhomb up to 100 microns (Bennet, 1980).

McKenzie et al. (1979/1980) proposed that the cyclic nature of the sediments resulted from global eustatic sea-level changes superimposed upon a tectonic constriction of the seaway between the Atlantic and Mediterranean. The stable isotope study of calcite components of the diatomaceous interbeds indicated normal marine conditions. Dolomite, on the other hand, gave oxygen-18 values from +4.0 to +8.5 o/oo. This range most likely signifies isotopic equilibrium with evaporated waters as normal marine water at normal bottom water temperatures could not result in the more positive values. The oxygen-isotope data requires that the pore waters or diagenetic solutions must have had variable but greater salinity than normal sea water. This is comparable with the modern Mediterranean, where excess evaporation over precipitation run-off from a silled basin produces dense surface waters which sink to form bottom waters. An "estuarine" circulation pattern in the early Messinian could have provided oxygen-18 enriched pore waters which participated in anoxic dolomite diagenesis.

McKenzie et al. (1979/1980) suggested that the dolomite in the Tripoli Formation was a product of dolomitization associated with evaporite deposition. They used a model whereby warm, oxygen-18 rich waters with high Mg:Ca ratios were formed in shallow marginal areas via gypsum precipitation and subsequently sank into the deeper basin where they dolomitized the calcium carbonate components of pelagic sediments. With the model derived from DSDP Legs 63 and 64 for deep-sea dolomite and the recognition of the inhibiting effect of sulfate ions on the formation of sedimentary dolomite (Baker and Kastner, 1981), we can reevaluate the origin of dolomite in the Tripoli Formation. We now propose that these dolostones had a similar deep-sea origin within anoxic sediments.

The stable isotope data for Tripoli dolomite are given in Table 1. The carbon-isotope values (0 to -22 o/oo) from dolomite samples represent a significant component of organically derived carbon dioxide, probably from the metabolic activity of sulfate-reducing bacteria in anoxic sediments. The extreme range is an indication that little or no originally deposited precursor carbonate was in the sediments. There are numerous reports emphasizing a possible euxinic character for the Tripoli Basin; fossil fish are perfectly preserved in some layers of laminated diatomites; sediments are rich in organic carbon and pyrite; faunal diversity is low (Ongnibin, 1957); Decima and Wezel 1973). From this sedimentological and geochemical data, we conclude that the Tripoli dolostones are a product of anoxic dolomite precipitated within the sulfate reducing layer of diatomaceous pelagic sediments.

Table 1: Stable Isotope Data for Tripoli Formation dolomite

Locality[1]	Sample No.	Lithology[2]	$\delta^{13}C$ °/oo PDB	$\delta^{18}O$ °/oo PDB
Calascibetta	S9	dolomite	-12.94	+3.53
	BO 25/40	marl	-4.58	+7.50
	BO 25/37	marl	-1.11	+7.45
	BO 25/32	dolomite	0.29	+6.82
	BO 25/17	marl	0.28	+7.97
Camastra	S26	marl	-1.47	+8.25
	BO 06/33	diatomite	-1.16	+7.25
	BO 06/31	dolomite	-2.18	+8.02
	X006/10	diatomite	1.66	+6.85
	X006/6	diatomite	-2.69	+5.42
Falconara	952	dolomite	-11.63	+4.41
	944	marl	-18.92	+5.12
Monte Capadarso	B068/01	dolomite	-19.20	+7.03
	CAP 3	marl	-20.78	+7.55
	4554	marl	-20.37	+7.47
	4555	dolomite	-21.94	+6.65
	S15	dolomite	-17.58	+6.70
	4556	marl	-11.49	+7.43
	S16	dolomite	-4.77	+8.43
	4558	dolomite	-15.61	+7.34
	4560	dolomite	-13.10	+8.51
Gibel Gabel	82-7	dolomite	-0.65	+7.62
	82-8	dolomite	-3.58	+6.78
Monte Giammora	Gi3	dolomite	-0.94	+5.13
Riesi	Rie9	dolomite	-10.58	+5.64
	Rie10	dolomite	-5.59	+6.66
	Rie11	marl	-3.14	+7.87
Licata	Li2	dolomite	-6.83	+6.41
	Li3	dolomite	-8.28	+6.72
	Li5	dolomite	-4.07	+7.36

1. Location of various locatities in Bennet (1980).
2. Carbonate component of samples is consistently greater than 75% dolomite and generally 100%.

Stable-isotope data for Tripoli Formation dolomites are plotted along with the Monterey Formation envelope and Gulf of California results (Fig. 4). The Tripoli data encompases the lower portion of the Monterey data in terms of carbon-13 values providing an excellent complement and contrast to the Gulf of California data. Thus, whereas the dolomites in the Tripoli Formation apparently formed in zones of sulfate reduction, those in the Guaymas Basin are forming in zones of methanogenesis.

An evaluation of sedimentary differences which would explain the differences in the diagenetic processes leads to the conclusion that sedimentation rates are a major controlling factor for the carbon-isotope signal. We suggest that the isotope record in the dolomite layers is also an indirect measure of productivity rates for diatom sedimentation in the two basins. From an extensive compilation of Tripoli profiles, Bennet (1980) estimated that an average diatomite horizon is 0.60 m thick and represents approximately 1500 years whilst an average marly interbed is 0.34 m and represents 13,500 years. The sedimentation rates for diatom and marl deposition were therefore 0.40 and 0.03 mm/yr, respectively. An overall rate for the Tripoli Formation was 0.06 mm/yr, which is considerably lower than the 0.5 to 1.5 mm/yr estimated for the Guaymas slope sites (Schrader, 1982).

Nucleation and growth of anoxic dolomite is apparently a slow process. The slower sedimentation rate in the Tripoli Formation allowed sulfate to diffuse into the sediments and continuously renewed a reservoir being depleted by sulfate reduction. Thus the Tripoli sediments remained in zones of sulfate reduction and did not enter zones of methanogenesis. This process was particularly enhanced during periods of marl deposition, with perhaps porous diatom-rich layers contributing carbonate ions.

Some support for a sedimentation rate control on carbon-isotope signatures is given by a measurement on Miocene dolomite offshore from the tip of Baja California (DSDP Site 475, Kelts and McKenzie, 1982, and Fig. 4). This sample has strongly negative carbon-13 and is associated with glauconite, organic carbon and phosphate rich sediments which indicate a reduced sedimentation rate, but dysaerobic or anaerobic sediments. Sulfate reduction not only removes an inhibitor to dolomitization but provides excess alkalinity which raises the supersaturation for dolomite. Other consequences of a sedimentation rate control would predict that the range of oxygen-18 values would be less for a negative-carbon dolomite as is seen in the narrow band of the Tripoli Formation (Fig. 4). This record corresponds closely to near bottom temperatures. Gradients across individual beds are less likely than in the case of more rapid burial which conveys a slowly accreting layer across varied geochemical zones.

CONCLUSIONS

In conclusion, the carbon-isotope composition of anoxic dolomites not only records the type of organic diagenesis occurring in the sediments (sulfate reduction or methanogenesis dominant) but can be used as a rough evaluation of productivity rates in overlying waters. High fertility and burial of organic matter in the Guaymas area moves a slowly nucleating dolomitic layer quickly along the sedimentary conveyor belt into the zone of methanogenesis, whereas lower sedimentation rates as in the Tripoli Formation, allow dolomite formation to continue in a zone of sulfate reduction. End-member values of the carbon-isotope ranges indicate that significant precursor carbonate was lacking in the sediment and are evidence of primary dolomite precipitation in a zone with high diagenetic potential.

This proposed relationship between carbon-isotope ratios and sedimentation rates could be tested on the dolomites of the Monterey Formation. A relationship between positive carbon-isotope values and zones of maximum burial of organic matter, and hence, correlation with petroleum source rock potential of a deep-marine sequence is predicted.

REFERENCES

Baker P. and Kastner, M. 1981. Constraints on the formation of sedimentary dolomite: Science, v. 213, p. 214-216.

Bramlette, W., 1946 The Monterey Formation of California and the Origin of its Siliceous Rocks: U.S. Geol. Survey Prof. Paper 212, 57 p.

Bellance, A., Calderone, S. and Neri, R., 1982/83, Evidenze geochimica e mineralogiche de episodi evaporitici nella sequenza diatomitica (Messiniano "pre-Evaporitico") di Suterna (Sicilia Centrale): Rend. Soc. Ital. Miner. Petrol., v. 38, p. 1271-1280.

Bennet, G. 1980. The sedimentology, diagenesis and paleo-oceanography of diatomites from the Miocene of Sicily: Unpubl. Ph.D. Thesis, Univ. of Durham, England, 223 pp.

Calvert, S., 1966. Origin of diatom-rich varved sediments from the Gulf of California: Jour. Geol., v. 76, p. 546-565.

Claypool, G.E. and Kaplan, I., 1974. The origin and distribution of methane in sediments, in I. Kaplan (ed.) Natural Gases in Marine Sediments: Plenum, New York p. 99-139.

Curray, J., Moore, D., and scientific party; 1982. Initial Reports DSDP, v. 64: U.S. Gov't Print. Office Wash. D.C. pt. 1 and 2, 1313 p.

Decima, A., and Wezel, F.C., 1973. Late Miocene evaporites of the Central Sicilian Basin, Sicily, in W.B.F. Ryan, K.J. Hsu et al. Initial Reports DSDP, v. 13 (2): U.S. Gov't Print Office, Wash. D.C., p. 1234-1239.

Deuser, W.G., 1970. Extreme C-13/C-12 variations in Quaternary dolomites from continental shelf: Earth Planet. Sci. Lett., v. 8, p. 118-124.

Einsele, G., 1982. Mass physical properties of Pliocene to Quaternary sediment in the Gulf of California, Deep Sea Drilling Project Leg 64, in J. Curray, D. Moore, et al., Initial Reports DSDP, v. 64: U.S. Gov't Print. Office, Wash. D.C., p. 529-542.

Friedman, I., and Murata, K.J., 1979. Origin of dolomite in Miocene Monterey Shale and related formation of the Temblor Range, California: Geochim. Coscmochim. Acta, v. 43, p. 1357-1365.

Galimov, E.M. and Simoneit, B.R.T., 1982. Geochemistry of interstitial gases in sedimentary deposits of the Gulf of California, Deep Sea Drilling Project Leg 64, in J. Curray, D. Moore, et al., Initial Reports DSDP, v. 64: U.S. Gov't Print, Office, Wash. D.C., p. 781-789.

Gieskes, J.M., Elderfield, H., Lawrence, J., Johnson, J., Meyers, B., and Campbell.A., 1982. Geochemistry of interstitial waters and sediments, Leg 64, Gulf of California, in J. Curray, D. Moore, et al., Initial Reports DSDP, v. 64: U.S. Gov't Print. Office, Wash. D.C., p. 675-694.

Goldhaber, M., and Kaplan, I.R., 1980. Mechanisms of sulfur incorporation and isotope fractionation during early diagenesis in sediments of the Gulf of California: Marine Chemistry, v. 9, p. 95-144.

Hein, J.R., O'Neil, J.R. and Jones, M.G., 1979. Origin of authigenic carbonates in sediment from the deep Bering Sea: Sedimentology, v. 26, p. 681-705.

Irwin, H., Curtis, C., and Coleman, M., 1977. Isotopic evidence for source of diagenetic carbonates formed during burial of organic rich sediments: Nature, v. 269, p. 209-213.

Kelts, K. and McKenzie, J. 1982. Diagenetic dolomite formation in Quaternary anoxic diatomaceous muds of Deep Sea Drilling Project Leg 64, Gulf of California, in J. Curray, D. Moore, et al. Initial Reports DSDP, v. 64: U.S. Gov't Print. Office, Wash. D.C., p. 553-570.

LeClaire, J. and Kelts, K., 1982. Calcium carbonate and organic carbon stratigraphy of late Quaternary laminated and homogeneous diatom oozes from the Guaymas slope, HPC site 480, Gulf of California, in J. Curray, D. Moore, et al., Initial Reports DSDP, v. 64: U.S. Gov't Print. Office, Wash. D.C., p. 1263-1278.

McKenzie, J.A., Jenkyns, H.C. and Bennet,G.G., 1979/80. Stable isotope study of the cyclic diatomite claystones from the Tripoli Formation, Sicily: a prelude to the Messinian Salinity Crisis: Palaeogeogr. Palaeoclimat. Palaeoecol., v. 29, p. 125-141.

Moore, D.G., 1973. Plate-edge deformation and crustal growth, Gulf of California structural province: Geol. Soc. Amer. Bull., v 84, p. 1883-1906.

Murata, K.J., Friedman, I., and Madsen,B.M., 1969. Isotopic composition of diagenetic carbonates in marine Miocene formations of California and Oregon: U.S. Geol. Surv. Prof. Paper 614-B, 24 p.

Ogniben, L., 1957. Petrografia della serie Solfitera sicilana e considerazioni geologiche relative: Mem. descr. Carta Geol. Ital., v. 35, 275 p.

Pisciotto, K.A., 1981. Review of secondary carbonates in the Monterey Formation, in R.E. Garrison et al. (eds.) The Monterey Formation and Related Siliceous Rocks of California: Pacific Section, Soc. Econ. Paleontol. Mineral. Spec. Publ., p. 273-283.

Pisciotto, K.A. and Mahoney, J.J., 1981. Isotopic survey of diagenetic carbonates, Deep Sea Drilling Project Leg 63, in R. Yeats, B. Haq., et al., Initial Reports DSDP, v. 63: U.S. Gov't Print. Office, Wash. D.C., p. 595-609.

Russell, K.L. et al., 1967. Marine dolomite of unusual isotopic composition. Science, v. 130, p. 1658-1659.

Schrader, H., et al. 1980. Laminated diatomaceous sediments from the Guayamas Basin slope (central Gulf of California): 250,000-year climate record: Science, v. 207, no. 4436, p. 1209.

Schrader, H., 1982. Diatom biostratigraphy and laminated diatomaceous sediments from the Gulf of California, Deep Sea Drilling Project Leg 64, in J. Curray D. Moore, et al., Initial Reports DSDP, v. 64: U.S. Gov't Print. Office, Wash. D.C., p. 973-982.

Simoneit, B., Summerhayes, C.P., and Meyers P.A., 1982. Sources, preservation, and maturation of organic matter in Pliocene and Quaternary sediments of the Gulf of California: a synthesis of organic geochemical studies from Deep Sea Drilling Project Leg 64, in J. Curray, D. Moore, et al., Initial Reports DSDP, v. 64: U.S. Gov't Print. Office, Wash. D.C., p. 939-954.

Schoell, M., 1982. Stable isotopic analyses of interstitial gases in Quaternary sediments from the Gulf of California, in J. Curray, D. Moore, et al., Initial Reports DSDP, v. 64: U.S. Gov't Print. Office, Wash. D.C., p. 815-818.

Wada, H., Niitsuma, N., Nagasawa, K., and Okada, H., 1982. Deep sea carbonate nodules from the Middle America Trench area off Mexico, Deep Sea Drilling Project Leg 66. in J.S. Watkins, J.C. Moore, et al., Initial Reports DSDP, v. 66: U.S. Gov't Print. Office, Wash. D.C., p. 453-474.

DOLOMITES IN ORGANIC-RICH MUDS OF THE PERU FOREARC BASINS: ANALOGUE TO THE MONTEREY FORMATION

LaVerne D. Kulm, Erwin Suess, and Todd M. Thornburg
College of Oceanography
Oregon State University
Corvallis, Oregon 97331

ABSTRACT

Dolomite from the Peru continental margin is intimately associated with Quaternary organic-rich mudstone deposits and their Neogene counterparts. The present-day upwelling circulation associated with the Peru Current and the regional pattern of organic matter and skeletal calcium carbonate in Holocene sediments suggest that basin tectonics exert a specific control on the distribution of sedimentary facies. This relationship persists throughout diagenetic alteration, after burial and dolomitization of sediments. Similar processes and conditions are believed to have played an analogous role in the development of the Monterey Formation.

Organic matter decomposes by bacterial fermentation in the rapidly accumulating carbonate-free mud facies of the subsiding Lima and Progresso basins, whereas in the calcareous mud of the vertically stable Trujillo Basin organic matter undergoes decomposition largely by microbial sulfate reduction. These differences in facies composition and organic matter degradation are recorded in the diagenetic dolomites.

Accordingly, the isotopic signatures and mineral assemblages of the dolomites in the Peru forearc basins display two distinctive regional groupings:

1. Lima Basin and Progresso Basin deposits are characterized by dolomites containing 46-52 mol-% Ca with isotopically heavy $\delta^{13}C$ values, which suggest a chemical environment of organic matter decomposition dominated by bacterial fermentation, while

2. Trujillo Basin deposits are characterized by Ca-rich dolomites containing 52-58 mol-% Ca with isotopically light $\delta^{13}C$ values which suggest decomposition largely by microbial sulfate reduction.

The different tectonic styles which ultimately control these geochemical characteristics in the margin basins are:

1. Rapid subsidence of the Lima Basin, which is attributed to tectonic erosion (i.e., subduction erosion) beneath the forearc, and sedimentation in oxygen-deficient bottom waters that are favorable to the preservation of organic-rich muds;

2. Neogene uplift of the upper slope Trujillo Basin, due to tectonic underplating, followed by vertical stability in an environment that is too agitated and oxygenated to allow organic-rich muds to accumulate; and

3. Impingement of the subducting Nazca Ridge on the Peru margin, which inhibits the structural formation of the West Pisco Basin and restricts the areal and stratigraphic accumulation of organic-rich muds.

During burial, certain stratal zones within the organic-rich mud sequences are pervasively cemented by micritic dolomite to form thin beds of hard, dense carbonate. Chemical and isotopic evidence indicate that a major source of carbonate for this cement is derived from the degradation of organic carbon. Foraminifers may be dissolved from the dolomite groundmass to form fabric-selective, biomoldic porosity.

Finally, tectonic deformation of the Peru forearc has fractured the dolomites to varying degrees, allowing a younger generation of fracture filling carbonate cement. Both moldic and fracture porosity could provide possible reservoirs for the associated organic-rich source rocks. If the indurated carbonates are subsequently exposed at the seafloor, they may be bored by organisms or encrusted with phosphorite.

INTRODUCTION

Dolomites in deep-sea sediments have been reported from many localities of the world's ocean. In continental margin settings, dolomite was recovered from the slopes or rises off Japan, Oregon, Baja California, the northeastern United States, western Africa, southern Africa, the Cariaco Basin on the Venezuelan shelf, and the Gulf of California (Russell, Deffeyes and Fowler, 1967; Deuser, 1970; Berger and von Rad, 1972; Hayes, Pimm and others, 1972; Edgar, Saunders and others, 1973; Arthur, von Huene and Adelseck, 1980; Matsumoto and Iijima, 1980; Okada, 1980; Pisciotto and Mahoney, 1981; Kelts and McKenzie, 1982). Dolomites occur on both convergent and passive margins in water depths ranging from a few hundred meters to several thousand meters; they are Cenomanian to Pleistocene in age. The dolomite recovered in cores and dredges usually occurs as euhedral grains, small nodules, or thin layers and appears to be limited in volume and geographic distribution. The Peru margin dolomites, however, extend for more than 1100 km along the upper slope (Kulm and others, 1981), where at present strong coastal upwelling and margin morphology combine to shape the deposition of a unique organic-rich sedimentary facies.

The association of dolomite with organic-rich terrigenous deposits has been noted in continental margin settings other than Peru. For example, dolomite of the Miocene Monterey Formation, which consists in part of finely laminated siliceous shale

in Garrison, Robert E.; Kastner, Miriam and Zenger, Donald H., eds., 1984, **Dolomites of The Monterey Formation and Other Organic-Rich Units:** Pacific Section S.E.P.M., Vol. 41, p. 29-47.

rich in organic matter, appears to have formed in anoxic basins similar to the Cariaco Basin on the Venezuelan shelf (Friedman and Murata, 1979) or possibly in a coastal upwelling environment similar to that found off Peru (Garrison and Douglas, 1981; Suess and Thiede, 1983). Numerous studies of dolomites and associated terrigenous lithologies in progress on the Monterey Formation and of lithologies recovered during DSDP legs 63 and 64 off Baja California (Pisciotto and Mahoney, 1981; Kelts and McKenzie, 1982), coupled with the results reported here on the Peru margin rocks will show more conclusively the similarities among these depositional environments and the mode of origin of diagenetic dolomites. Basically, the Peru dolomite lithologies, which are interbedded with organic-rich terrigenous sediments deposited beneath the coastal upwelling regime, are formed as a result of diagenetic transformation of organic matter. The occurrence of "organic dolomites" in Peru forearc basins provides a modern analogue to ancient rocks with similar lithological and environmental affinities. Specifically, we wish to address the relationship between basin tectonic history, rate and composition of sediment supply, and the type of organic dolomites that ultimately form off Peru.

PERU FOREARC BASINS

Several interconnecting forearc basins of various dimensions characterize the Peruvian continental margin (Fig. 1; Thornburg and Kulm, 1981). The basins of the continental shelf (Sechura, Salaverry, and East Pisco) are floored by block-faulted Paleozoic or Precambrain metamorphic terrane. A horst-like structural high on the outer shelf trends roughly parallel to the Andean strike and limits the shelf basins seaward. North of 6°S and south of 14°S, the shelf basins rise above sea level to form continental alluvial basins, and the outer shelf high (OSH) emerges to form a low range of coastal hills. The metamorphic pre-Andean massif extends some distance seaward beneath the Trujillo and Lima basins on the upper continental slope where it is believed to interface with the subduction complex of variable width (Kulm and others, 1982).

East of the OSH, the offshore Cenozoic stratigraphy on the continental shelf is unsampled and known only geophysically. However, the shelf basins have been studied where they outcrop subaerially. In the north, the Sechura Basin contains up to 3000 m of post-Eocene, marine to brackish water shale and argillaceous sandstone, with interbedded tuffaceous units (Travis, Gonzalez and Pardo, 1976; Fig. 1). Glauconite, fish remains, diatomite, and phosphate beds are especially prominent in strata of Miocene age, indicating that upwelling and high productivity were important during the development of this basin. In southern Peru, Miocene-age strate of the East Pisco Basin are similar, and characterized by diatomite, fossiliferous shale, and phosphate beds (Travis, Gonzales and Pardo, 1976). High productivity waters have influenced Peru margin sedimentation not only at present, but also well into the Neogene.

This study is mostly concerned with sedimentation in the upper slope Lima and Trujillo basins. Diagenetic dolomites recovered from these basins suggest that organic-rich deposits, associated with upwelling, have accumulated during Quaternary, Neogene, and possibly Paleogene times on the upper continental slope.

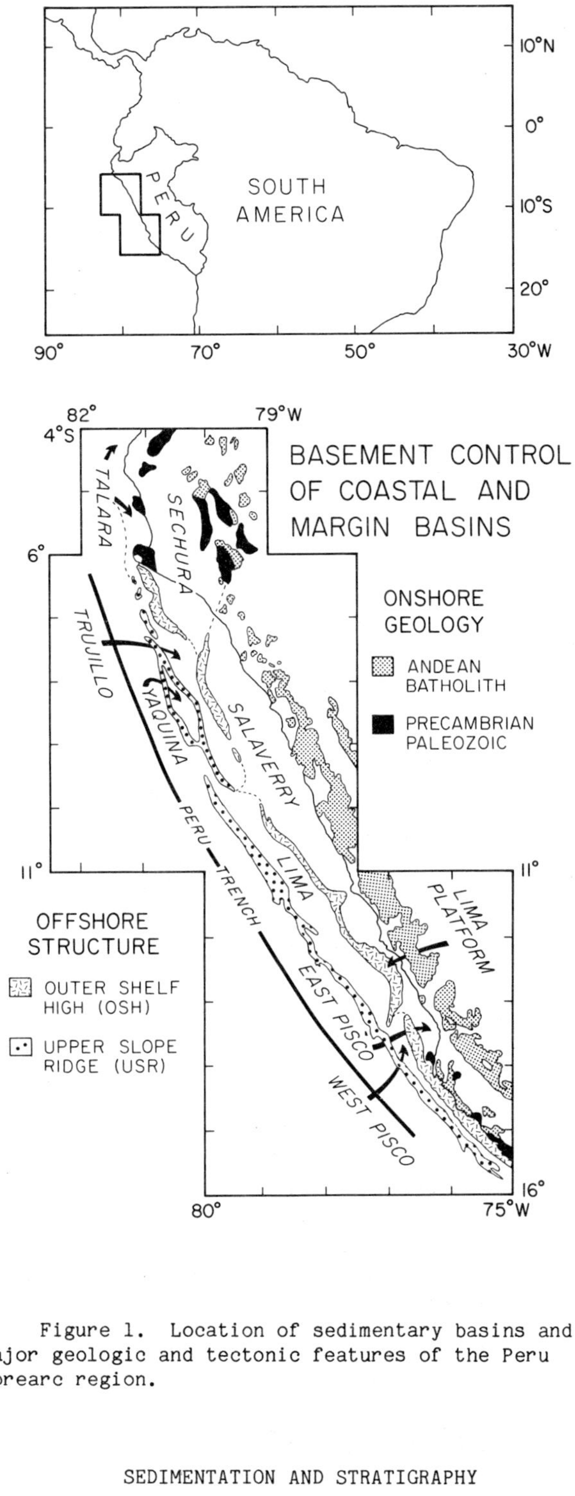

Figure 1. Location of sedimentary basins and major geologic and tectonic features of the Peru forearc region.

SEDIMENTATION AND STRATIGRAPHY

Holocene Sedimentary Facies

Organic-Rich Mud Facies

Coastal upwelling off Peru generates some of the highest biological production measured in coastal waters of the world (Zuta and Guillen, 1970). The present-day centers of productivity are located between latitudes 7-8°S, 11-12°S, and 14-16°S (Reimers, 1982). The organic-rich mud facies (>>5 % C-org) is especially prominent in the upper slope mud lens that covers the landward flank of the Lima Basin (Fig. 2). The mud lens extends from shelf depths of 50 m down

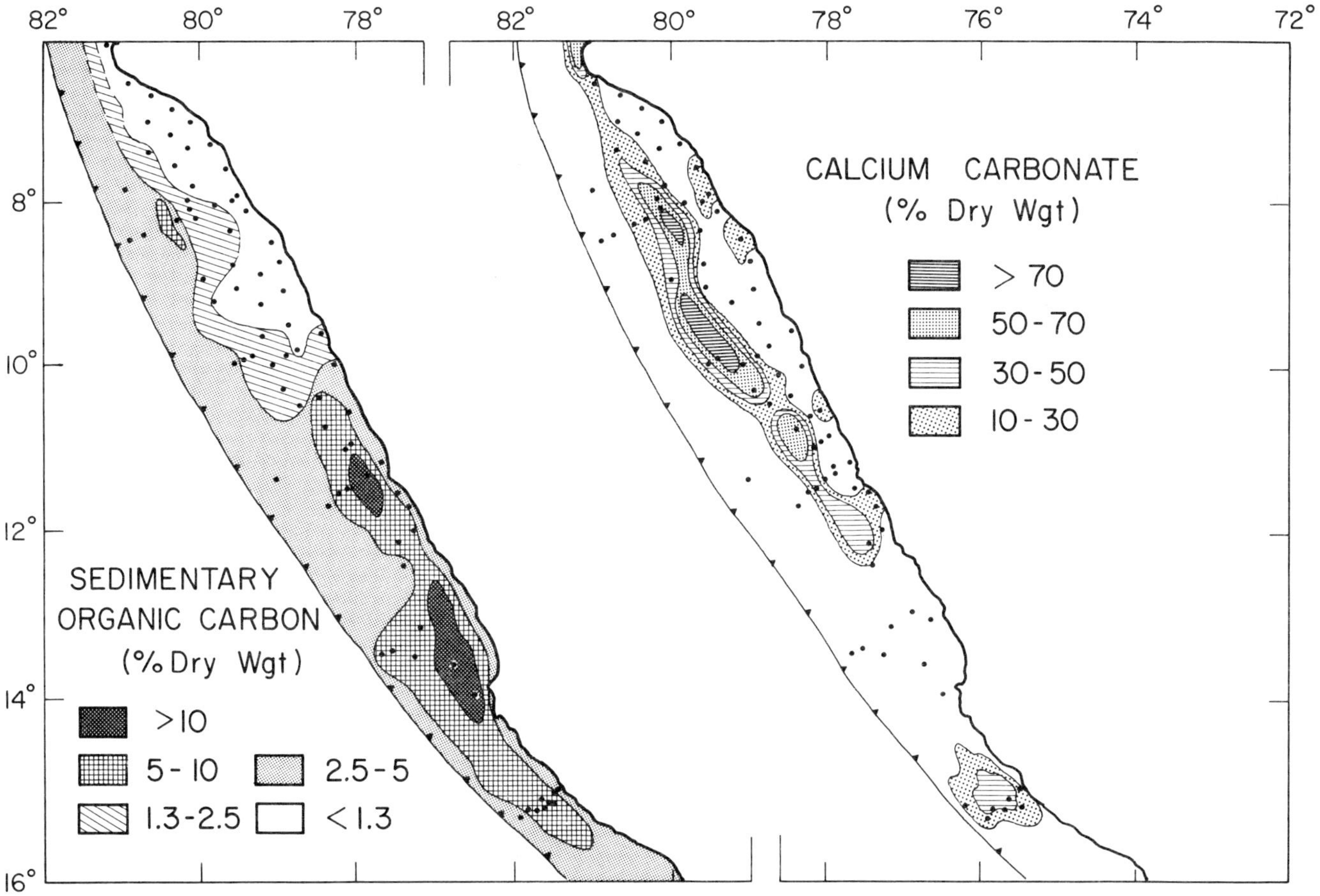

Figure 2. Left: Distribution of organic-rich mud facies as reflected by the organic carbon content of surface sediments. Maximum concentrations fall on the landward flank of the Lima Basin on the upper continental slope. Right: Distribution of biogenic sand/calcareous mud facies as shown by the calcium carbonate content of the surface sediments. Maximum concentrations are coincident with the axial trend of the Trujillo Basin on the outer shelf and upper slope. Data from Reimers and Suess (1983) and Suess, Kulm and Killingley (in press).

to slope depths of 680 m, pinching out in both landward and seaward directions. The depositional axis is located at water depths of from 160 to 370 m, where as much as 38 m of Quaternary sediment has accumulated.

The organic-rich mud lens contains a virtually complete record of all the components derived from this classic upwelling regime. It consists chiefly of terrigenous silt and clay, diatoms (opal), and organic carbon with subordinate consistuents such as calcium carbonate, phosphorite, glauconite, and fish debris. Opaline silica and organic carbon may contribute as much as 40% by weight of the total mud deposit, and terrigenous detrital material a similar amount. However, because of the low specific gravity of the fibrous organic matter and porous opaline silica tests, the sediment consists largely of biogenic material by volume (Scheidegger and Krissek, 1983), producing a siliceous mud facies.

The mud lens is characterized by >60% clay-sized material which produces an anomalously fine-grained sediment compared to the hemipelagic facies found farther seaward on the adjacent slope. This fine-grained material is contained in abundant fecal pellets and aggregates of clay flocs produced from ingestion by zooplankton and nekton living in the overlying water column (Krissek, Scheidegger and Kulm, 1980; Scheidegger and Krissek, 1983; Reimers and Suess, 1983). These pellets of clay-sized particles and organic matter fall to the sea floor at faster settling rates than unaggregated clay particles.

Benthic foraminifera that inhabit the organic-rich mud facies include Bolivina humilus nudi, which is typically found in the oxygen minimum zone associated with coastal upwelling. Off Peru, the oxygen minimum (<0.3 ml/ℓ) zone is centered at about 400 m water depth in the north (5°S), and becomes shallower (100 m) and more intense toward the south (15°S). Uvigernia species, which are more typical of higher contents of oxygen, also occur in these sediments indicating that the bottom waters undergo fluctuating oxygen conditions even where the organic-rich sediments are accumulating (Reimers, 1982).

Biogenic Sand Facies

A marked angular unconformity occurs in the shallow subsurface of the Trujillo Basin, apparently separating Quaternary from underlying Neogene age deposits. It is especially prominent between 8 and 10.5°S latitude. The biogenic sand facies is best developed above this erosional surface on the outer continental shelf and upper slope.

The biogenic sand facies is a residual lag deposit characterized by a high bulk calcium carbonate content (Fig. 2). This facies consists mainly of benthic foraminifera, fish debris, phosphorite, minor glauconite, and relatively low contents of organic carbon (<3.3%; Fig. 2). These deposits are largely sand-sized material, well sorted, somewhat oxidized, and may be partially cemented by phosphorite and glauconite. They are similar to relict deposits found on other continental margins where there is little or no modern sedimentation. Seaward of these relict materials (>500 m water depth), fine-grained sediments accumulate with slightly higher organic carbon content. However, the organic-rich mud facies is patchy and more poorly developed than in the Lima Basin to the south.

These well-sorted biogenic sands are apparently produced by the winnowing of silts and clays from pre-existing organic-rich hemipelagic sediments by uni-directional bottom currents (contour currents) that sweep the outer continental shelf and upper slope. Benthic foraminifera that inhabit these sands include the *Uvigerina* species that are characteristic of an oxygenated environment, indicating that the oxygen minimum zone is not so well developed and persistent in this calcareous facies as in the organic-rich mud facies. During normal upwelling conditions, a strong poleward flowing undercurrent reaches its highest velocities (2 to 24 cm/sec) in the north from 5 to 10°S, and lowest velocities (0.6 to 8.0 cm/sec) in the south (15°S) at water depths of about 50 to 150 m. Current velocities are also stronger to the north than to the south at deeper water depths (200 to 800 m). During El Niño events, the undercurrent velocities increased dramatically up to 36 cm/sec at 10°S (Smith, 1983) where the biogenic sand facies is most prominent. These accelerated currents averaged from 25 to 30 cm/sec over an 8- to 64-day period, which is ample time to winnow the underlying deposits.

No modern surface depositional environment hypersaline or supratidal has been found on the Peru margin that is conducive to the formation of dense micritic dolomites, but these and related carbonate lithologies are common in the Pleistocene and Neogene deposits of the upper slope basins. It is apparent that the dolomite facies is strictly a subsurface, diagenetic phenomenon that results from pervasive, isochemical, carbonate cementation of an initial terrigenous/biogenic mud.

Cenozoic Carbonates

Pleistocene and Neogene Carbonates

Carbonate rock, together with phosphorite, calcareous siltstone, siltstone, claystone, and glauconite, have been recovered by dredges from several rock outcrops along more than 1000 km of the Peru continental margin between 3°30'S and 13°37'S latitude (Figs. 3,4; Kulm and others, 1981). These lithologies are middle Miocene to middle Pleistocene in age and occur in water depths ranging from 200 to 3800 m. The carbonates are classified as dolomicrite, brecciated dolomicrite, micrite, and glauconitic micrite. The bulk of the samples were obtained from strata exposed in upper slope forearc basins. Both undisturbed and strongly faulted sections were sampled. These older deposits are overlain by the Quaternary organic-rich muds or residual biogenic sands, as described above, or they may crop out directly on the sea floor.

Lima Basin

The Pliocene to middle Pleistocene deposits dredged from the Lima Basin consist chiefly of dolomicrite and calcareous siltstone with minor micrite and brecciated dolomicrite samples (Fig. 3). Calcareous siltstone is the most weakly indurated, while the dolomicrite is the most strongly indurated lithology (Kulm and others, 1981). The least indurated calcareous siltstones have a rather high organic carbon content (2-9%) compared to most indurated micrite and dolomicrite (<1-2%). Foraminifers, diatoms, and radiolarians occur in all lithologies, with diatoms typical of coastal upwelling being especially prominent in the less indurated and in some indurated deposits (Kulm and others, 1981). In some barren lithified samples, there are molds of diatoms and foraminifers indicating dissolution after lithification by carbonate cementation. Fish debris and microfossils comprise as much as 11% of some dolomicrites and much higher percentages of the calcareous siltstone and siltstone. The abundant diatoms and fish scales indicate that primary productivity played a similarly important role when these sediments were deposited as it does today. The terrigenous components of these rocks consist of quartz, feldspar, and heavy minerals. All of the biogenic and terrigenous constituents found in the dolomicrites and associated terrigenous siltstones are the same as those found in the overlying Holocene organic-rich mud facies.

Benthic foraminifera typically found in the oxygen minimum zone of the organic-rich mud lens (species of the *Bolivina* assemblage) also occur in the Neogene siltstones and dolomicrites of the Lima Basin, indicating a similar depositional environment. In general, benthic foraminiferal assemblages of the Pliocene and Pleistocene rocks closely resemble the modern assemblages from the outermost continental shelf and uppermost slope between water depths of 100 to 300 m (Kulm and others, 1981; Resig, 1981). Because the dolomites were dredged from depths ranging from 837 to 2263 m and paleodepths indicate a much shallower depositional environment, subsidence rates from 275 to 500 m/m.y. are calculated for the Lima Basin during the past 4 m.y. (Fig. 5; Kulm and others, 1981). The highest rates of subsidence occurred during the Pleistocene.

From a stratigraphic point of view, these Pliocene to middle Pleistocene dolomites were recovered from the upper 1.3 km of a 2.0-km thick stratigraphic section in the Lima Basin (Fig. 5). The Pleistocene portion of the section is estimated to be between 400 and 500 m thick based upon multichannel seismic reflection records and assuming that the DR-series (DRAG) rocks are representive of this section (Fig. 3). Diatom biostratigraphy shows that the youngest dolomicrites are middle Pleistocene (about 1.0 m.y.) in age. While there is no direct evidence for the abundance or stratigraphic distribution of dolomite in the upper slope basins, a velocity analysis of multichannel seismic data in the Lima Basin compared to carbonate velocities of the dredge samples shows that dolomite lithologies comprise no more than 10% and more likley 1-2% of the entire section. For example, the composite velocity of the Lima Basin section is 2.0 km/s, but our dredged dolomicrites yielded compressional velocities of from 4.5 to 6.6 km/s. An analysis of downhole sonic logs from industry wells suggests that a similar situation exists in the Oligocene of the Trujillo Basin, where dense carbonate lithologies occur as thin localized beds.

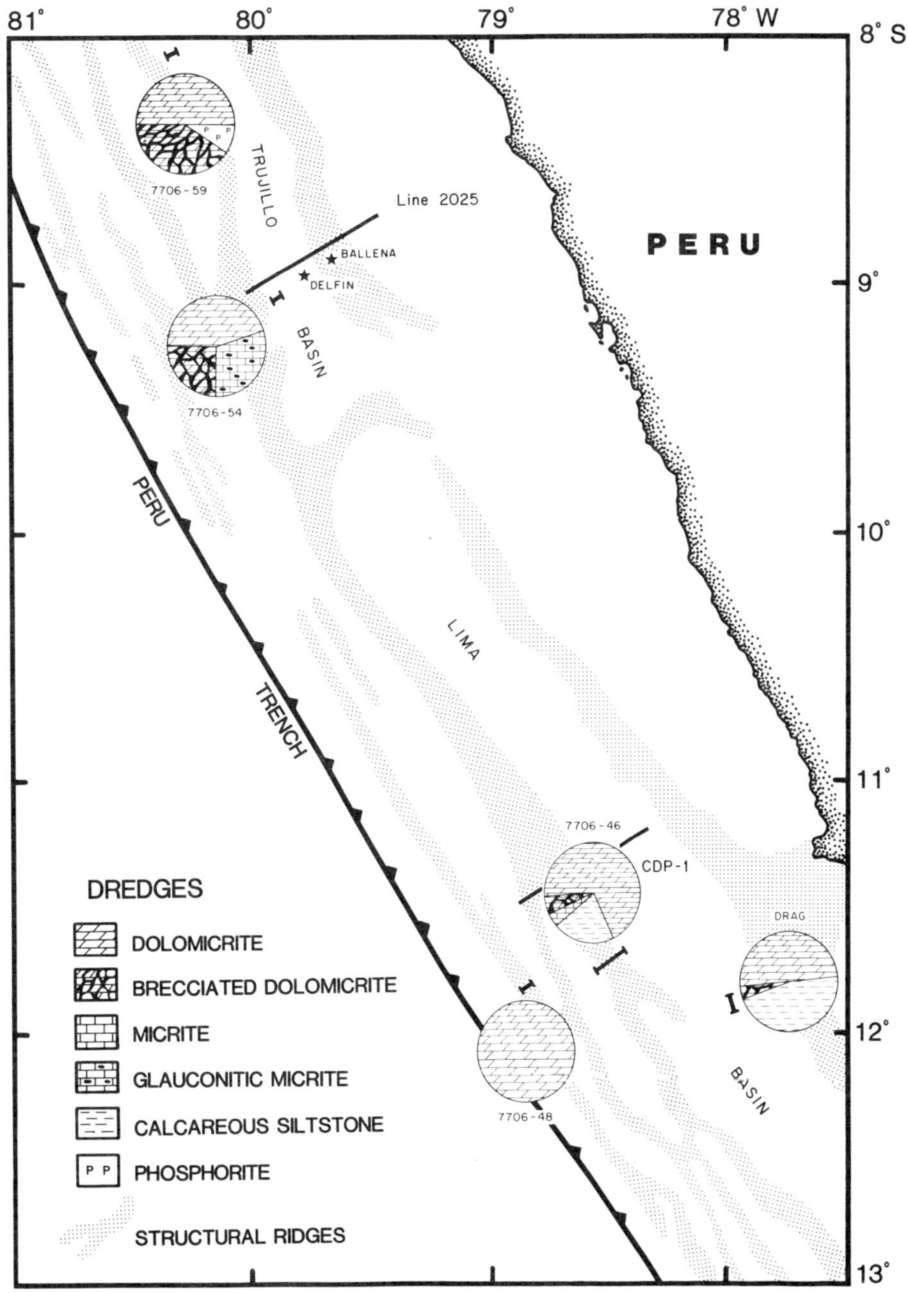

Figure 3. Relative abundance of Neogene and Quaternary carbonate lithologies dredged from the Lima and Trujillo basins. Note locations of multichannel seismic lines 2025 and CDP-1 (Figs. 5,6).

The dredged dolomites occur as slabs, concretions or irregular blocks, providing further evidence for their stratigraphic localization. However, the dredged brecciated dolomicrites can be large, massively cemented and highly indurated blocks more than one meter in diameter.

Trujillo Basin

Carbonate rocks dredged from the Trujillo Basin are characterized by dolomicrite, brecciated dolomicrite, glauconitic micrite, micrite, and phosphorite (Fig. 3). Late Miocene to Pleistocene dolomicrites dominate the dredged lithologies and no terrigenous lithologies were recovered. Phosphorite occurs as nodules, in a carbonate matrix, or as a veneer over other lithologies. All lithologies are indurated to highly indurated due to pervasive carbonate cementation. The carbonates contain less than 1% organic carbon. Microfossils are present, but they are generally scarce and cemented in the matrix or have been dissolved and occur only as molds within the indurated rock.

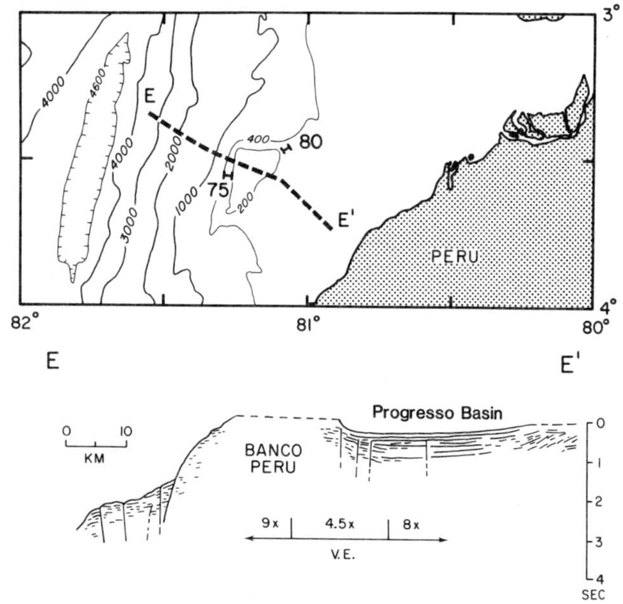

Figure 4. Location of dredge samples (top) and single-channel seismic section over Banco del Peru; all lithologies dredged from the bank are dolomicrite.

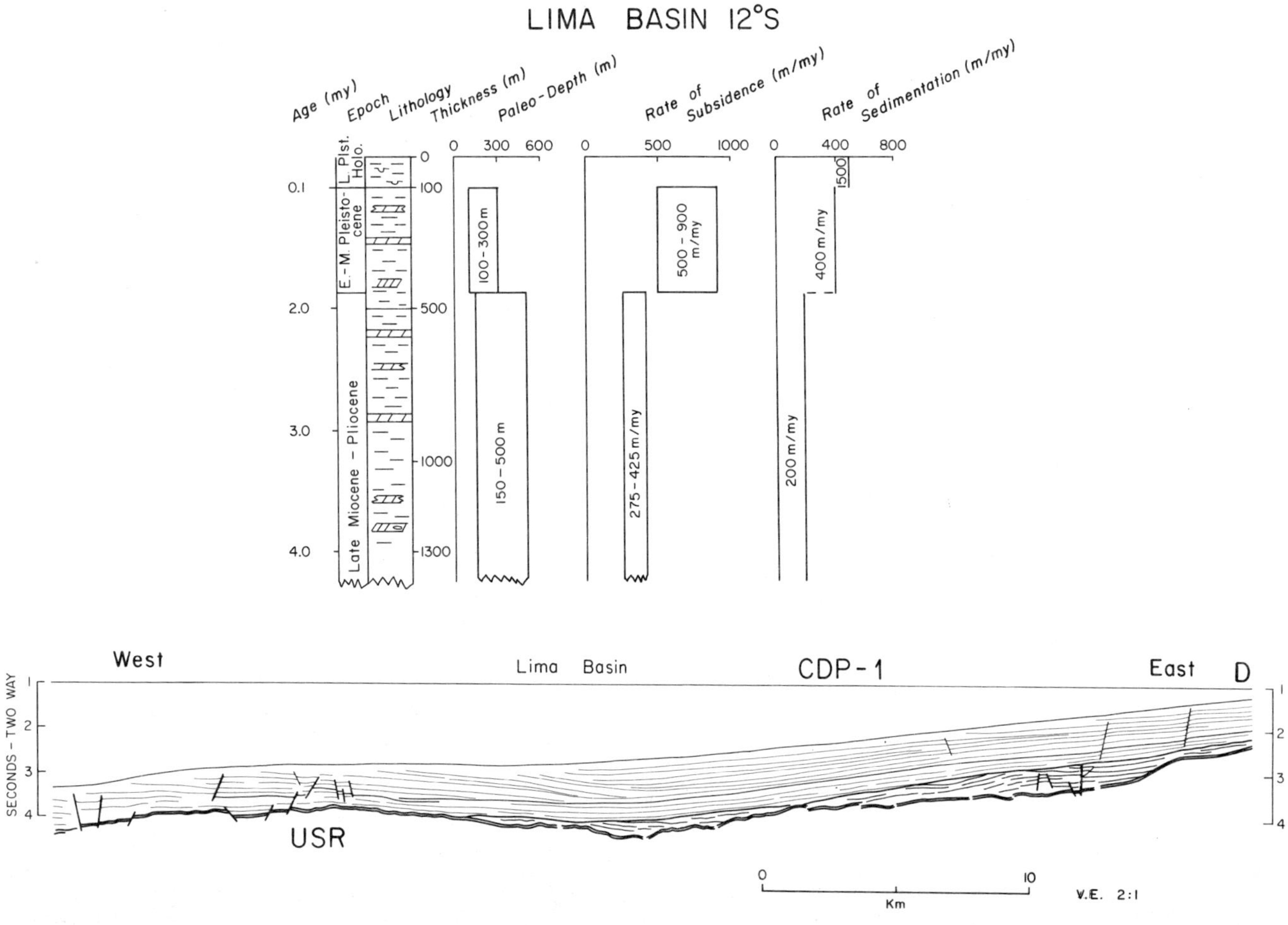

Figure 5. Top: Generalized stratigraphic section for Lima Basin constructed from dredged lithologies (this study). Bottom: Multichannel seismic section (Thornburg, in press). Rates of subsidence calculated from paleodepth information, age data, and stratigraphic thicknesses in a multichannel depth section. See Figure 3 for location of seismic section CDP-1.

Glauconitic micrite is concentrated in the rocks of dredge 7706-54 and phosphorite in dredge 7706-59 (Fig. 3). Glauconitic rocks contain 30-50% angular to rounded glauconite which is green to yellow-green in color. The grains are up to 2 mm in diameter and are often enclosed by coarse carbonate cement consisting primarily of calcite with subordinate dolomite that occasionally replaces the glauconite (Kulm and others, 1981). Pyrite is a common inclusion in the glauconite. Collophane may comprise more than 4% of the rock. Some glauconitic micrites contain pebble and sand-sized phosphatic material, giving the appearance of a conglomerate. Terrigenous minerals are generally minor constituents of the rocks.

Benthic foraminifers indicate that the Neogene-Quaternary Trujillo rocks were deposited in outer shelf to upper slope water depths (upper bathyal), identical to the depths from which the rocks were dredged. Little vertical movement is thus implied for the Trujillo Basin since late Miocene time (Fig. 6). The large amounts of authigenic material (glauconite and phosphorite) and minor terrigenous material found in these dredged rocks indicate deposition in an environment that receives little detrital material and where input from coastal upwelling is prevalent. Phosphatic crusts, oölitic coatings of phosphate, rounded glauconite grains, and sparry calcite suggest agitation and current activity in the depositional environment. All lines of evidence imply a close analogy with the modern day setting in the region where the dredge materials were recovered (8-10°S).

The residual biogenic sand facies covers the indurated carbonates (Figs. 2,3) and is itself partly cemented. Apparently the region has been characterized by a similar depositional environment from late Miocene to Holocene time. Thinly layered and laminated dolomicrites occur with the glauconitic micrites and phosphorites, suggesting periods of higher rates of sedimentation than those associated with the authigenic mineral formation and residual biogenic sand accumulation. However, microfossils (largely foraminifers) are concentrated in thin layers in the dolomicrites. These were derived from the muddy groundmass, which suggests winnowing of contemporaneous organic-rich muds to form a lag deposit similar to the residual biogenic sand. Variability in the subsidence, sediment supply, or contour current activity could produce biogenic sand layers of varying thickness, as seen in the dolomicrites. The present biogenic sand facies is a few centimeters to a few tens of centimeters thick, suggesting a somewhat longer duration or higher intensity of winnowing than seen in the older rocks obtained in the region (Suess, 1981; Reimers and Suess, 1983). Alternatively, it may be that the organic-rich mud facies is preferentially cemented to form dolomicrite, while the biogenic sand facies remains less indurated, less resistant in outcrop, and less susceptible to dredge sampling.

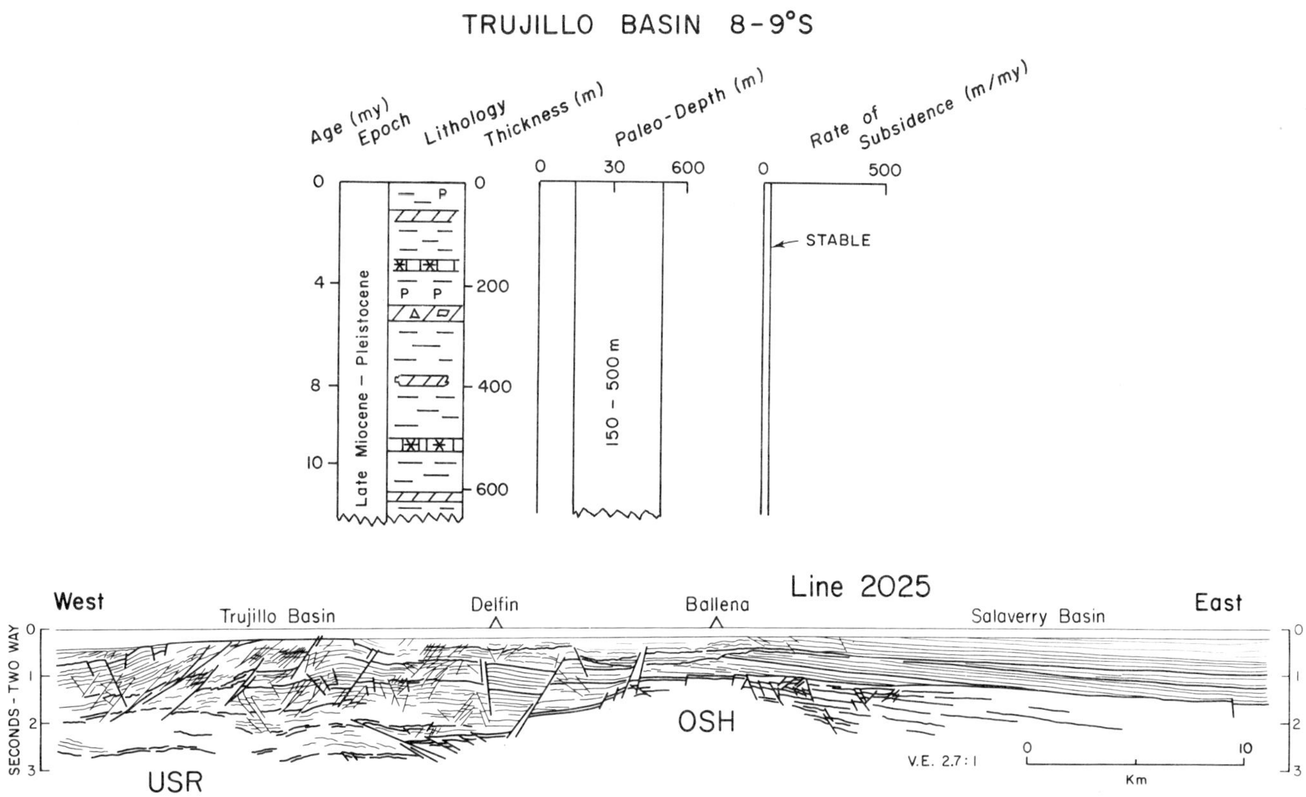

Figure 6. Top: Generalized stratigraphic section (this study) for neogene and Quaternary portion of Trujillo Basin. Bottom: Multichannel seismic section (Thornburg, in press). See Figure 5 for explanation. Note location of Delfin and Ballena wells; see also Figure 3 for location of wells in relationship to dredge samples. Outer Shelf High (OSH) and Upper Slope Ridge (USR) are indicated (Fig. 1).

Banco del Peru

Banco del Peru is an elevated horst-like feature situated near the boundary between the Talara Basin to the south and the Progresso Basin to the north (Fig. 4). Carbonate lithologies dredged from rock outcrops at two sites on the bank consist of massive, medium to dark gray dolomicrite of uniform composition. These indurated lithologies apparently range in age from middle to late Miocene based upon strontium isotopic ratios, but may be Pliocene to Holocene in age based upon microfossils (Kulm and others, 1981). No paleoenvironmental or paleodepth information are currently available for these carbonates. The proximity of Banco del Peru to a structural basin containing more than 5.5 km of sediment immediately to the east of the bank (Travis, Gonzales and Pardo, 1976) suggests that carbonate lithologies may be present in this basin.

Paleogene Lithologies

Upper Eocene to Oligocene terrigenous marine deposits with minor carbonate were recovered in two industry drill holes that penetrated Paleozoic or Precambrian metamorphic basement in the Trujillo Basin (Figs. 3,6; Thornburg and Kulm, 1981; Kulm, Thornburg and Dang, in press; Thornburg, in press). These older, intensely faulted deposits underlie the late Cenozoic lithologies described above. The Delfin well, situated on the landward hinge of the basin, recovered a thin basal sandstone with about 300 m of overlying Upper Eocene fossiliferous siltstone and claystone followed by 750 m of claystone, siltstone, limestone and calcareous shale (Thornburg and Kulm, 1981). Higher up in the section very hard, dense microcrystalline dolomites and dolomitic limestones are commonly interbedded with the claystones and siltstones. These Paleogene carbonates may be comparable in origin to those recovered in the dredges from the seaward margin of the Trujillo Basin on the upper slope. Graded sandstones and siltstones (probably turbidites) become common upsection, indicating a possible upper slope environment with access to terrigenous sources during middle to late Oligocene time (Thornburg and Kulm, 1981; Kulm, Thornburg and Dang, in press). These lithologies are capped by fine- to coarse-grained sand, the youngest unit drilled. Farther landward, the Ballena well recovered younger deposits consisting of a thin basal sand and an Oligocene claystone, siltstone, and sandstone sequence; graded beds and dolomitic limestone may be present as well. This section is capped by fine to coarse sand similar to the Delfin well (Thornburg and Kulm, 1981; Kulm, Thornburg and Dang, in press). The Miocene and younger lithologies obtained in the dredges from the Trujillo Basin were not recovered in these two wells. Truncated seismic reflectors on the sea floor in the vicinity of the wells show that a portion of the younger deposits have been stripped from the section by subaerial erosion (Fig. 6). The coarse sands at the top of the Paleogene section in both wells also signify pronounced uplift of the region following Oligocene turbidite deposition.

MINERALOGY, CHEMISTRY AND STABLE ISOTOPE COMPOSITIONS OF CARBONATES

Methods

The major lithologies obtained from the three basins along the Peru continental margin are dolomicrite, brecciated dolomicrite, micrite, laminated micrite, phosphorite, calcareous siltstone, siltstone, and claystone (Fig. 3) as described above. Two typical lithologies and rock fabrics from each dredge haul are illustrated in Figures 7 and 8. For the analyses discernible rock textures were subsampled. Thus, where possible, groundmass was separated from different generations of cement, and sometimes variations in texture and porosity of the groundmass allowed even further subdivision of the samples.

The intention was to obtain homogenous phases of dolomite, calcite, and apatite for isotopic, mineralogical, and chemical analyses. The mineralogy of each subsample was qualitatively ascertained by standard X-ray diffraction technique (Chave, 1952; Murata, Friedman and Madsen, 1969) and quantified from analyses of the soluble constituents (Suess Killingley and Kulm, in prep.). The method of estimating mineralogy is briefly outlined in the next section but is described in detail by these authors.

The carbonate and phosphate phases were dissolved in acid. They comprised 85-100% of the rocks and only where glauconite was present did the insoluble fraction constitute more than 15% (for representative results see Table 1). The leachate was used to determine phosphate colorimetrically and calcium, magnesium, and iron by standard atomic absorption spectroscopy. Carbonate and organic carbon were determined on separate sample splits by thermal conductivity detection of carbon dioxide after acid digestion and dichromate/sulfuric acid oxidation, respectively (Weliky, Suess and Ungerer, 1983). Oxygen and carbon isotope mass ratios were obtained by J. S. Killingley at Scripps Institution of Oceanography, La Jolla. All analyses of the selected data set reported here are with reference to carbon dioxide from a calcite standard relative to PDB. The oxygen isotope ratios were not corrected for the fractionation between dolomite and calcite undergoing acid decomposition as is sometimes customary (Pisciotto, 1981).

Table 1 shows the estimated mineralogy of the selected eight rock samples illustrated in Figures 7 and 8 and, in addition, the results of two rocks dredged from Banco del Peru (7706-75-1 and 7706-80-9) which represent the deposits from the Progresso Basin. Four of the rocks are from the Lima Basin (7706-46-3, -91 and DR-9, -14) and four are from the Trujillo Basin (7706-54-5, -6 and 7706-59-11, -28A). Subsequent figures illustrating mineralogy and isotopic characteristics, are based on a total of 92 analyses from all 65 samples reported by Suess, Killingley and Kulm (in prep.).

Carbonate Mineralogy

Several episodes of dolomite formation can usually be discerned in the sedimentary rock fabric (Figs. 7,8). During the first stage, the groundmass is lithified by dolomite cement preserving much of the original sedimentary structure. This is followed by one or more events of dolomite precipitation evident in fillings of fractures of brecciated rocks or along bedding planes. These later vein and fracture fillings sometimes consist of magnesian calcite but usually also contain dolomite. The final precipitate around most dredged rock fragments is a veneer of apatite or fillings of bore holes by apatite-cemented glauconitic sands. This most recent event is not related to the early diagenesis of continental margin facies but rather results from exposure of the dredged rock fragments to the present bottom waters.

Table 1. Stable isotope and chemical and mineralogical characteristics of representative dolomite rocks from Lima Basin (46- and DR-series), Trujillo Basin (54- and 59-series) and Progresso Basin (75- and 80-series). $\delta^{18}O$ values (by J.S. Killingley, SIO, La Jolla) are with respect to PDB calcite standard and are not corrected for fractionation between dolomite and calcite; insoluble residues are calculated from the sum of the soluble constituents and represent a maximum estimate because of compounded analytical undertainties, soluble constituents not analyzed for, and sorbed/occluded gases and water; organic carbon was determined by wet oxidation after removal of carbonate carbon (Weliky and others, 1983); mineralogies were calculated from total soluble constituents and dolomite, calcite, and apatite compositions; Ca in dolomite and Mg in calcite were estimated from XRD analyses. A brief outline of the procedure is presented in the text and Table 2, original data and details are in Suess, Killingley and Kulm (in prep.).

		Stable Isotopes		Residue	Organics	Mineralogy				
Sample	Designation	Oxygen $\delta^{18}O$ (‰PDB)	Carbon $\delta^{13}C$	Insol. (wt-%)	C-org (wt-%)	Dolomite (wt-%)	Calcite (wt-%)	Apatite (wt-%)	Ca in Dolomite (mol-%)	Mg in Calcite (mol-%)
46-3	groundmass	5.77±.02	13.15±.18	4	0.31	81	14	1	47.5	13.4
	cement									
	gray	5.54±.22	17.29±.41	4	0.31	84	12	0	47.5	12.7
	white	6.10±.20	17.20±.10	5	0.42	78	14	0	47.7	14.5
46-91	groundmass									
	porous	5.09±.20	3.92±.28	10	0.50	80	9	1	50.5	8.2
	dense	4.78±.07	3.52±.04	11	0.51	78	10	1	49.0	10.4
54-5	groundmass									
	lamin.	5.53±.02	-12.40±.02	6	1.05	86	5	2	54.5	6.3
	homog.	5.47±.18	-13.30±.30	7	(1.15)	88	8	0	54.5	7.0
	cement	5.76±.19	-14.39±.52	1	0.10	88	11	0	53.8	7.0
54-6	groundmass									
	porous	6.44	-12.47	7	0.62	72	19	2	54.3	8.9
	dense	6.28	-10.55	9	0.81	68	20	2	53.0	8.9
59-1	groundmass									
	homog.	5.17±.09	-8.83±.05	47*	0.11	49	3	0	55.8	4.4
	pebbly	2.47±.33	-2.61±.60	24*	1.16	3	11	63	50	0
	cement									
	white	5.23±.01	-7.61±.60	0	0.12	86	13	0	55.5	5.2
	gray	5.37±.13	-13.41±.07	0	0.09	92	8	0	54.8	6.3
59-28A	groundmass	5.64±.67	-7.73±.05	2	0.66	91	1	0	55.3	7.0
	cement									
	homog.	5.62±.09	-8.04±.10	0	0.12	93	8	0	55.8	5.2
	pebbly	1.11±.07	-2.72±.10	14	1.00	0	7	71	--	1.4
D-2	groundmass									
	porous	5.74	12.57	11	0.68	80	8	0	53.0	7.0
	dense	5.84	10.57	14	1.10	77	8	0	52.5	7.0
D-9	groundmass	6.59	10.14	14	0.66	80	4	0	51.3	7.0
75-1	groundmass	7.00±.10	8.02±.06	21†	1.49	64	11	4	48.0	7.0
80-9	groundmass	5.92	1.82	25†	2.33	55	17	1	49.5	7.0

*Glauconite
†Clay and sulfides

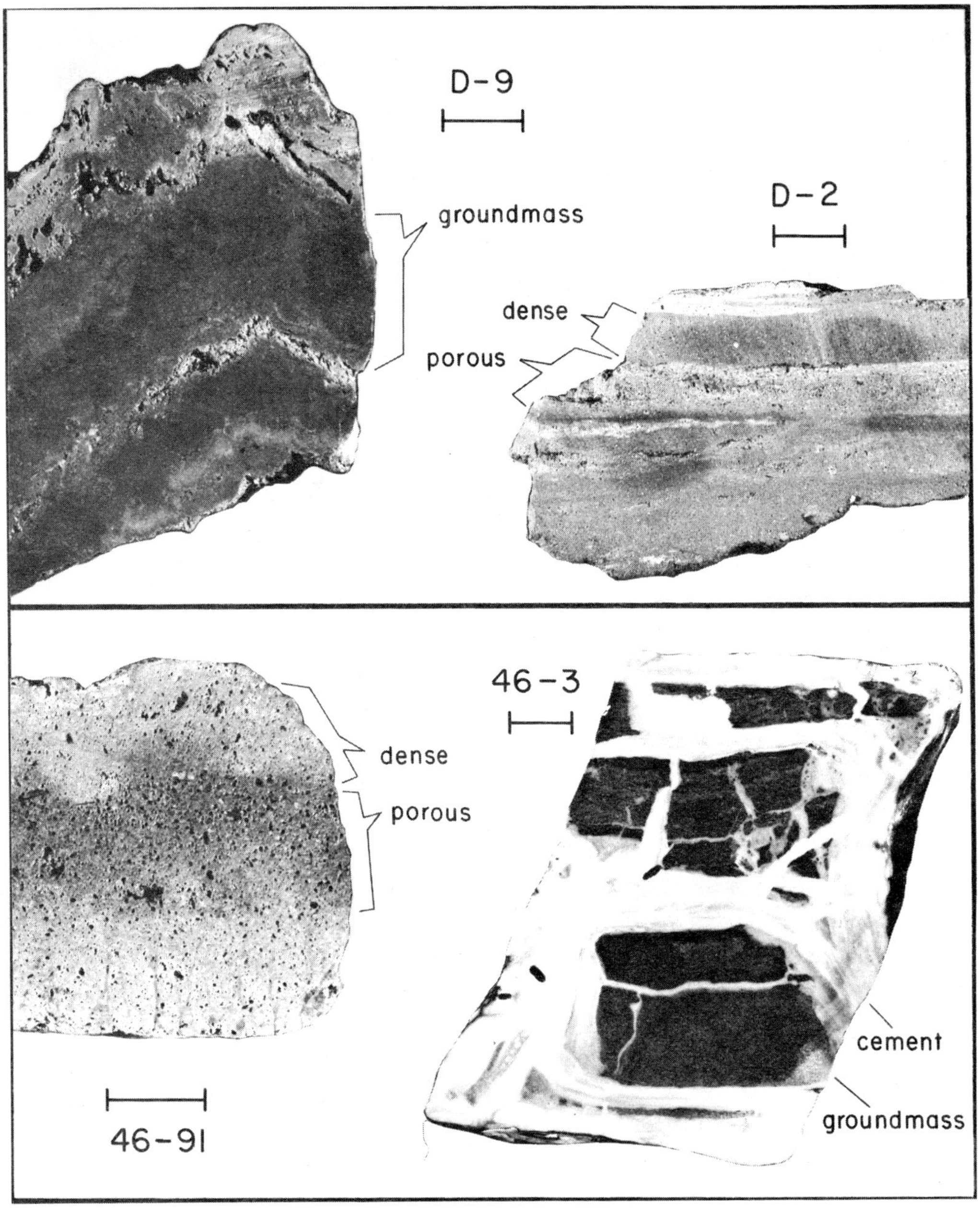

Figure 7. Polished sections of typical dolomite rock fabrics from the Lima Basin (DR-series and 7706-46 dredge series) showing porous and dense groundmass and different generations of cement, which were isolated and separately analyzed. For results see Tables 1 and 2. DR-9 = dense dolomicrite; DR-2 = laminated dolomicrite; 46-91 = dolomicrite with molds of foraminifers; 46-3 = brecciated dolomicrite with two generations of cement. Scale = 1 cm.

The dolomite-cemented groundmass and different generations of fracture fillings vary in chemical composition from 46-58 mol-% calcium with a distinct regional grouping (Fig. 9). Both calcium-deficient and calcium-enriched dolomites were recovered in the Lima Basin dredge hauls (7706-46-series and DR-series) and off Banco del Peru (Progresso Basin; 7706-80 and 7706-75), and the most calcium-enriched dolomite mineral phases characterize the Trujillo Basin (7706-54 and 7706-59). In both Ca-enriched and Ca-deficient dolomites, the combined Ca+Mg provide >99 mol-% of all cations. Thus, there appears to be negligible substitution of Fe (ankerite) into the carbonate lattice. This is inferred not only from X-ray diffraction techniques, but also from analysis of the acid-soluble cations.

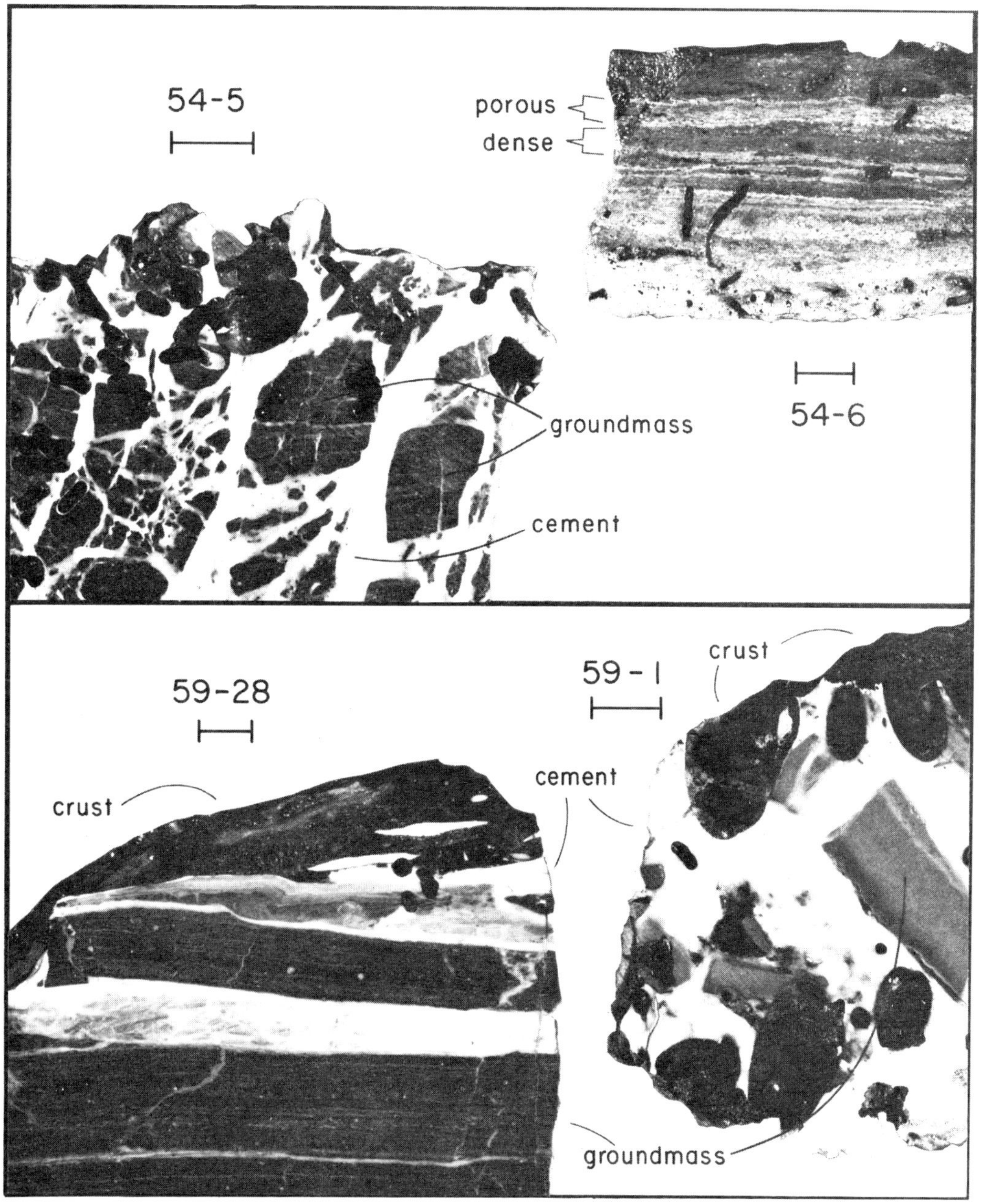

Figure 8. Polished sections of typical dolomite rock fabrics from the Trujillo Basin (7706-54 and 7706-59 dredge series); groundmass and cement as in Figure 7; in addition apatitic crusts either as veneer or fillings of borings are indicated. For results see Table 1. 54-5 = brecciated dolomicrite; 54-6 = finely laminated dolomicrite; 59-28A = finely laminated dolomicrite with sparry cement and apatite veneer; 59-1 = conglomeratic dolomicrite with angular groundmass fragments, sparry cement, and apatite-cemented, glauconite-filled borings. Scale = 1 cm.

Estimation of the mineralogy is based on partitioning the total soluble fraction (Table 2, sample 46-3) between fluorapatite

$$Ca_{10}(PO_4)_{5.5}(CO_3)_{0.5}F_{2.5};$$

(Suess, Killingley and Kulm, in prep.), magnesian calcite (13 mol-% $MgCO_3$ from XRD) and Ca-deficient dolomite (47 mol-% $CaCO_3$ from XRD). The total soluble fraction represents to within 1% the carbonates and phosphates as indicated by the anion-cation balance.

The rather low soluble iron in the leachate appears to be related to the insoluble residue fraction (Fig. 10) and not to the carbonates. Moreover, the low contents of <2000 ppm iron in general are insufficient to compensate for the Ca-deficiency of some of the dolomites, particularly as illustrated for sample 46-3 (Table 2). The subsitution of excess Mg over Ca in the dolomite lattice appears to be an unusual and diagnostic feature of some of the Lima Basin dolomites and awaits confirmation by microprobe analyses.

Table 2. Acid soluble constituents from three subsamples of Lima Basin rocks. Data sets of this kind were used to ascertain the possible Fe content of dolomites and estimate the total rock mineralogy of each sample as listed in Table 1. The estimate involves (1) cation-anion balance to evaluate whether or not only carbonate and phosphates were dissolved and Fe is a substitute in the carbonate lattice; (2) calcite and dolomite characterization by XRD to obtain respective mol-% $MgCO_3$ and $CaCO_3$, see also Table 1; (3) partitioning of soluble constituents between dolomite, calcite and apatite using three simultaneous equations; (4) recalculating mineral estimates to include residue fraction.

	Sample 7706-46-3		
	Groundmass	Gray Cement	White Cement
	(soluble equivalents*/100 g rock)		
Ca	1.099	1.093	1.051
Mg	0.988	0.989	0.925
Fe	0.0017	0.0008	0.0003
CO_3	2.043	2.082	2.085
PO_4	0.013	0.005	0.005
F†	0.002	<0.001	<0.001
Σ cations	2.088	2.083	1.977
Σ anions	2.058	2.088	2.091
	(estimated weight-%)		
Calcite	15	12	15
Dolomite	84	88	85
Apatite	1	<1	<1

*Equivalents = (wt-% · valency)/formula weight.
†Calculated from phosphate content and apatite composition as listed in text.

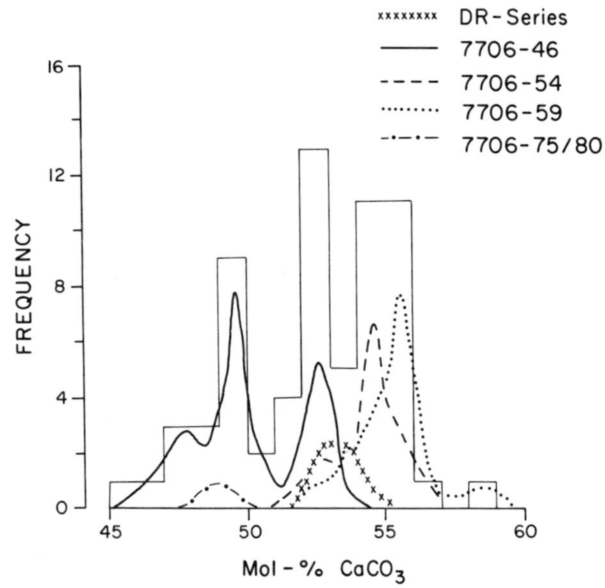

Figure 9. Frequency distribution of chemical compositions from 67 dolomite samples of the Peru continental margin. Three regional groupings are evident: Ca-deficient dolomites (46-50 mol-% $CaCO_3$) of the late Miocene Lima Basin (7706-46) and Progresso Basin (7706-75/-80), Ca-enriched dolomites (52-58 mol-% $CaCO_3$) of The Trujillo Basin (7706-54 and 7706-59) and dolomites with intermediate Ca-contents (50-52 mol-% $CaCO_3$ of the Pliocene-Pleistocene Lima Basin (DR-series and 7706-46).

Stable Carbon Isotope Characteristics

The most obvious isotopic signal of the Peru margin dolomites and calcites is in the large range of $\delta^{13}C$ values from +20‰ to -20‰ relative to the PDB standard (Fig. 11). The carbon isotope signal shows a regional grouping similar to that observed for the chemistry of the dolomites between basins along the margin, i.e., isotopically heavy dolomites were recovered from the Lima and Progresso basins and isotopically light dolomites from the Trujillo Basin. Both variables, mineralogy and stable carbon isotope mass ratios, therefore, help to define each of the series of dredged rocks uniquely (Fig. 11). The light and heavy $\delta^{13}C$ values identify the carbonate ions as end members derived from organic matter by two separate decomposition processes, sulfate and carbonate reduction, respectively, and hence the term "organic dolomites" (Spotts and Silverman, 1966; Deuser, 1970; Claypool and Kaplan, 1974; Irwin, Curtis and Coleman, 1977; Irwin, 1980; Dickson and Coleman, 1980; Matsumoto and Iijima, 1980; Reitsema, 1980; Pisciotto, 1981; Kelts and McKenzie, 1982; and many others).

Microbial sulfate reducers generate isotopically light metabolic carbon dioxide in accordance with the carbon isotope signal of the organic substrate (i.e., particulate organic carbon = ~-20‰ PDB; methane = ~-60‰ PDB. Thereby they may induce precipitation or replacement by solid carbonate phases with strongly negative $\delta^{13}C$ characteristics. The actual range of $\delta^{13}C$ of the resulting carbonates depends on the substrate signals, the relative contributions from the organic carbon reservoirs of different substrates, and the carbon, if any, from the inorganic bicarbonate pool (Suess, Kulm and Killingley, in press). The key to formation of dolomites rather than calcites or siderites under conditions of microbial sulfate reduction is the lowered concentration of sulfate ions below that of normal seawater (Baker and Kastner, 1981).

The diagnostic features for the organic dolomites of the Trujillo Basin resulting from SO_4 reduction are: $\delta^{13}C$ characteristics between -18 to -8‰ PDB, excess Ca of between 52-58 mol-% $CaCO_3$ (Fig. 11) and discrete iron sulfide minerals. We will discuss later why dolomite with excess Ca (i.e., 52-58 mol-%) might form during this process in the Trujillo Basin and suggest where the Mg- and Ca-sources are to be found.

Carbon dioxide reduction coupled to fermentation is responsible for the formation of isotopically "heavy" carbonates. Again, the SO_4-free environment promotes dolomite precipitation (Baker and Kastner, 1981). It appears overwhelmingly that "heavy" (+20 to +4‰ PDB) residual carbon dioxide from microbial carbonate reduction is incorporated into the dolomite phases of the Lima and Progresso basins (Fig. 11). It remains unclear if and why some of these dolomites acquire an excess of Mg (i.e., 46-50 mol-% Ca), but we will suggest below different possible sources of cations.

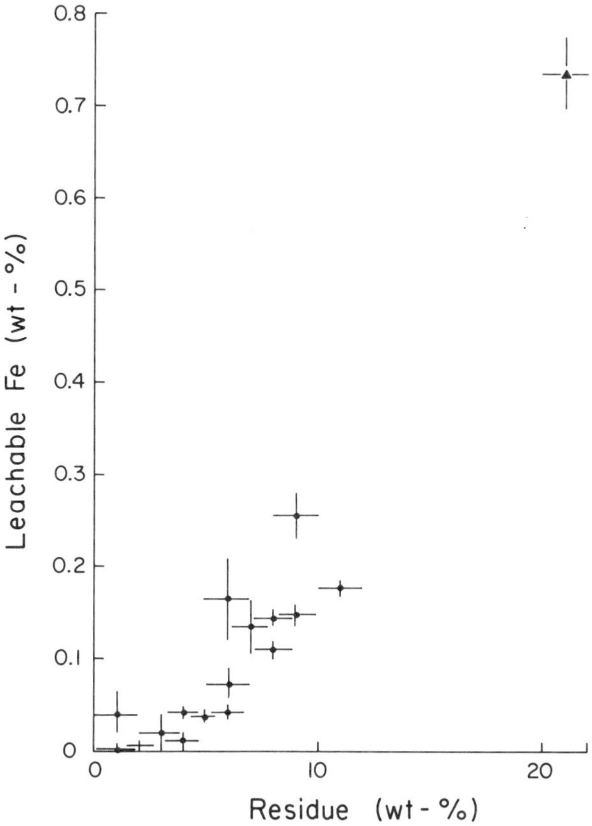

Figure 10. Low leachable iron in dolomite rocks appears to be associated with insoluble residue rather than with carbonate fraction. Data from Suess, Killingley and Kulm (in prep.); (+) = 46-3 (▲) = 75-1 as listed in Table 1.

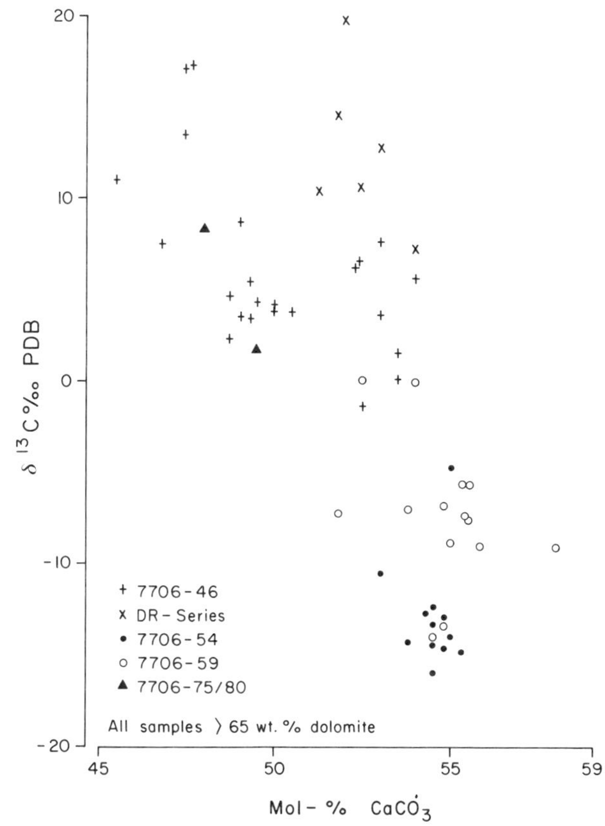

Figure 11. Changes in stable carbon isotope mass ratios and the mineralogy of dolomites from the Peru continental margin uniquely define samples from each basin or sub-basin; excluded are samples with >35 wt-% apatite and/or calcite. Regional groupings according to dredge sites are: late Miocene Lima Basin (+) and Progresso Basin (▲), Plio-Pleistocene Lima Basin (x and +), and Trujillo Basin (o and •).

Oxygen Isotope Characteristics

The $\delta^{18}O$ signature of the Peru dolomites varies only between +7.0 and +5‰ PDB (Fig. 12). Exceptions are the apatitic samples where $\delta^{18}O$ values of the structural carbonate are significantly lower (+1 to +4‰ PDB). A narrow range of oxygen isotope values has also been observed in numerous studies on similar dolomites from other continental margin sites (Deuser, 1970; Pisciotto, 1981; Pisciotto and Mahoney, 1981; Kelts and McKenzie, 1982). On average published data of the Monterey dolomites and those of the Gulf of California are "lighter" than the Peru dolomites by about 2-3‰ PDB. Irwin (1980) reports still "lighter" values for diagenetic dolomites from the Jurassic Kimmeridge Clay Formation, ranging from -1.6 to -4.5‰ PDB. At this point it seems likely that $\delta^{18}O$ of the Peru margin dolomites is a function of both temperature and $\delta^{18}O(H_2O)$ of the precipitating fluid.

The above authors have discussed the oxygen isotope record of diagenetic dolomites as a potential indicator of temperatures of formation. The $\delta^{18}O$ values of diagenetic dolomites from the Monterey, the Gulf of California, and the continental margin of Baja are consistent with temperatures of between 20° and 70°C, assuming $\delta^{18}O$ values of the interstitial fluids of between 0 and 1.5‰ SMOW.

Kastner (1982) reports on diagenetic calcites that formed in close proximity to hot basaltic sill intrusions in sediments of the Gulf of California. She calculates isotopic equilibrium at temperatures up to 130°C, with $\delta^{18}O$ of the pore fluids ranging from 1.96 to -0.05‰ SMOW. Irwin (1980), on the other hand, assumes an initial pore fluid composition of $\delta^{18}O(H_2O)$ = -4.9‰ SMOW in order to arrive at temperatures of formation between 15-30°C for the Kimmeridge Clay dolomites. Such ^{16}O-enriched waters are extremely rare in the marine environment; it is possible that the Jurassic dolomites of the Kimmeridge Clay actually formed at higher temperatures or have undergone exchange of oxygen isotopes with meteoric waters.

Extending these considerations to the Peru margin, it becomes clear that the rather heavy $\delta^{18}O$ values (+7 to +5‰ PDB) yield very low temperatures of formation, between 2° and 12°C calculated on the basis of the equation used by Pisciotto and Mahoney (1981), and assuming $\delta^{18}O$-depleted interstitial fluids ($\delta^{18}O$ = -1.0‰ SMOW). As is evident from these calculations, the temperatures of formation can only be assessed accurately if the $\delta^{18}O$ of the pore fluids at the time of cementation is known. Without

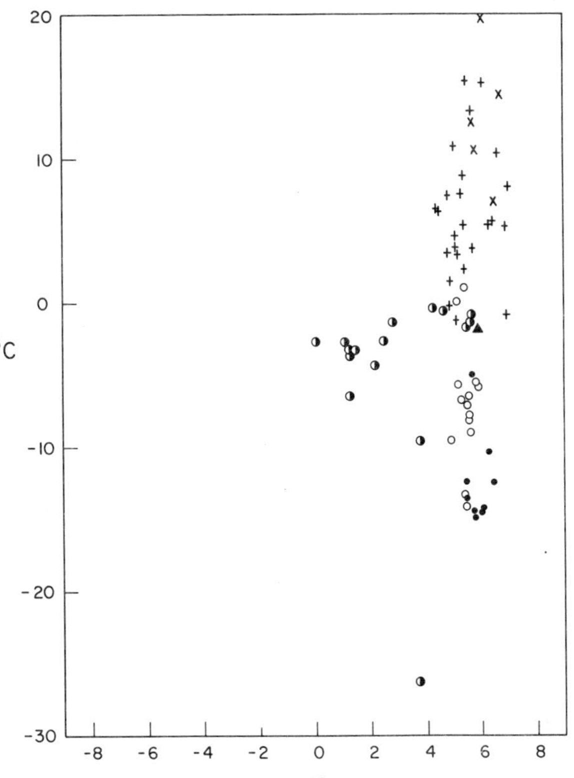

Figure 12. Oxygen and carbon isotope characteristics of Peru margin dolomite and apatite (◑) samples. The large range in $\delta^{13}C$ values reflects different mechanisms of carbonate generation as discussed in the text, and the narrow, rather "heavy" $\delta^{18}O$ signal (with the exception of most apatites which are "lighter") indicates a low temperature of formation between 2-12°C. Published data from the Monterey dolomites (Pisciotto, 1981; Pisciotto and Mahoney, 1981; Roehl, 1981) and the Kimmeridge Clay dolomites (Irwin, 1980) are significantly "lighter" and indicate higher temperatures of formation. Scales are in ‰; measurements by J. S. Killingley, Scripps Institution of Oceanography. Symbols same as in Figure 11.

this information we can only speculate that dolomite formation along the Peru continental margin took place at temperatures sufficiently low that thermal decarboxylation of organic matter may be excluded as a mechanism for carbonate generation. This in turn supports the earlier contention that only microbial sulfate and/or carbonate reductions are significant processes in generating carbonate for dolomite formation.

DISCUSSION

Facies Development, Preservation/Destruction, and Diagenetic Modification in the Peru Forearc Basins

Tectonics, sediment supply, and oceanographic-climatic setting control the accumulation of organic-rich shales and associated calcareous mudstones and dolomites in the Peru forearc basins. These processes often interact to produce a particular facies composition, contribute to the diagenetic modification of the facies and, in some cases, remove deposits from the geologic record. While all processes are important in facies development and preservation, the temporal tectonic processes associated with the ongoing development of the subduction zone are the most instrumental in the evolution and modification of the depositional basin.

Tectonic Processes

Vertical movements on the order of 2000 m have been documented in the subsiding Lima Basin during Pliocene-Pleistocene time (Fig. 5; Kulm and others, 1981). This rapid subsidence has created an ideal basinal environment for the accumulation and preservation of the late Pleistocene organic-rich mud facies and its diagenetic equivalents. The Trujillo Basin has experienced at most a few hundred meters of subsidence since late Miocene time, allowing the accumulation of only a thin Neogene section (Fig. 6; Thornburg, in press). Deposition in this basin has also been inhibited by the strong poleward flowing undercurrent that tends to rework and winnow the sediments on the seafloor, producing a biogenic sand or calcareous mud facies.

Recent geological and geophysical studies of modern subduction zones have shown that major subduction-related processes can induce rapid and substantial vertical movements in the forearc region (Kulm and Fowler, 1974; Schweller and Kulm, 1978; Scholl and others, 1980; von Huene and others, 1982; Moore and others, 1982a, 1982b; von Huene and Lee, 1982). Two large-scale subduction processes, subduction erosion (i.e., removal of material from the underneath side and/or leading edge of the overriding South American plate by tectonic erosion) and underplating (i.e., addition of material to underneath side of the overriding plate), have apparently caused the subsidence and uplift, respectively, of the crystalline Paleozoic-Precambrian basement of the Peru forearc basins.

The post-Oligocene uplift of the outer shelf high (OSH) near 9°S is most likely related to underplating of the arc massif with low-density material, causing the forearc region to isostatically rise. Peru Trench turbidites are a probable source of low density material for the underplating process. The variable amounts of sediment cover over the crystalline OSH and the local emergence of this feature as islands suggest that the rates of vertical tectonism vary along the forearc in response to differential underplating and along strike changes in basement structure.

Rapid Neogene subsidence (275-500 m/m.y.) of the Lima basin is probably related to tectonic erosion beneath the arc massif by the subducting Nazca plate. Sediment loading and thermal contraction are not viable subsidence mechanisms in a forearc with only minimal to moderate (up to 4 km) sediment loading on ancient continental crust. Geophysical data show that the accretionary prism (subduction complex) is small (7-35 km wide), considering the 150 m.y. subduction history for the South American arc. Not only has the bulk of the incoming sediments been subducted, a substantial amount of crystalline forearc material must have been tectonically eroded during Andean convergence as well (Hussong and others, 1976; Hussong and Wipperman, 1981; Kulm and others, 1982).

Tectonic erosion can accentuate either forearc basin formation or basin destruction, depending upon where material is removed from the margin. Tectonic erosion has removed most of the subduction complex from the leading edge of the forearc near 15°S, largely as a result of the Nazca Ridge impinging on the margin, and has severely restricted the formation of the West Pisco Basin (Fig. 1; Thornburg and Kulm, 1981). In this case, tectonic control is particularly noteworthy because the intense upwelling centers off Pisco should produce a facies that is richest in organic carbon over the entire Peru margin, but conditions are not favorable for its preservation. The convergence of large, buoyant features such as the Nazca Ridge with a continental arc, therefore, has a profound influence on the size, geometry, and facies development of sedimentary basins in the forearc region.

Sediment Supply and Climate

Sedimentation in Peru forearc basins is controlled by the tectonic processes just described and the climatic conditions in the continental drainage basins. These factors will determine whether the terrigenous clastic sediments supplied to the offshore basins consist largely of hemipelagic or turbidite deposits. Off Peru, the OSH has created a structural barrier to the dispersal of clastics away from the shelf by turbidity currents (Figs. 1,5,6). This is evidenced by the hemipelagic silts and clays found in the Quaternary mud lens and in the Neogene rocks dredged from the upper slope basins. However, in some areas it has been more effective as a topographic barrier than in others depending upon its relative elevation with time. Turbidites are found in the deposits of the Ballena well, suggesting that deeper basinal conditions once prevailed where now we find a crystalline horst with truncated overlying strata. The frequency of turbidites also increases upsection at the expense of carbonate interbeds in the Oligocene of the Delfin well. Shallow-water sands cap both the Ballena and Delfin sections and thus provide evidence of post-Oligocene uplift of the OSH.

The input of suspended terrigenous materials to the continental margin by Peruvian coastal rivers ranges from 1.2 to 5.0 x 10^6 tons/y. (Zuta and Guillen, 1970). Because of the extremely arid climate of the Peruvian coastal plain, this fluvial input is smaller by a factor of 3 to 10 than that estimated for the Oregon margin in the temperate Northeast Pacific (Krissek and Scheidegger, 1983). It is because of terrigenous dilution that an organic-rich mud facies is not well developed off Oregon, even though favorable upwelling conditions are present. The suspended sediment input from Peruvian rivers decreases from the north to the south with some local fluctuations, thereby establishing a regional gradient of terrigenous sediment supply to the margin. Large amounts of sediment were once supplied to the Peru margin by the Amazon River which flowed into the Pacific Ocean in the vicinity of the Gulf of Guayaquil. However, the Amazon's discharge shifted to the Atlantic Ocean in middle Miocene time and dramatically reduced the amount of sediment supplied to the northern Peruvian margin (Jenks, 1956). This decrease in terrigenous input correspondingly increases the organic matter content of the upwelling sediments deposited in the forearc basins (Krissek and Scheidegger, 1983).

Sediment supply to the basins and dispersal by turbidity currents are important determinants in the development of the upwelling facies, because the clastics will mask and dilute the organic-rich sediments. Eventually turbidites may overcome hemipelagic sedimentation and terminate the upwelling-related facies as demonstrated in the Monterey Formation (Pisciotto and Garrison, 1981). The arid climate, the outer shelf structural barrier (the crystalline OSH), and the subsiding basins of the continental shelf presently limit the amount of bottom-transported clastic sediment supplied and deposited in upper-slope forearc basins. It is therefore a very special set of sedimentary, climatic, and structural conditions that combine to enhance deposition and preservation of the organic-rich facies of the Peru forearc basins.

Oceanographic Setting

The coastal upwelling regime is essential for high primary productivity and the supply of organic matter for the Quaternary organic-rich mud facies. The centers of upwelling and productivity vary in intensity along the margin (Fig. 2), and accordingly, the composition and thickness of the upwelling facies may vary from one basin to another or within a single basin. When localized high organic inputs are combined with minimal terrigenous sediment supply (either due to low overall rates of input or due to trapping of clastics by upslope structural barriers), then an organic-rich sediment will be supplied to the seafloor. If these sediments are allowed to accumulate in a rapidly subsiding basin with an oxygen minimum zone (e.g., the Lima Basin), then the organic-rich mud facies will likely be preserved and incorporated into the basin's stratigraphy.

But even under such ideal conditions, a portion of the Quaternary upwelling facies may be removed and sorted by the action of contour currents that sweep the margin. Interestingly, this reworking of the primary upwelling facies is accentuated by the poleward flowing undercurrent of an eastern boundary coastal upwelling regime. The episodes of strongest reworking appear to be associated with periodic El Niño conditions.

Diagenetic Imprint

The type of sedimentary facies and hence the composition of the associated diagenetic dolomites appears to be related to the subsidence history of the forearc basins of the Peru continental margin. Burial rates of sediments are high and preservation of organic matter is enhanced in the Lima Basin because of rapid subsidence and high input from biological productivity (Fig. 2). The Quaternary organic-rich mud (>>5 wt-% carbon) is low in skeletal calcium carbonate, and the middle Pleistocene to Pliocene dolomites resulting from diagenesis of this margin facies have from 45-52 mol-% Ca and "heavy" $\delta^{13}C$ values. The rapid accumulation rates, which are controlled to a large extent by the rapid subsidence rates, produce a relatively thick stratigraphic section. This together with high organic carbon contents exhausts the dissolved sulfate of the pore water close to the sediment-water interface, and the dominating diagenetic reaction in the sediment column changes from sulfate reduction to bacterial fermentation/carbonate reduction.

On the other hand, the Quaternary calcareous muds and biogenic sands of the Trujillo Basin have generally less than 2 wt-% organic carbon. Burial rates of sediments are low because of tectonic stability of the basement, and continuous reworking of the sediments by bottom currents. Preservation of organic matter is diminished under oxygenated conditions. These factors combine to inhibit the accumulation of an organic-rich mud facies in the

Trujillo Basin, and the middle Miocene to Pleistocene diagenetic carbonates are characterized by Ca-rich dolomite (52-58 mol-%) and "light" $\delta^{13}C$ isotope values. This is in accord with microbial sulfate reduction being the dominant decomposition mechanism of organic matter because of lowered sulfate reduction rates.

Linking the stable carbon isotope signal of the diagenetic dolomites to either carbonate reduction/ fermentation or sulfate reduction is not entirely new; however, the geochemical results from the dolomites of the Peru forearc basins suggest a slightly more detailed, though tentative, interpretation.

Ca-to-Mg Ratio: The preferential incorporation of Ca over Mg in the Trujillo Basin dolomites might reflect a contribution of Ca from skeletal carbonates, including aragonite, low- and high-magnesian calcite, pre-concentrated in the reworked biogenic sand facies. The reverse relationship, i.e., preferential incorporation of Mg over Ca in some of the Lima Basin dolomites, could be related to the low content of skeletal carbonate in the primary, organic-rich mud facies. Terrigenous clays may provide a possible source of excess Mg. This type of ionic substitution, however, needs to be confirmed independently. In addition to the carbonate derived from organic decomposition, the dissolution of biogenic calcareous tests could release to the pore waters a significant amount of skeletal-derived carbonate ions. The isotopic composition of skeletal carbon is between +2 and -2°/₀₀ PDB.

Diagenetic dolomite on the Peru margin, as in the other regions of "organic" dolomites, can be thought of as a phase incorporating components from three different sources. Figure 13 schematically illustrates these sources. Source A would be dominated by skeletal calcium carbonate with Ca >> Mg and $\delta^{13}C$ = -20 ± 2°/₀₀. Source B would be dominated by sedimentary organic matter undergoing sulfate reduction (slow rates of subsidence, <150 m/m.y.), some lesser proportion of skeletal calcium carbonate, and terrigenous matter, possibly rich in Mg, such that

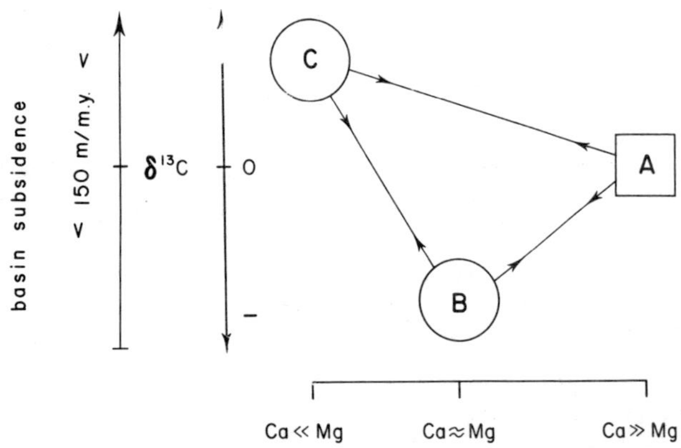

Figure 13. Qualitative environmental model relating subsidence history and sediment sources to the type of dolomites forming in the Peru forearc basins. A = biogenous skeletal material, B and C = sedimentary organic matter with possibly Mg-rich terrigenous material; these sources provide the starting material for essentially isochemical dolomitization. See text for discussion.

Ca = Mg and $\delta^{13}C$ = -20 ± 2°/₀₀. Source C would be exclusively of organic matter undergoing fermentation/ carbonate reduction (rapid rates of subsidence, >150 m/m.y.), and would contain Mg-rich terrigenous matter; hereby Ca << Mg and $\delta^{13}C$ = +25 ± 2°/₀₀ PDB. These sources adequately describe the contrasting sedimentary facies of the Trujillo Basin and Lima Basin and they readily explain the mineralogic and isotopic variability of dolomites found in the Peru forearc region. This qualitative environmental model does not address the chemical question of the actual mechanism of dolomite formation, i.e., whether through direct precipiation or replacement of precursors. However, it does provide a hypothesis worth testing by identifying possible sources of Ca and Mg and evaluating the geochemical mass balance of certain continental margin basins for isochemical dolomitization as also suggested by Baker and Kastner (1981). Such evaluations are presently being undertaken for the Peru forearc basins.

Oxygen Isotopes and Temperature of Formation: It is likely that the $\delta^{18}O$ of the Peru margin dolomites records the temperature of formation, as discussed previously. However, any such interpretation depends on accurate knowledge of the oxygen isotope composition of the interstitial water in contact with the dolomites at the time of formation. This knowledge can only rarely be obtained and certainly is lost from the rocks dredged from the Peru forearc basins. A large number of oxygen isotope analyses from pore waters of other continental margin sediments, some of which also contain diagenetic dolomites, is available (Kastner, 1982) and can be used to constrain the range of possible temperatures of formation. More than 95% of $\delta^{18}O$ (H_2O) values of pore fluids fall in the range of +1.5 to -2.0°/₀₀ SMOW which yield temperatures for the Peru dolomites between 2 and 12°C. The significance of this low temperature estimate implies that dolomitization occurred very early and at shallow burial.

Post-Depositional Deformation of Upwelling Deposits

As noted by Redwine (1981) in his review of the Monterey Formation, fractured dolomitic rocks have a substantial reservoir capacity for petroleum, and are associated with organic-rich source beds. Many of these reservoirs are the result of repeated periods of rock dilation followed by natural hydraulic fracturing, all of which may be produced by episodic tectonic compression. Both multichannel seismic reflection records (Fig. 6) and dredge samples of brecciated dolomicrite (Fig. 3,7,8) collected in the Peru forearc basins suggest a history that includes compressional tectonics. Multiple episodes of compression are documented by swarms of reverse faults that truncate at several different unconformities in seismic section. Comparing the seismic records in Figures 5 and 6, it is evident that the conformable, synclinal sequence in the Lima Basin is much less disturbed than the shattered and broken stratigraphy of the Trujillo Basin. However, Trujillo is an older basin that developed dominantly during Paleogene time, while the Lima Basin section accumulated largely during late Neogene time (Thornburg and Kulm, 1981; Thornburg, in press). It is clear that both basins have undergone different tectonic histories even though they lie adjacent to one another. The Lima Basin has been subsiding thoughout most of its history, but even its deposits have been punctuated by minor periods of compression.

Late Miocene to late Pliocene dolomicrites from the Trujillo Basin exhibit a matrix lithology that is strongly brecciated along subparallel fracture planes and infilled with younger carbonate cement of late Pliocene to early Pleistocene age (Fig. 8, sample 54-5) (Kulm and others, 1981; Thornburg and Kulm, 1981). The fracture orientation and granulated grains in this section suggest a compressive origin for these rocks. Brecciaed rocks were also dredged from the Lima Basin, but they are a minor part of the recovered lithologies.

Basinal Distribution of Isotopic Signatures

Because the Peru dolomites display distinctive basinal groupings of isotopic signatures and mineral assemblages, we attempted a similar comparison for the Monterey Formation. The Peru dolomites display a regional clustering of isotopic and mineralogic composition that corresponds directly to the distribution of sedimentary forearc basins. By comparison, Roehl (1981) noted a geographic distribution of carbon isotope values of dolomites in the surficial outcrops of the Miocene Monterey Formation which he attributed to certain tectonic districts that are separated by the Nacimiento fault, a major strike-slip fault. Dolomites on one side of the fault (south and west) are isotopically "light," whereas those on the opposite side of the fault (north and east) are isotopically "heavy." This surface distribution of light isotopic values continues into the subsurface in the Santa Maria Basin.

In Peru, diagenetic dolomites appear to form preferentially in upper slope water depths. However, regional along-strike variations in isotopic composition are evident, such that "heavy" carbon isotope values typify the Lima Basin, and "light" values the Trujillo Basin. If the Peru forearc was subsequently subjected to strike-slip tectonism in its orogenic history, then the isotope provinces that are presently developed along strike in the upper slope basins might be brought to lie laterally adjacent, similar to the situation that Roehl describes.

CONCLUSIONS

Several factors, including tectonic history, sediment supply and oceanographic-climatic setting, are responsible for the development, diagenetic modification and fate (preservation vs. destruction) of the Cenozoic sedimentary facies of the Peru forearc basins. The interaction of these processes influences the composition and areal distribution of the sedimentary facies and the type of diagenetic dolomite that forms in a given basin and among basins.

1. Two prominent late Pleistocene to Holocene sedimentary facies, organic-rich mud and biogenic calcareous sand/mud, are accumulating on the outer continental shelf and upper slope in the Lima Basin and Trujillo Basin, respectively. The organic-rich muds contain abundant fish debris, diatoms, and benthic foraminiferal assemblages derived from Peru's classic upwelling regime.

2. Late Miocene to middle Pleistocene dolomites contain similar upwelling constituents, but these are pervasively bound by carbonate cement to form indurated dolomicrite. The dolomites are interbedded with mudstones, and apparently constitute no more than 1-2% of the stratigraphic section.

3. Neogene and Quaternary dolomites found in Peru forearc basins are characterized by distinctive stable isotope signatures and mineral assemblages. Dolomites in the rapidly subsiding Lima Basin exhibit isotopically heavy $\delta^{13}C$ values and 46-52 mol-% Ca, whereas dolomites in the slowly subsiding or stable basins Trujillo Basin exhibit light $\delta^{13}C$ values and 52-58 mol-% Ca. We suggest that the subsidence history of the forearc basins during deposition may exert a strong influence on the type of sedimentary facies that is initially deposited. A critical rate of subsidence appears to be 150 m/m.y. At rates slower than this, a calcareous sand/mud facies may preferentially develop, yielding dolomites with isotopically "light" carbon and Ca-rich mineralogy. At faster subsidence rates, an organic-rich mud facies will be favored to accumulate, and "heavy" carbon isotopes and Ca-poor mineralogies may characterize the diagenetic dolomites.

4. Stable oxygen isotopes suggest that the Peru dolomites form at lower temperatures than the Monterey Formation dolomites. This indicates that dolomitization occurred very early and at a shallow depth of burial.

ACKNOWLEDGEMENTS

This research was supported by the National Science Foundation through grant OCE 8118006 to Oregon State University. We gratefully acknowledge the encouragement and constructive reviews by the editors of this volume towards publication of the article. We thank C. A. Ungerer for excellent laboratory assistance and M. J. Armbrust for expert help in preparing the various versions of this manuscript.

REFERENCES CITED

Arthur, M.A., von Huene, R. and Adelseck, C.G., Jr., 1980, Sedimentary evolution of the Japan forearc region off northern Honshu, Legs 56, 57, DSDP: Initial Reports of the DSDP, v. 56-57, part 1, p. 521-568.

Baker, P.A. and Kastner, M., 1981, Constraints on the formation of sedimentary dolomites: Science, v. 213, p. 214-216.

Berger, W.H. and von Rad, U., 1972, Cretaceous and Cenozoic sediments from the Atlantic Ocean N. 26: Initial Reports of the DSDP, v. 14, p. 787-954.

Chave, K., 1952, A solid solution between calcite and dolomite: J. Geol., v. 60, p. 190-192.

Claypool, C.E. and Kaplan, I.R., 1974, The origin and distribution of methane in marine sediments, in Kaplan, I.R., ed., Natural Gases in Marine Sediments: Plenum Press, New York, p. 99-140.

Deuser, W.G., 1970, Extreme $^{13}C/^{12}C$ variations in Quaternary dolomites from the continental shelf: Earth Planet. Sci. Lett., v. 8, p. 118-124.

Dickson, J.A.D. and Coleman, M.A., 1980, Changes in carbon and oxygen isotope composition during limestone diagenesis: Sedimentology, v. 27, p. 107-118.

Edgar, N.T., Saunders, J.B., et al., 1973, Initial Reports of the DSDP, v. 15, U.S. Government Printing Office, Washington, D.C., 1137 p.

Friedman, I., and Murata, K.J., 1979, Origin of dolomite in Miocene Monterey shale and related formations in the Temblor Range, California: Geochim. Cosmochim. Acta, v. 43, p. 1357-1365.

Garrison, R.E. and Douglas, R.G., eds., 1981, The Monterey Formation and Related Siliceous Rocks of California: Soc. Econ. Paleontologists and Mineralogists, 327 p.

Hayes, D.C., Pimm, A.C., et al., 1972, Initial Reports of the DSDP, v. 14: U.S. Government Printing Office, Washington, D.C., 975 p.

Hussong, D.M. and Wipperman, C.K., 1981, Vertical movement and tectonic erosion of the continental wall of the Peru-Chile Trench near 11°30'S latitude: Geol. Soc. America Mem. 154, p. 509-524.

Hussong, D. M., Edwards, P.B., Johnson, S.H., Campbell, J.F. and Sutton, G. H., 1976, Crustal structure of the Peru-Chile Trench: 8°S-12°S latitude: Am. Geophys. Union Monogr. 19, p. 71-86.

Irwin, H., 1980, Early diagenetic carbonate precipitation and pore fluid migration in the Kimmeridge Clay of Dorset, England: Sedimentology, v. 27, p. 577-591.

Irwin, H., Curtis, C.D. and Coleman, M., 1977, Isotopic evidence for source of diagenetic carbonates formed during burial of organic-rich sediments: Nature, v. 269, p 209-213.

Jenks, W.J., 1956, Handbook of South American Geology; An Explanation of the Geologic Map of South America: Geol. Soc. America Mem. 65, 378 p.

Kastner, M., 1982, Evidence for two distinct hydrothermal systems in the Guaymas Basin: Initial Reports of the DSDP, v. 64, part 2, p. 1143-1157.

Kelts, K. and McKenzie, J.A., 1982, Diagenetic dolomite formation in Quaternary anoxic diatomaceous muds of Deep Sea Drilling Project, Leg 64, Gulf of California: Initial Reports of the DSDP, v. 64, part 2, p. 553-570.

Krissek, L.A. and Scheidegger, K.F., 1983, Environmental controls on sediment texture and composition in low oxygen zones off Peru and Oregon: in Thiede, J. and Suess, E., eds., Coastal Upwelling: Its Sediment Record, Part B, Plenum Press, New York, p. 163-180.

Krissek, L.A., Scheidegger, K.F. and Kulm, L.D., 1980, Surface sediments of the Peru-Chile continental margin and the Nazca plate: Textural and geochemical characteristics and their controls: Geol. Soc. America Bull., v. 91, p. 321-331.

Kulm, L.D. and Fowler, G.A., 1974, Oregon continental margin structure and stratigraphy: A test of the imbricate thrust model, in Burk, C.A. and Drake, C.L., eds., The Geology of Continental Margins: Springer-Verlag, New York, p. 261-284.

Kulm, L.D., Schrader, H., Resig, J.M., Thornburg, T.M., Masias, A. and Johnson, L., 1981, Late Cenozoic carbonates on the Peru continental margin: Lithostratigraphy, biostratigraphy, and tectonic history: Geol. Soc. America Mem. 154, p. 469-508.

Kulm, L.D., Resig, J.M., Thornburg, T.M. and Schrader, H.-J., 1982, Cenozoic structure, stratigraphy and tectonics of the central Peru forearc, in Leggett, J.K., ed., Trench and Forearc Sedimentation and Tectonics on Modern and Ancient Active Plate Margins: Geol. Soc. London Spec. Pub. 10, p. 151-169.

Kulm, L.D., Thornburg, T.M. and Dang, S., in press, Drill hole stratigraphy, dredge lithologies, and sample locations, in Hussong, D.M. et al., eds, Atlas of the Ocean Margin Drilling Program, Peru Continental Margin, Region VI: Joint Oceanographic Institutions, Inc., Marine Science International, Woods Hole, MA, 1 map sheet with text.

Matsumoto, R. and Iijima, A., 1980, Carbonate diagenesis in cores from Sites 438 and 439 off northeast Honshu, northwest Pacific, Leg 57, Deep Sea Drilling Project: Initial Reports of the DSDP, v. 56-57, part 2, p, 1117-1127.

Moore, J.C., Watkins, J.S., Shipley, T.H., McMillen, K., Bachman, S.B. and Lundberg, N., 1982a, Geology and tectonic evolution of a juvenile accretionary terrane along a truncated convergent margin: Synthesis of results from Leg 66 of the Deep Sea Drilling Project, southern Mexico: Geol. Soc. America Bull., v. 93, 847-861.

Moore, J.C., Biju-Duval, B., Bergen, J.A., Blackington, G., Claypool, G.E., Cowan, D.S., Duennebier, F., Guerra, R.T., Hemleben, C.H.J., Hussong, D., Marlow, M.S., Natland, J.H., Pudsey, C.J., Renz, G.W., Tardy, M., Willis, M.E., Wilson, D. and Wright, A.A., 1982b, Offscraping and underthrusting of sediment at the deformation front of the Barbados Ridge: Deep Sea Drilling Project 78A: Geol. Soc. America Bull., v. 93, p. 1065-1077.

Murata, K.J., Friedman, I. and Madsen, B.M., 1969, Isotopic composition of diagenetic carbonates in the marine Miocene formations of California and Oregon: Geol. Surv. Prof. Paper 614, p. 824.

Okada, H., 1980, Pebbles and carbonate nodules from Deep Sea Drilling Project Leg 56 cores: Initial Reports of the DSDP, v. 56-57, part 2, p. 1089-1105.

Pisciotto, K.A., 1981, Review of secondary carbonates in the Monterey Formation California, in Garrison, R.E. and Douglas, R.G., eds., The Monterey Formation and Related Siliceous Rocks of California: Soc. Econ. Paleontologists and Mineralogists, p. 273-284.

Pisciotto, K.A. and Mahoney, J.J., 1981, Isotopic survey of diagenetic carbonates, Deep Sea Drilling Project, Leg 63: Initial Reports of the DSDP, v. 63, p. 595-610.

Pisciotto, K.A. and Garrison, R.E., 1981, Lithofacies and depositional environments of the Monterey Formation, California, in Garrison, R.E. and Douglas, R.G., eds., The Monterey Formation and Related Siliceous Rocks of California: Soc. Econ. Paleontologists and Mineralogists, p. 97-122.

Redwine, L.E., 1981, Hypothesis combining dilation, natural hydraulic fracturing, and dolomitization to explain petroleum reservoirs in Monterey shale, Santa Maria area, California, in Garrison, R.E. and Douglas, R.G., eds., The Monterey Formation and Related Siliceous Rocks of California: Soc. Econ. Paleontologists and Mineralogists, p. 221-248.

Reimers, C.E., 1982, Sedimentary organic matter: distribution and alteration processes in the coastal upwelling region off Peru: Ph.D. Thesis, Oregon State Univ., Corvallis, 219 p.

Reimers, C.E. and Suess, E., 1983, Spatial and temporal patterns of organic matter accumulation on the Peru continental margin, in Thiede, J. and Suess, E., eds., Coastal Upwelling: Its Sediment Record, Part B: Plenum Press, New York, p. 311-346.

Reitsema, R.H., 1980, Dolomite and nahcolite formation in organic rich sediments: isotopically heavy carbonates: Geochim. Cosmochim. Acta, v. 44, p. 2045-2050.

Resig, J.M., 1981, Biogeography of benthic foraminifera of the northern Nazca plate and adjacent continental margin: Geol. Soc. America Mem. 154, p. 619-666.

Roehl, P.O., 1981, Dilation brecciation--A proposed mechanism of fracturing, petroleum expulsion and dolomitization in the Monterey Formation, California, in Garrison, R.E. and Douglas, R.G., eds., The Monterey Formation and Related Siliceous Rocks of California: Soc. Econ. Paleontologists and Mineralogists, p. 285-316.

Russell, K.L., Deffeyes, K.S. and Fowler, G.A., 1967, Marine dolomites of unusual isotopic compositions: Science v. 155, p. 189-191.

Scheidegger, K.F. and Krissek, L.A., 1983, Zooplankton and nekton: natural barriers to the seaward transport of suspended terrigenous particles off Peru, in Suess, E. and Thiede, J., eds., Coastal Upwelling: Its Sediment Record, Part A: Plenum Press, New York, p. 303-333.

Scholl, D.W., von Huene, R., Vallier, T.L. and Howell, D.G., 1980, Sedimentary masses and concepts about tectonic processes at underthrust ocean margins: Geology, v. 8, p. 564-568.

Schweller, W.J. and Kulm, L.D., 1978, Extensional rupture of oceanic crust in the Chile Trench: Marine Geology, v. 28, p. 271-291.

Smith, R. L., 1983, Peru coastal currents during El Niño: 1976 and 1982: Science, v. 221, p. 1397-1399.

Spotts, J.H. and Silverman, S.R., 1966, Organic dolomite from Point Fermin, California, in Zenger, D.H. and Mazzullo, S.J., eds., Dolomitization: Hutchinson Ross Pub. Co., Stroudsburg, PA, p. 339-349.

Suess, E., 1981, Phosphate regeneration from sediments of the Peru continental margin by dissolution of fish debris: Geochim. Cosmochim. Acta, v. 45, p. 577-588.

Suess, E., and Thiede, J., 1983 (eds.), Coastal Upwelling: Its Sediment Record, Part A: Plenum Press, New York, 604 p.

Suess, E., Killingley, J.S. and Kulm, L.D., in prep., Organic dolomites of the Peru continental margin.

Suess, E., Kulm, L.D. and Killingley, J.S., in press, Coastal upwelling and a history of organic-rich mudstone deposition off Peru, in Brooks, J. and Fleet, A., eds., Marine Petroleum Source Rocks, A special publication of the Geological Society: Blackwell Publ. Ltd., London.

Thornburg, T.M., in press, Multichannel seismic interpretations--shelf and upper slope basins, in Hussong, D.M. et al., eds., Atlas of the Ocean Margin Drilling Program, Peru Continental Margin, Region VI: Joint Oceanographic Institutions, Inc., Marine Science International, Woods Hole, MA, 1 map sheet with text.

Thornburg, T. and Kulm, L.D., 1981, Sedimentary basins of the Peru continental margin: Structure, stratigraphy, and Ceonzoic tectonics from 6°S to 16°S latitude: Geol. Soc. America Mem. 154, p. 393-422.

Travis, R.B., Gonzales, G. and Pardo, A., 1976, Hydrocarbon potential of coastal basins of Peru: Amer. Assoc. Petroleum Geol. Mem 25, p. 331-338.

von Huene, R. and Lee, H., 1982, The possible significance of pore fluid presssures in subduction zones: American Assoc. Petroleum Geol. Mem. 34, p. 781-792.

von Huene, R., Langseth, M., Nasu, N. and Okada, H., 1982, A summary of Cenozoic tectonic history along the IPOD Japen Trench transect: Geol. Soc. America Bull., v. 93, p. 829-846.

Weliky, K., Suess, E. and Ungerer, C.A., 1983, Problems with accurate carbon measurements in marine sediments and particulate matter in seawater: A new approach: Limnol. Oceanogr., v. 28, p. 1252-1259.

Zuta, S. and Guillen, O., 1970, Oceanografia de las aquas costeras del Peru: Institute de Mar Peru Boletin, v. 2, no. 5, p. 157-324.

DIAGENETIC ASPECTS, SANDHOLDT MEMBER, MIOCENE MONTEREY FORMATION, SANTA LUCIA MOUNTAINS, CALIFORNIA: IMPLICATIONS FOR DEPOSITIONAL AND BURIAL ENVIRONMENTS

MERTZ, K.A. JR., Earth Science Board, University of California, Santa Cruz, California 95064

ABSTRACT

X-ray and scanning electron microscope analyses on Sandholdt hemipelagic shales reveal the presence of authigenic apatite, iron-rich dolomite, pyrite, kaolinite, quartz, and opal-CT (rare). Petrographic relationships suggest that apatite and pyrite formed early, followed by dolomite, kaolinite, opal-CT, and, finally, by quartz. Analyses demonstrate that: (1) apatite and dolomite commonly occur together, but whereas dolomite is most abundant in highly laminated, organic-rich rocks, apatite becomes increasingly abundant in moderately to faintly laminated rocks, and even occurs in massive-bioturbated, organic-poor samples; (2) dolomite occurs by replacement of host shales, apparently along beds that are somewhat more calcite-rich, and relatively less detrital-rich, than interbedded shales. Dolomite bodies themselves are organic-poor (TOC $< 1\%$) and contain more oxygen-rich, hydrogen-poor organic matter than surrounding organic-rich rocks; and (3) *in situ* apatite occurs as a replacement product and authigenic precipitate, often associated with pyrite and bacterial mat structures commonly found in shales and interbedded dolomite bodies.

Isotopic analyses on dolomites reveal "heavy" carbon-13 and systematically variable oxygen-18 values. It appears that these dolomites formed in response to bacterial breakdown of organic matter, primarily methanogenesis by CO_2 reduction, over a range of burial temperatures and depths (approximately 12 to 33°C; 100-450 meters). Carbon-13 isotopic analyses on *in situ* apatite concretions and pelletal phosphorites suggest early formation by direct authigenic precipitation or replacement of precursor carbonate, under sulphate reducing conditions. Depletion of SO_4^{2-} and increased alkalinity (HCO_3^-) resulting from these reactions may serve to facilitate subsequent dolomite formation. Dolomite, however, does not appear to be a prerequisite for apatite precipitation. Consistently "light" oxygen-18 values from PO_4^{3-} and CO_3^{2-} in apatite are enigmatic, but probably reflect diagenetic temperatures (approx. 27-40°C).

Mineralogical and organic geochemical data put clear constraints on Sandholdt burial history. Maximum temperatures and burial depths range from 80 to 105°C and 2400-3300+ meters. The unit is characterized (in decreasing order of abundance) by the association of: (1) well-laminated, organic- and dolomite-rich "methane-zone" rocks deposited under anaerobic to dysaerobic conditions; (2) faintly to moderately laminated (low to moderate organic matter) apatite-rich, pyritic rocks suggestive of formation under low-oxygen (sub-oxic) conditions in sulphate reducing environments; and (3) organic-poor, bioturbated rocks deposited under more oxic conditions. These associated lithotypes (deposited at rates up to 400 m/my) suggest variability in the position and intensity of an oxygen minimum zone that affected deposition throughout Sandholdt time.

INTRODUCTION

Hemipelagic, deepwater sediments reflecting the onset of rapid subsidence and basin formation in late Early Miocene time (17-16 m.y.b.p.) are widespread in the Santa Lucia Mountains of the central Salinian Block, California (Fig. 1). Originally termed the Sandholdt Formation by Thorup (1941), these rocks were later incorporated into the Monterey Formation as the Sandholdt member (Durham, 1974). In California, the Monterey Formation represents the basinal facies of a circum-Pacific Neogene stratigraphic sequence including Oligocene non-marine to shallow marine deposits, Miocene basinal facies, and late Miocene-Pliocene shallow marine deposits (Ingle, 1981). The Monterey can be roughly divided into a lower calcareous shale facies, a transitional calcareous-phosphatic facies, and an upper siliceous facies (Pisciotto, 1978; Pisciotto and Garrison, 1981). In the Santa Lucia Mountains, the Sandholdt member encompasses only the calcareous-phosphatic transitional facies.

Basin formation occurred in response to Neogene movement of the Salinian Block (thought by some to be one of California's numerous "exotic" terranes: McWilliams and Howell, 1982; Page, 1982) along the San Andreas transform system following the collision of the East Pacific Rise with the North American-Farallon trench system (Atwater, 1970; Blake *et al.*, 1978; Graham, 1979). Dextral shear led to formation of an anastomosing system of strike-slip faults in a broad wrench tectonic zone, resulting in formation of en-echelon and pull-apart basins, a complex system of adjacent uplifts and crustal sags similar to the present-day California Borderland (Crowell, 1974; Page, 1981; Mann *et al.*, 1983; Douglas, 1981).

in Garrison, Robert E.; Kastner, Miriam and Zenger, Donald H., eds., 1984, **Dolomites of The Monterey Formation and Other Organic-Rich Units:** Pacific Section S.E.P.M., Vol. 41, p. 49-73.

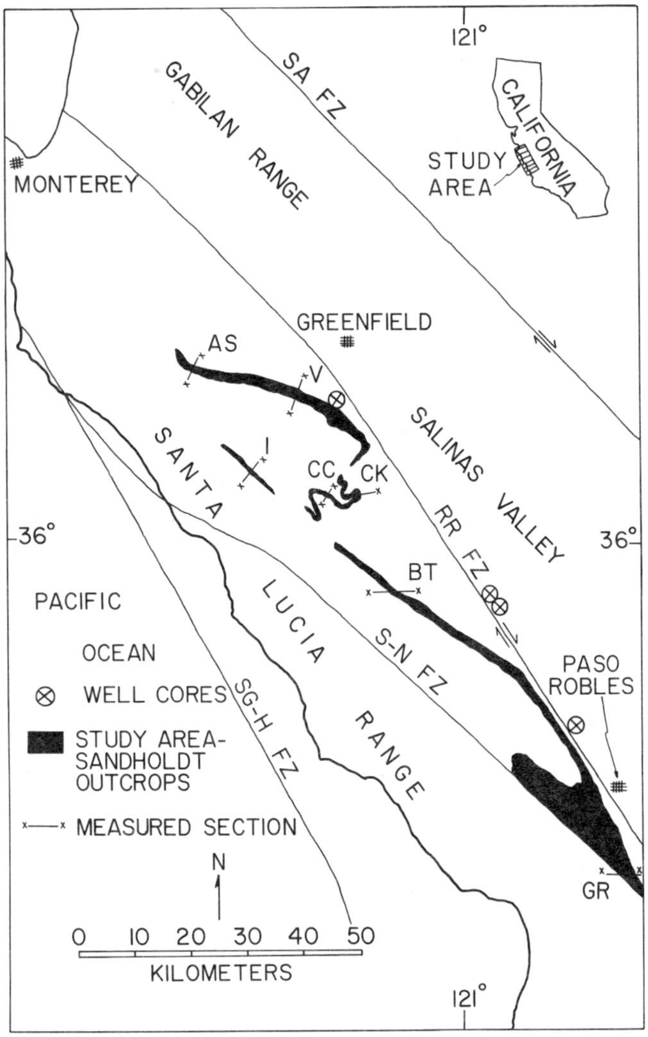

Fig. 1: Location map of study area showing outcrop and well core localities and extent of Sandholdt surface outcrop in the Santa Lucia Mountains. AS = Arroyo Seco; V = Vaqueros Canyon; I = Indians Ranch; CK = Cosio Knob; CC = Coleman Canyon; BT = Bear Trap Road; GR = Graves Creek; RR FZ = Reliz-Rinconada fault zone; SG-H FZ = San Gregorio-Hosgri fault zone; S-N FZ = Sur-Nacimiento fault zone; SA FZ = San Andreas fault zone.

Formation of starved basins is reflected in an abrupt shift from Oligocene-Early Miocene shelf and shelf-break deposits (Ingersoll and Graham, 1983), to bathyl hemipelagic shales of the Sandholdt Member. Deposition commenced in late Saucesian time and ranged through the Relizian and Luisian benthonic foraminiferal stages (Graham, 1976). Nannofossil biostratigraphy indicates that Sandholdt deposition was essentially synchronous throughout the study area (Fig. 1), and ranges from the upper part of the H. ampliaperta zone (≈17-16 m.y.b.p.) through the C. miopelagicus subzone of the D. exilis zone (C. H. Ellis, written comm., 1983). Planktonic foraminifera (G. Keller, written comm., 1983) suggest an early Middle Miocene age for all sections studied.

Sandholdt deposition spans a critical Middle Miocene climatic transition from a non-glacial to a glacial world (Pisciotto and Garrison, 1981; Ingle, 1981; Savin et al., 1981), straddling a brief early Middle Miocene high stand of sea level characterized by relatively warmer bottom water temperature (Keller and Barron, 1983; Woodruff et al., 1981). Organic-rich, phosphatic Sandholdt shales are thought to reflect increased upwelling and productivity and development of low oxygen to anoxic conditions in marginal basins, in response to initiation of Middle Miocene climatic cooling and intensified oceanic and atmospheric circulation (Ingle, 1981). Oxygen-minimum conditions are conducive to the preservation of abundant, hydrogen-rich organic matter and lamination (Demaison and Moore, 1980; Pisciotto and Garrison, 1981). Additionally, anoxic conditions within these hemipelagic sediments offer ideal conditions for the growth of authigenic apatite, dolomite, and pyrite, all commonly observed in Sandholdt shales. This paper focuses on selected authigenic mineral phases in the Sandholdt, their burial history, and how they may reflect the original depositional environment.

METHODOLOGY

Sandholdt samples were analyzed for bulk mineralogy by standard X-ray diffraction (XRD) techniques. Chemical (X-ray Fluorescence-XRF) analyses were performed using an AXF automated X-ray wavelength spectrometer (Phillips) following standard preparation by heating and fusing whole-rock samples until all organic-matter and CO_2 were driven off. XRD relative peak intensities, and XRF oxide values were corrected and interpreted using methods of Isaacs (1980).

Scanning electron microscopy (SEM) was performed using a Nanolab SEM. Chemical determinations of SEM images were made by qualitative X-ray microanalysis using an energy-dispersive spectrometer (Goldstein et al., 1981).

Organic geochemical data was obtained by standard total organic carbon (TOC) determinations and ROCK-EVAL pyrolysis techniques (Milner, 1982; Tissot and Welte, 1978; Espitalie et al., 1977; G. Demaison, written comm., 1983). Analyses were performed courtesy of Chevron Overseas Petroleum Co., and SOHIO Petroleum, both of San Francisco.

Carbon-13 and oxygen-18 isotopic analyses on CO_3^{2-} in carbonate fluorapatite and dolomite were run courtesy of M. Kastner, Scripps Institute of Oceanography, and M. Arthur, University of Rhode Island. Dolomite and apatite samples were "cleaned" to remove calcite by, respectively, a 4.7 pH solution of 0.1 molar Na acetate plus 0.1 molar acetic acid (M. Kastner, oral comm., 1982), and Silverman' solution (Silverman et al., 1952). All samples were confirmed calcite-free by XRD. Minor amounts of silica were ignored. CO_2 extraction was performed by standard vacuum techniques outlined by McCrea (1950) and Craig (1957). Isotopic results are reported in per mil relative to the PDB standard using the delta notation:

$$\delta^{13}C \text{ or } \delta^{18}O = [(R_{sample}/R_{std}) - 1] \times 10^3$$

where R is C^{13}/C^{12} or O^{18}/O^{16}. Temperature calculations using dolomite oxygen-18 values are based on the proto-dolomite-water fractionation expression of Fritz and Smith (1970). Equilibrating pore waters were assumed to have a value of 0 $^{o}/oo$ SMOW (Pisciotto and Mahoney, 1981) near the sediment-water interface, but decreasing by approximately 0.3 $^{o}/oo$ for each 100 meters of burial depth (Pisciotto, 1981b; Lawrence et al., 1975).

Oxygen-18 isotopic analyses from PO_4^{3-} in apatite were conducted courtesy of Y. Kolodny at The Hebrew University in Jerusalem. Standard methods for precipitation of $BiPO_4$ and extraction of CO_2 by fluorination were used (Tudge, 1960; Shemesh et al., in press). Temperature calculations based on oxygen-18 values are based on the revised scale derived by Longinelli and Nuti (1968; 1973), where $T°C = 111.4 - 4.3\ (\delta p - \delta w)$. δp is the $\delta^{18}O$ of oxygen in PO_4^{3-} from solid phosphate and δw is the $\delta^{18}O$ of equilibrating pore waters.

OBSERVATIONS AND DATA

Lithotypes

The most common Sandholdt lithologies (Fig. 2A,B) are hemipelagic, nannofossil- and foraminifera-rich siliceous shales (calcareous-siliceous shales of Isaacs, 1981a). X-ray diffraction and chemical analyses indicate that these rocks are variably argillaceous, containing from 15-50% aluminosilicates by weight. Detrital quartz ranges from 5-15 weight percent (by point count and XRF estimates (see Isaacs, 1980)), diagenetic quartz from 10-50%, and calcite from 4-35 weight percent. The rocks are commonly organic-rich (total organic carbon values up to 7.1%), and contain variable amounts of authigenic apatite, dolomite, pyrite, and kaolinite. Other constituents include reworked foraminifera, fish bones, compacted aggregates of siliciclastic and biogenic fragments (possibly of fecal origin), and clays including mixed layer illite/smectites, some illite, and kaolinite. Locally important, but not discussed here in detail, are interbedded sediment gravity flow deposits including pelletal phosphorite and glauconite, and arkosic pebbly sandstones, most of which appear to be grainflows, and coarse-grained turbidites.

A striking feature in Sandholdt rocks is the high variability in the degree of lamination (Figs. 2, 3). Samples range from highly laminated shales to nearly massive (homogenized) calcareous mudstones (Fig. 2B). In laminated samples, depositional laminae range from 0.25 mm to 2.0 mm in thickness, averaging near 1.0 mm. Thin, wavy, organic-rich zones commonly parallel lamination (Fig. 3A) and, as a result, visually accentuate the degree of lamination in the rock. These organic-rich bands are similar in texture to bacterial "mat" structures reported off Peru today (Reimers, 1982), and for the Monterey Formation (Williams and Reimers, 1983). The "mats" are thought to represent bacterial sheaths and filaments binding the sediment into a cohesive layer. The result is a porous, "spongy," organic-rich mat running parallel to lamination. In highly laminated Sandholdt shales, organic "mats" are well developed with few disturbances or discontinuities (Fig. 3A,B,D; 2C). SEM analyses (Fig. 2E) reveal the porous, spongy texture of these organic mats, and post-compactional remnants of bacterial sheaths and veils (Fig. 2E; Mertz, in prep.). Well laminated shales are clearly laminated in outcrop (Fig. 2A), and display fissility. Moderately to faintly laminated samples generally lack fissility, and show increasing signs of disruption ("mottling") of lamination by both micro- and macro-bioturbation (Fig. 2B; 3C). Depositional lamina, and organic rich mats are still evident in hand specimen and thin-section, but commonly pinch out, or are clearly disturbed by activity of a burrowing infauna, possibly polychaete worms (Fig. 2D; Demaison, G., oral. comm., 1982). Massive, homogenized mudstones, volumetrically less important, are pervasively bioturbated and lamination is obliterated (Fig. 2B). Variability in the degree of lamination exists both basinwide, and within given sections often on a bed-to-bed basis.

Organic Geochemistry

Variability in lamination is clearly reflected in organic geochemical data (TOC and ROCK-EVAL pyrolosis). These are summarized on a Van-Krevelen type plot (Fig. 4; see also Mertz et al., 1983; Tissot and Welte, 1978). Highly laminated samples are typically organic-rich ($4 < TOC < 7\%$), and are characterized by high hydrogen and low oxygen indices. These shales are excellent potential oil source rocks. Moderately and faintly laminated rocks (Fig. 4) are moderately organic-rich to organic-poor ($0.5 < TOC < 4\%$) and are less hydrogen-rich and more oxygen-rich than highly laminated samples. Massive, bioturbated rocks are organic-poor (<0.5% TOC), and whatever kerogen is left tends to be very oxygen-rich. Organic matter in dolomite bodies rarely exceeds 1.0% TOC, and is distinctly hydrogen-poor and oxygen-rich.

Typically, a Van Krevelen diagram (Fig. 4) reflects variable kerogen sources (Tissot et al., 1974; Tissot and Welte, 1978). Path I indicates lipid-rich, algal kerogens of marine origin. Path II includes "mixed" kerogens of marine origin. Path III reflects a significant input of humic-rich, terrestrial organic carbon. The paths themselves indicate increasing "maturity" during burial where type I kerogens are oil-prone, type II are oil and gas prone, and type III are gas prone. Figure 4 suggests a very different trend for the Sandholdt depicting a strong inverse relationship between hydrogen index and oxygen index values. This may indicate increasing oxidation and bacterial breakdown of organic matter following deposition (Bein and Sandler, 1983), and increasing degree of micro- and macro-bioturbation (Pratt and Claypool, 1983), in samples with higher oxygen indices.

T-MAX values (the temperature of maximum hydrocarbon generation during ROCK-EVAL pyrolysis) and vitrinite reflectance data (Ro) are shown in Table I. T-MAX values average 416°C and range from 308° to 438°C. Vitrinite reflectance values average 0.40% and range from 0.31 to 0.59%. The

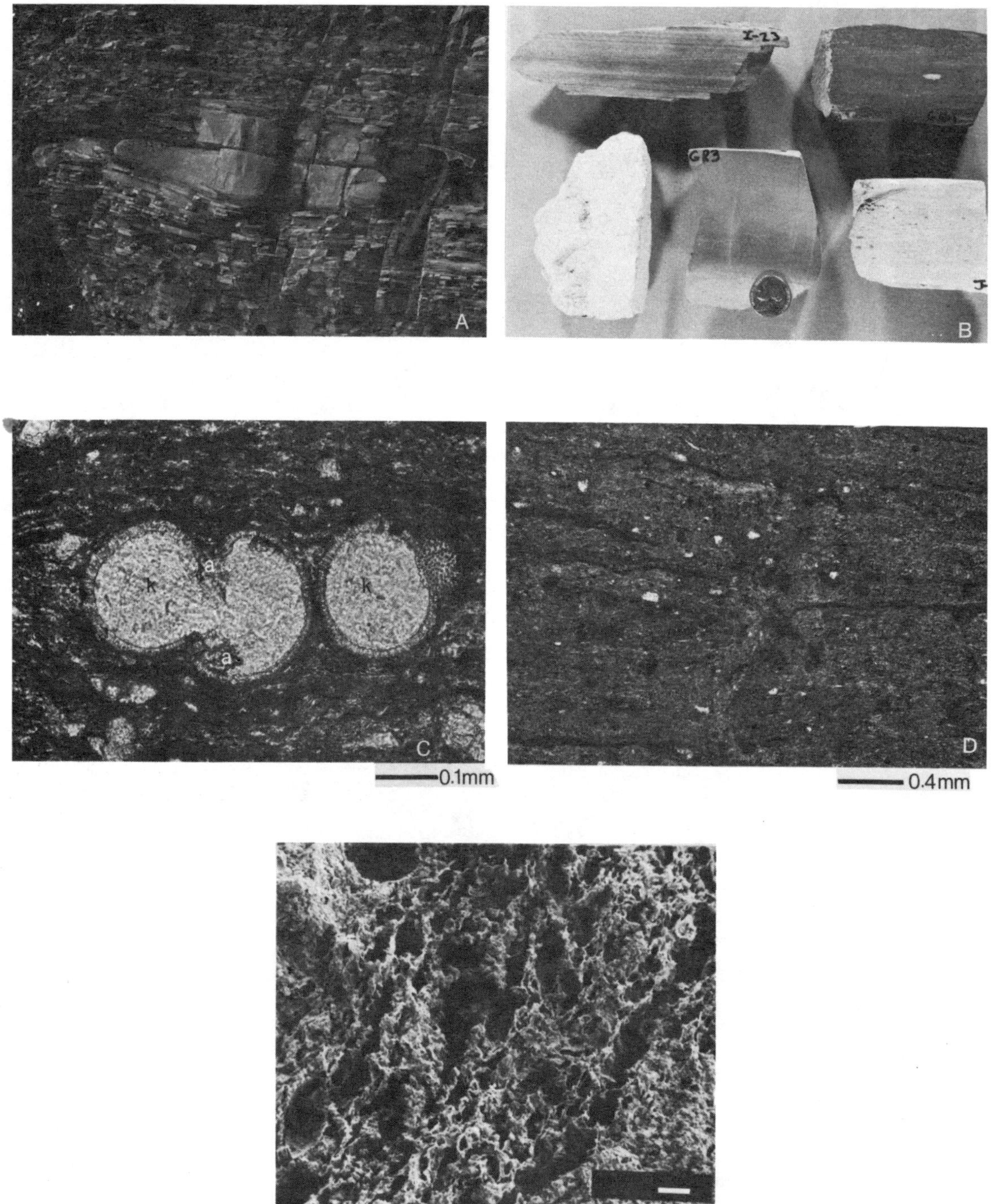

Fig. 2: (A) Outcrop photograph showing lensoid dolomite body enclosed in highly laminated, organic-rich hemipelagic shale; (B) Hand specimens of typical Sandholdt lithotypes. Clockwise from upper left: highly laminated shale; moderately laminated shale (note small apatite concretion); faintly laminated ("mottled") mudstones (2 examples); and massive-bioturbated mudstone (note burrows approximately 1.0 cm across); (C) Thin-section photomicrograph of foraminiferal test in laminated shale. Kaolinite (K) fills test chambers. A small amount of previously formed apatite (a) is visible on the side of the test (see also Fig. 6A,B); (D) Moderately laminated shale laterally equivalent to lensoid dolomite body. Note distinct disruption of lamination by small burrow, possibly a worm trace; and (E) SEM photomicrograph of organic-rich, wavy bands visible in figures 2C, D and Fig. 3A, B, D. Note porous spongy texture, and small "veils" and filaments binding sedimentary constituents. Texture is strikingly similar to organic-rich, bacterial "mat" zones reported off Peru (Reimers, 1982) and for the Monterey Formation (Williams and Reimers, 1983). Scale in microns (white bar).

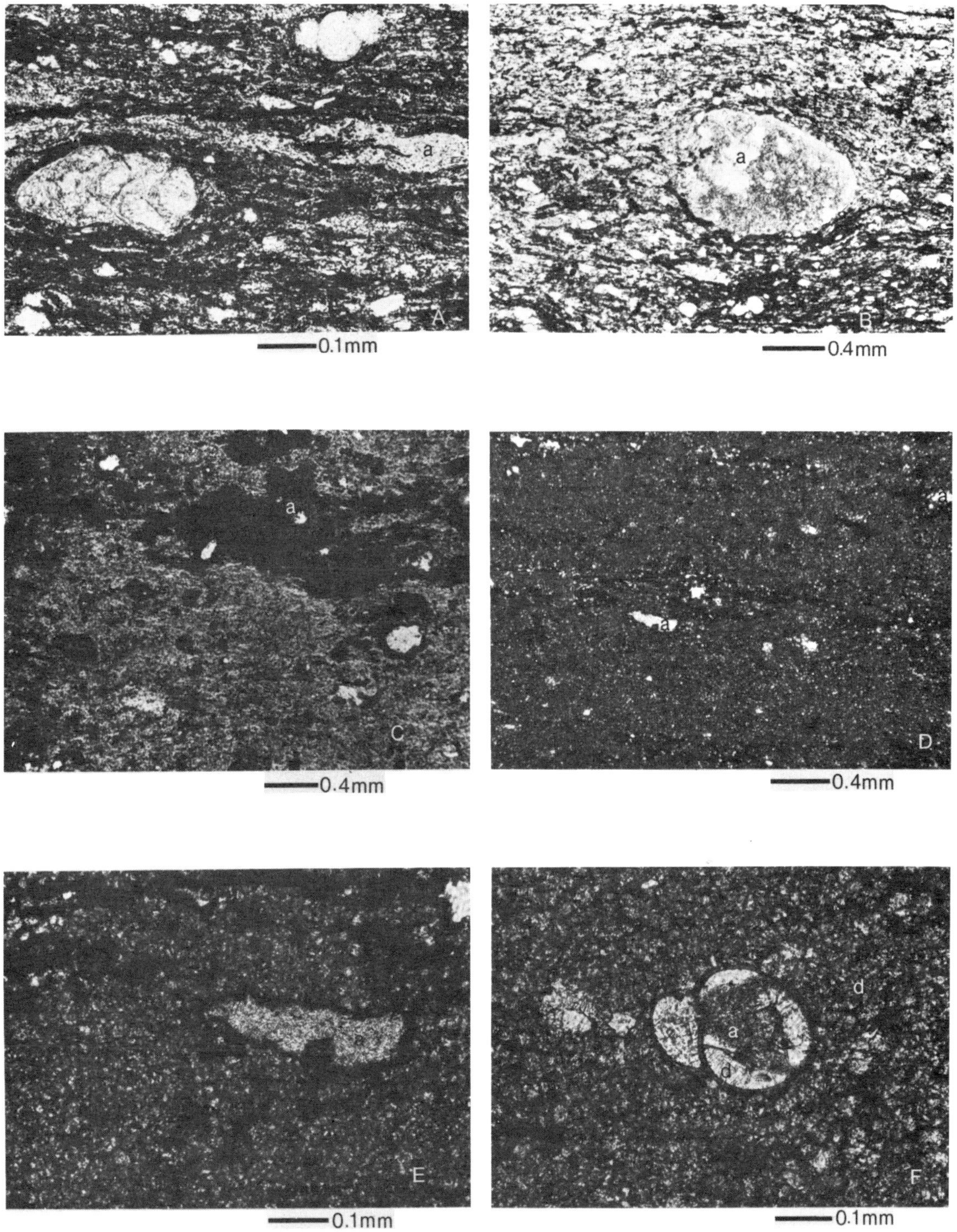

Fig. 3: Thin-section photomicrographs (scale in millimeters): (A) Example of highly laminated, organic-rich shale. Note foraminifera, thin organic-rich zones (dark stringers), and small apatite lens (a) with lamina wrapped around it; (B) *in situ* apatite (a) concretion in moderately laminated shale. Note "wrapping" or bending of lamination around apatite; (C) Faintly laminated, "mottled," sample. Lamination is highly disturbed. Note apatite blebs (a) associated with organic-rich zones, probably old bacterial mat zones; (D) Overview of "sugary" microcrystalline dolomite. Note faint lamination (relict organic-rich mat zones), and previously formed apatite concretions (a) enveloped by authigenic dolomite; (E) Detail of "sugary" dolomite texture, apatite lens, and laminated, organic-rich zone. Apatite is often associated with these zones both in shales and dolomite; (F) Dolomitized foraminiferal test in dolomite matrix. Apatite (a) formed early, later surrounded by large dolomite rhombs (d) in chamber.

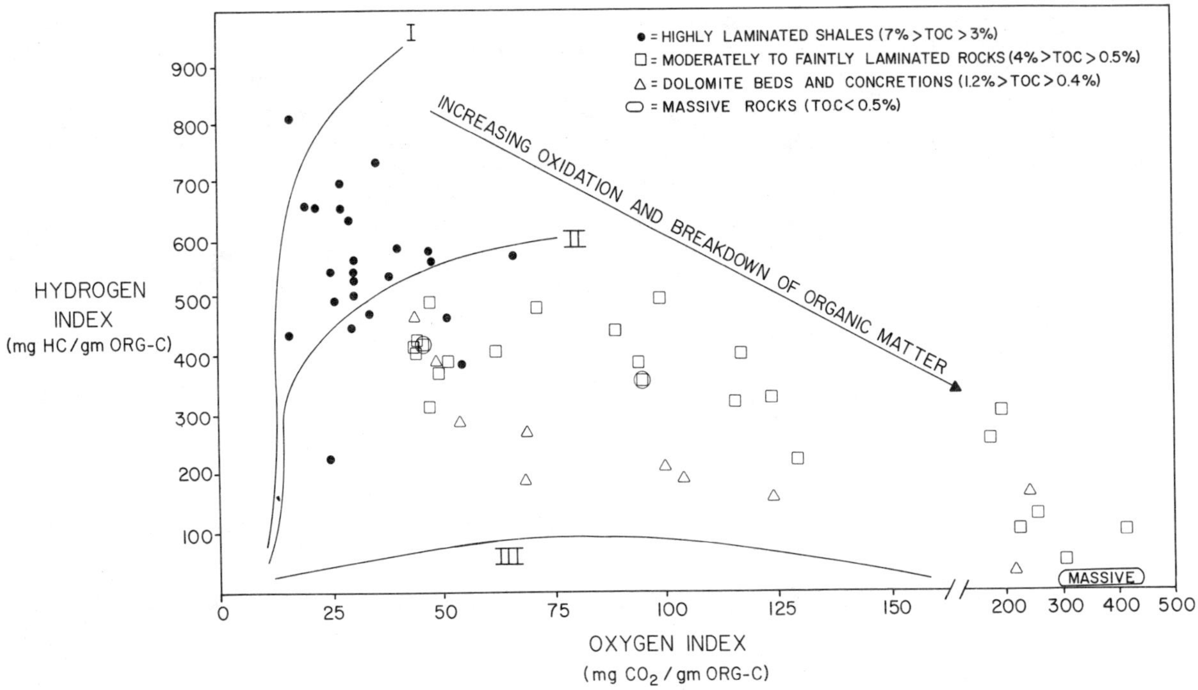

Fig. 4: Organic geochemical data from ROCK-EVAL pyrolysis on Sandholdt shales and dolomite bodies depicted on a Van Krevelen-type diagram (Tissot et al., 1974). Hydrogen index is derived by dividing S_2 (mg hydrocarbons/gm of rock) by TOC. The oxygen index is S_3 (mg CO_2/gm rock) divided by TOC. Circled values are from shales lateral to lensoid dolomite bodies. Massive rocks include eight samples. For discussion see text.

productivity ratio, S_1/S_1+S_2 (S_1 represents the "free" hydrocarbons in the sample and S_2 = mg of hydrocarbons/gm of rock derived from kerogen "cracking") is consistently low, rarely exceeding 0.04. This, in addition to T-MAX and vitrinite reflectance data, suggests that the majority of Sandholdt samples have not entered the main phase of oil generation (G. Demaison, oral and written comm., 1983).

Authigenic Phases

Apatite

Apatite (carbonate fluorapatite-$Ca_5(PO_4, CO_3, SO_4)_3F$--see Table II) occurs in most Sandholdt rocks either as in situ concretions, and stringers in shales ($0.2 < P_2O_5 < 5\%$) or as phosphatic grains in reworked pelletal phosphorites (P_2O_5 approx. 18%). In shale (Fig. 3A,B,C,E,F) apatite occurs by replacing calcareous material (usually foraminifera), or, more commonly, as small (.05 mm to 3.0 cm) in situ authigenic micro-concretions that formed prior to any significant compaction as evidenced by the distinct "wrapping" of host shale lamina around concretions (Fig. 3B). Apatite concretions are nearly pure carbonate fluorapatite (only minor calcite and/or silica), and are commonly associated with organic-rich mat structures (Figs. 3A,B,C). Small clusters of apatite crystallites also form in micro-pore space, often foraminiferal test chambers which they may share with other authigenic phases as pyrite, kaolinite, or dolomite (Fig. 3F; 2C; 6A,B). Occasionally, concretions are nucleated around variable material including siliciclastic grains, fish bones, and foraminiferal tests.

Apatite occurs in all rock types, but varies inversely with organic content and degree of lamination (Fig. 5A), being most common in moderately to faintly laminated (moderate to low TOC) samples. Apatite even occurs in massive rocks. Relative to dolomite (Fig. 5B), which is consistently more abundant in highly laminated, organic-rich rocks (Fig. 5C), apatite becomes the more significant authigenic phase of the two in samples that are more poorly laminated and contain less organic carbon. Most dolomite-rich rocks contain less than 0.5% P_2O_5, while dolomite is insignificant in many apatite-rich samples. In outcrop, apatite-rich sections contain few discrete dolomite bodies, whereas in sections dominated by laminated, organic-rich shales dolomitized zones are abundant.

Petrographic and isotopic evidence suggests that apatite forms during earliest diagenesis, before most other phases. Apatite appears to form contemporaneously with pyrite as suggested by X-ray analyses on SEM images (Table II), which reveal that clusters of apatite crystallites often contain iron sulphides, and pyrite framboids occasionally contain detectable phosphate. These relationships are not visually apparent in SEM, but are inferred from data in Table II which suggests a close association between apatite and iron sulphides. Apatite generally pre-dates dolomite, usually occurring as discrete concretions enveloped by later dolomite (Fig. 3D,E), or crystallites in

Table I

T-MAX (°C) Data from ROCK-EVAL Pyrolysis; and Vitrinite Reflectance Values from Selected Sandholdt Shales.

Locality*	No. of Samples	T-MAX(°C) Ave.	T-MAX(°C) Range	Sample #	Vitrinite Reflectance R_o (%)
AS:	22	419°	(413-425°)	AS-4	0.38
V:	11	413°	(400-433°)	GR-8	0.39
I:	4	423°	(407-430°)	GR-9	0.33
GR:	12	409°	(398-422°)	GR-10	0.42
C-14:	2	424°	(421-426°)	GR-11	0.35
C-233:	2	439°	(436-442°)	I-11	0.54
F-99:	2	414°	(408-421°)	I-13	0.48
	N=55	$\bar{x}$ = 417°C		I-21	0.40
				C-233-7552	0.59
				C-14-4830	0.31
				F-99-3645	0.35
				F-99-3675	0.34
					$\bar{x}$=0.40%

*AS = Arroyo Seco; V = Vaqueros Canyon; GR = Graves Creek; I = Indians Ranch. C-14, C-233, and F-99 are well cores (see Figs. 1, 12).

Table II

Qualitative Element Identification: X-Ray microanalysis (EDS) of Scanning Electron microscope images

Mineral:	Apatite	Dolomite	Pyrite	Kaolinite	Quartz	Calcite	Illite/smectite(?)
Habit:	concretions	Isolated Rhombs and Microcrystalline matrix	Framboidal	Authigenic "Books"	Euhedral Hexagonal Pyramids	Foram test walls and Chambers	Distinctly banded (layered) authigenic clay
Elements:	P Ca S (Fe)*	Ca Mg Fe	Fe S (P)* (Ca)*	Al Si	Si	Ca (Fe)**	K Al Si Fe (Mg) (Ca)

* = Fe in apatite, and P and Ca in pyrite, probably reflect presence of iron sulphide with apatite crystallites, and small amounts of apatite in some pyrite framboids.

( ) = small, but clearly detectable peaks

** = occasional; most samples are "pure" low-Mg calcite.

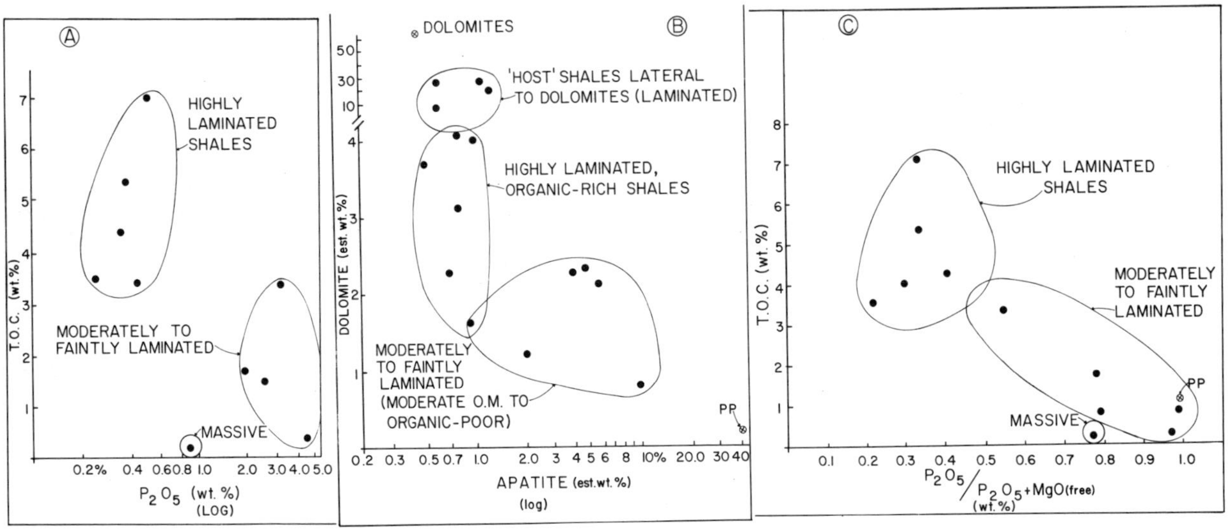

Fig. 5: Relationship between apatite, dolomite and organic-carbon (expressed as TOC) in Sandholdt shales and mudstones: (A) Plot of TOC values verus weight percent P_2O_5 determined by whole rock XRF analyses (concretions of "pure" carbonate-fluorapatite were carefully avoided). % P_2O_5 is assumed to primarily reflect authigenic apatite. Organically bound P is probably a relatively small constituent based on known TOC values, and the fact that in most marine organic matter there is only one "P" for each 106 carbons (Redfield et al., 1963). (B) Plot of estimated weight % dolomite versus weight % apatite (log scale) in varying Sandholdt lithologies. Estimates are based on chemical (XRF) and XRD data using methods of Isaacs (1980). "PP" = pelletal phosphorites. Note distinct increase in apatite (in Fig. 6A and B) in moderately to faintly laminated, moderately organic-rich shales; (C) Plot of organic-richness (as TOC) versus the ratio of P_2O_5/P_2O_5 + MgO. This is meant to show, for varying lithotypes, the relative abundance of apatite compared to dolomite. Apatite is expressed as wt % P_2O_5. Weight % MgO, after an approximate correction for Mg^{2+} in aluminosilicates (Isaacs, 1980), is assumed to be roughly indicative of dolomite contents. In organic-rich rocks dolomite is the more significant phase. In samples with lower TOC values, apatite becomes increasingly more abundant relative to dolomite.

foraminiferal tests surrounded by later-forming dolomite rhombs (Fig. 3F). Finally, apatite appears to form prior to authigenic kaolinite (Fig. 2C; Fig. 6A,B).

Isotope analyses on CO_3^{2-} and PO_4^{3-} in apatite are summarized in Fig. 8A,B and Table III. Carbon-13 values are generally "light" ranging from +3 ‰ to -12 ‰ PDB, suggesting an early diagenetic origin for these apatites (see discussion below). Oxygen-18 values in both CO_3^{2-} and PO_4^{3-} are consistently "light" (depleted in δ-18). These data fall within the range of values reported for other phosphatic material (Benmore et al., 1983; Shemesh et al., in press), and contrast markedly (Fig. 8B) with isotopic values of associated dolomites.

Pyrite

Confirmed by XRD and X-ray analysis of SEM images (Fig. 6C; Table I), pyrite occurs in small, but consistent amounts in all Sandholdt sections, and in all rock types with the exception of highly bioturbated, massive samples. Pyrite framboids (or an iron-sulphide precursor) appear to form very early, commonly associated with apatite (see above), and usually before authigenic dolomite (Fig. 7A), and kaolinite (Fig. 6D,E). Early formation is consistent with accepted notions that pyrite is a common byproduct of sulphate reducing reactions near the sediment-water interface under anaerobic conditions (Kaplan, 1983; Claypool and Kaplan, 1974; Altschuler et al., 1983).

Dolomite

Authigenic dolomite occurs as discrete concretions and lensoid beds (Fig. 2A), or as disseminated rhombs in interbedded shales and shales lateral to dolomite bodies. Dolomites contain significant amounts of iron as evidenced by XRF and X-ray data on SEM images (Table II). Relict lamination is usually visible in hand specimen and thin section (Fig. 3D,E). Lamination is commonly paralleled by wavy, irregular organic-rich zones thought, perhaps, to be bacterial mat structures. These organic-rich zones commonly contain apatite (Fig. 3D,E), and later authigenic phases such as as kaolinite (Fig. 7D).

Dolomite bodies appear to form by replacement of laminated host shales (Pisciotto and Garrison, 1981; Hennessy and Knauth, 1983). Relict lamination clearly runs from laterally equivalent shales through the dolomitized zone. Constituents of the host shale, including replaced foraminifera, siliciclastic grains, organic-rich "mat" laminations, and fish debris, are found "floating" or dispersed in a matrix of sugary, crystalline dolomite, suggesting that displacive growth (Roehl, 1981) took place as well as replacement (Fig. 2A; 3D,E,F; 6F). This, along with differential compaction, explains why dolomite bodies as

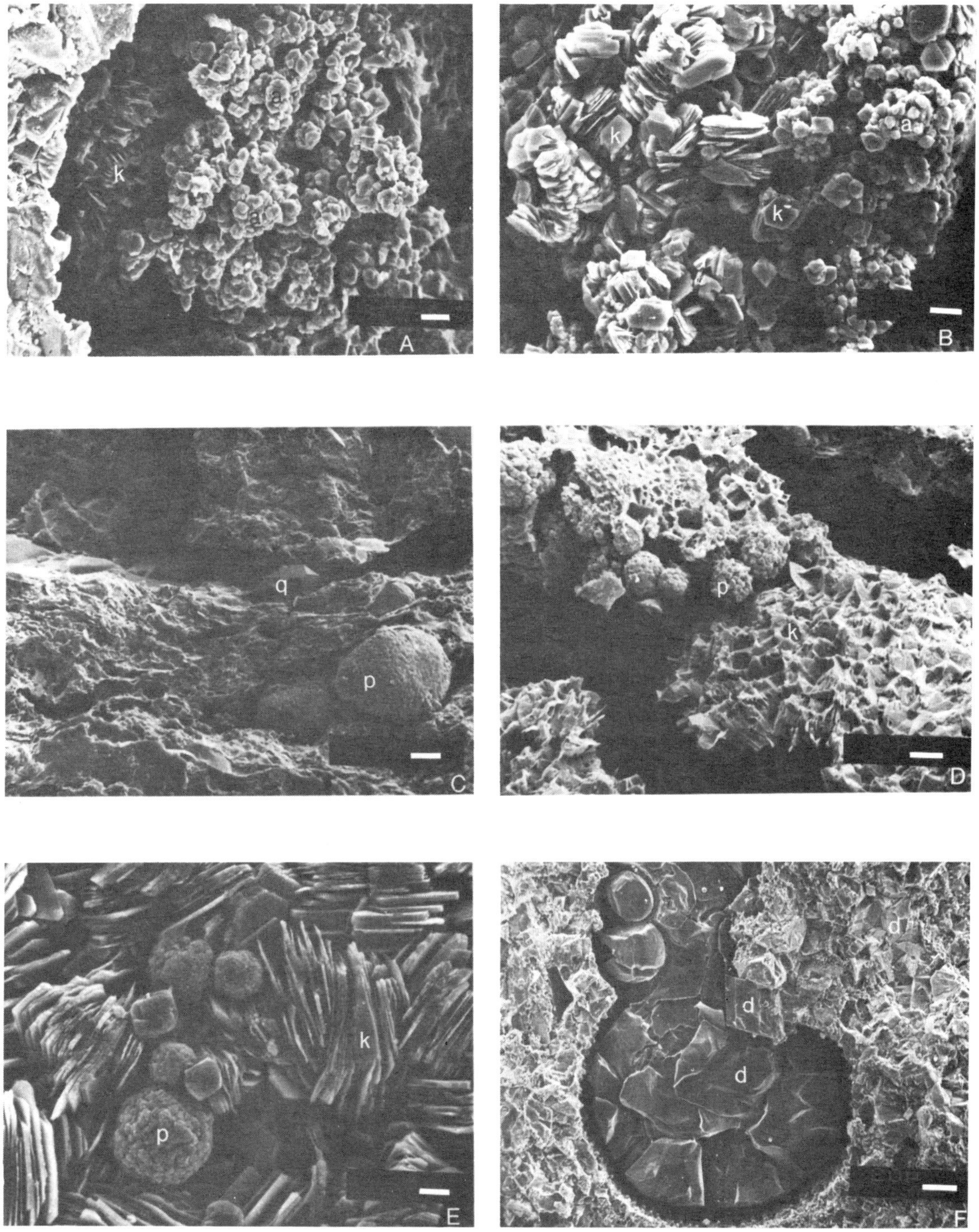

Fig. 6: Scanning electron micrographs. Scale in microns (white bars). (A) and (B) Apatite (a) and kaolinite (k) in two foraminiferal test chambers. Kaolinite, filling test chamber, post-dates formation of apatite which is shown as it commonly occurs, along fringes of chambers and test walls (see also Fig. 2C); (C) Pyrite framboids (p) and authigenic quartz crystal (q) forming in "wavy," porous mat lamination zone; (D) Pyrite (p), and kaolinite (k) in foraminiferal test. Pyrite formed early along fringes of test followed by infilling of foraminiferal cavity by authigenic kaolinite; (E) Detail of authigenic kaolinite (note "book" structure) completely filling a foraminiferal test chamber. Pyrite framboids (p) formed earlier near edge of chamber and are seen "enveloped" by void-filling kaolinite. (F) Typical SEM view of completely dolomitized foraminifera in matrix of microcrystalline dolomite.

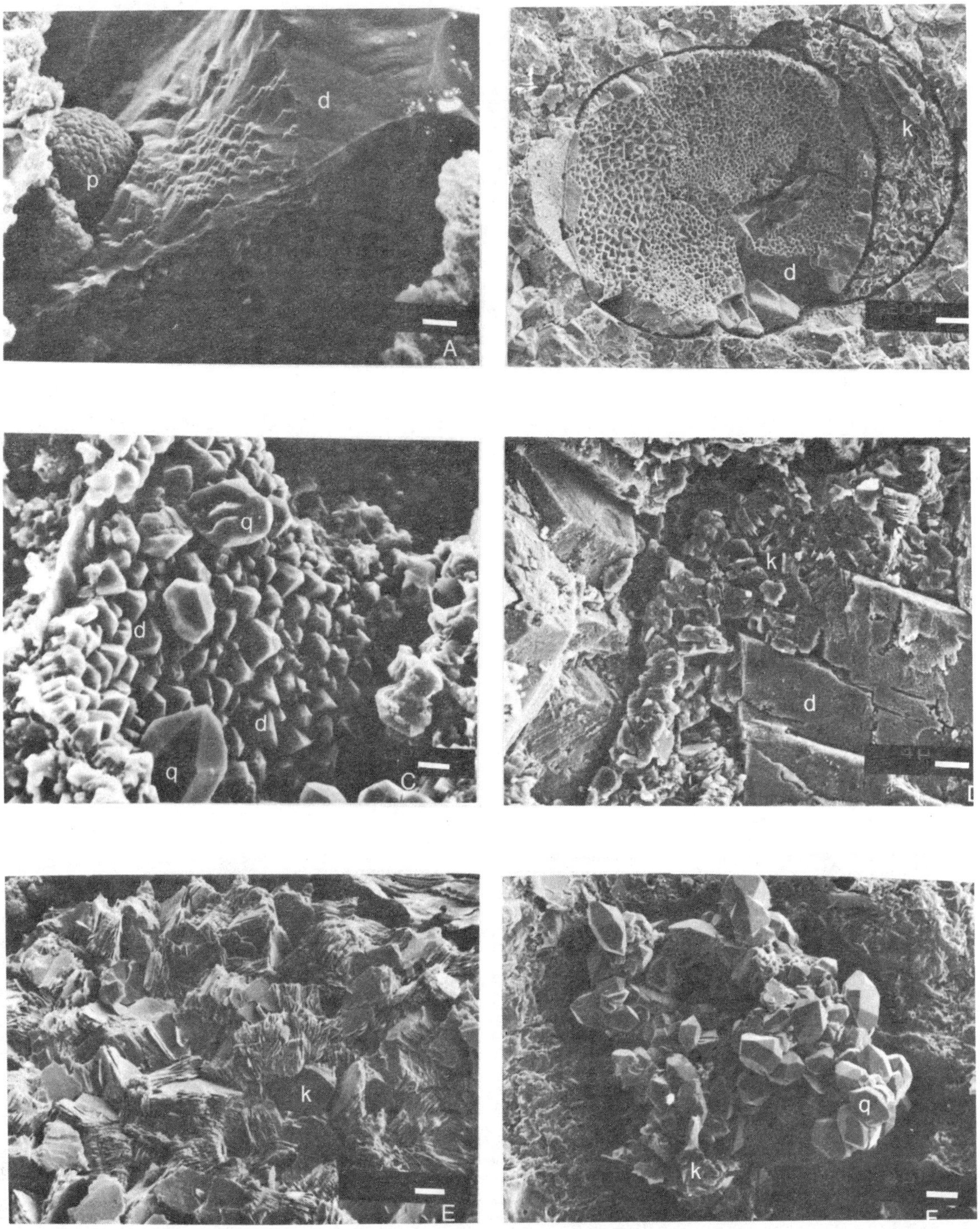

Fig. 7: Scanning electron photomicrographs. Scale in microns (white bars). (A) Dolomite (d) surrounding and enclosing previously formed pyrite (p) in foraminiferal test; (B) Dolomite (d) and kaolinite (k) each filling a separate chamber in a foraminiferal test; (C) dolomite (d) replacing a foraminiferal test wall, with later quartz crystals (q) forming within the chamber; (D) Dolomite (d) and kaolinite (k) in a dolomitized bed. Euhedral dolomite rhombs pre-date authigenic kaolinite "books" shown filling interstices between rhombs; (E) Excellent example of authigenic kaolinite infilling a foraminiferal test chamber (note distinct "book" structure); (F) Cluster of authigenic quartz crystals (q) infilling test chamber. Note kaolinite (k) at bottom of quartz cluster.

Table III
Isotopic Data: Authigenic Dolomite and Apatite

A. Dolomite

Sample	Lithology	$\delta^{13}C_{PDB}$	$\delta^{18}O_{PDB}$	Calculated Temp* (°C)
AS-13	Dolomite bed or concretion	+6.88°/oo	-3.68°/oo	50.1°
AS-25	"	+10.42	-1.16	35.8
AS-19	"	-1.94	-2.73	44.5
AS-106-T	"	+4.75	-5.04	
AS-106-B	"	+8.50	-1.77	
AS-104-M	"	+9.32	+0.18	28.9
AS-104-C	"	+11.77	+2.31	18.9
AS-104-T	"	+7.89	+1.40	23.1
AS-101-B	"	+1.09	-2.00	40.4
AS-101-T	"	+3.92	-1.18	35.9
AS-101-M	"	+7.59	+0.41	27.8
AS-102-B	"	+6.07	+0.15	29.1
AS-102-M	"	+6.54	+0.52	27.2
AS-102-T	"	+6.15	+2.23	20.9
028-T	"	+10.30	(+8.98?)	(-7.0?)
028-B	"	+9.49	+0.71	26.4
AS-25-T	"	+9.84	-1.21	36.1
AS-25-UM	"	+10.89	+1.41	23.1
AS-25-LM	"	+8.31	-1.88	39.7
AS-16-M	"	+6.90	-3.00	46.1
AS-16-B	"	+7.39	-2.67	44.2
V-15	"	+11.34	-2.16	41.3
V-8	"	+16.66	+0.47	27.5
I-19	"	-4.16	-6.35	67.8
I-7	"	-6.16	-6.41	68.3
GR-18	"	+5.99	-1.26	36.4

*Temperatures were calculated on the basis of $\delta^{18}O$ data using the protodolomite-water temperature expression of Fritz and Smith (1970): $1000 \ln \alpha = 2.78 \times 10^6 T^{-2} + .11$, where T = temperature in degrees Kelvin, and $\alpha = 1000 + \delta^{18}O$ dolomite/$1000 + \delta^{18}O$ water. Estimated depths and temperatures of formation are depicted in Figure 8C. Bases for these values are discussed in the text and figure caption.

Localities (see Fig. 1 or 12):
AS = Arroyo Seco; V = Vaqueros Canyon; I = Indians Ranch; GR = Graves Creek; J = Bear Trap Road (Sam Jones Canyon); CK = Cosio Knob.

B. Apatite--CO_3^{2-}

Sample	Lithology	$\delta^{13}C_{PBD}$ in CO_3^{2-}	$\delta^{18}O_{PDB}$ in CO_3^{2-}
AS-15	apatite micro-concretion	-3.00°/oo	-10.61°/oo
AS-22	"	-2.61	-10.28
AS-26	"	-1.83	-10.73
AS-37	"	-2.33	-11.09
V-9a	"	+3.05	-7.89
V-14	"	-0.85	-3.26
J-6	"	-6.58	-7.85
J-3	Pelletal phosphorite	-11.82	-8.99
I-3	"	+1.49	-11.07
I-12	"	-1.89	-10.18
I-7	Apatite concretion	-6.51	-7.34

Table III (continued)

C. Apatite--PO_4^{3-}

Sample	Lithology	$\delta^{18}O_{SMOW}$ in PO_4^{3-}
AS-15	Apatite concretion	+16.5 °/oo SMOW
SC-41	"	+15.4
V-14	"	+15.4
I-2	"	+14.6
AS-10	"	+15.1
I-13	"	+14.7
AS-22	"	+16.6
J-6	"	+17.9
V-26	"	+17.3
AS-45	"	+17.1
AS-46	"	+16.6
AS-49	"	+16.3
V-9a	"	+17.3
AS-42	Pelletal phosphorite	+16.7
I-3	"	+15.7
I-12	"	+14.9
AS-31	"	+14.95
CK-11	"	+18.17
KM-81-M-4	"	+17.70

measured in the field are approximately 1.5 to 4 times (ave. = 2.6x) as thick as replaced, laterally equivalent shales.

Dolomite often occurs with small amounts of apatite (Figs. 5B,C), and is most common in organic-rich, highly laminated shales, becoming less significant in more poorly laminated, organic-poor rocks. In laminated shales, dolomitized beds (up to 55-60 weight percent dolomite) occur regularly every three to five meters. Interbedded shales rarely contain more than 3-10% dolomite, which occurs mostly as scattered rhombs or replaced foraminiferal tests.

Preliminary XRD and XRF data from shale beds laterally equivalent to dolomite bodies suggest that these beds are less detrital rich (Fig. 9A), and somewhat more calcite-rich (Fig. 9B) than interbedded shales without dolomite bodies. Additionally, organic matter in dolomites is notably less hydrogen-rich and more oxygen-rich (Fig. 4) than organic-rich interbedded rocks. TOC values of dolomites range from 0.5 to 1.2%, and average 4 to 6 times lower than interbedded shales. Evidence for micro-bioturbation is found in many shales lateral to dolomite bodies and in dolomite bodies themselves, usually taking the form of small burrows (possibly made by worms) disturbing lamination (Fig. 2D). This may, in part, explain lower organic matter concentrations.

Most dolomite appears to form later than apatite (Fig. 3D,E,F), completely enveloping previously formed apatite concretions in dolomitized shales, or apatite occurring in foraminiferal test chambers. Dolomite clearly post-dates pyrite formation, enveloping and enclosing framboids (Fig. 7A). Dolomite-kaolinite relationships are less clear, but it appears that much kaolinite is contemporaneous with and/or post-dates dolomite formation (Fig. 7B,D). Often, one foraminiferal test chamber will be dolomitized, while another will be completely filled with kaolinite (Fig. 7B). Lastly, dolomite seems to form before the appearance of diagenetic quartz (Fig. 8C).

Isotopic analyses on dolomites are shown in figure 8B,C and Table III. Carbon-13 values are consistently "heavy," averaging +8 °/oo to +9 °/oo PDB, but ranging up to +16 °/oo PDB. Oxygen-18 values vary from near +2 °/oo to -6 °/oo PDB, probably a reflection of varying burial temperatures (see Methodology). Samples were taken through dolomite beds and concretions, on a line perpendicular to bedding, with use of a paleomagnetic core drill. The relationship between $\delta^{13}C$ and $\delta^{18}O$ from cores, and individual samples from hand specimens, is depicted in Figure 8C. Implications of $\delta^{18}O$ values for burial temperatures, and $\delta^{13}C$ values as indicators of diagenetic environments (Pisciotto and Mahoney, 1981; Irwin et al., 1977) will be discussed in greater detail below.

Kaolinite

Kaolinite is confirmed by XRD, and occurs as an authigenic mineral in both thin-section (Fig. 2C) and SEM (Fig. 6D, E; 7E). X-ray analyses of SEM images (Table II) likewise suggest the presence of kaolinite. It occurs as distinct "book" structures of aggregated crystals which have low birefringence, and are optically negative, commonly filling micro-pore space (especially test-chambers). It is most common in detrital-rich rocks where aluminosilicates (other clays and K-feldspar) provided a good source of required Al and

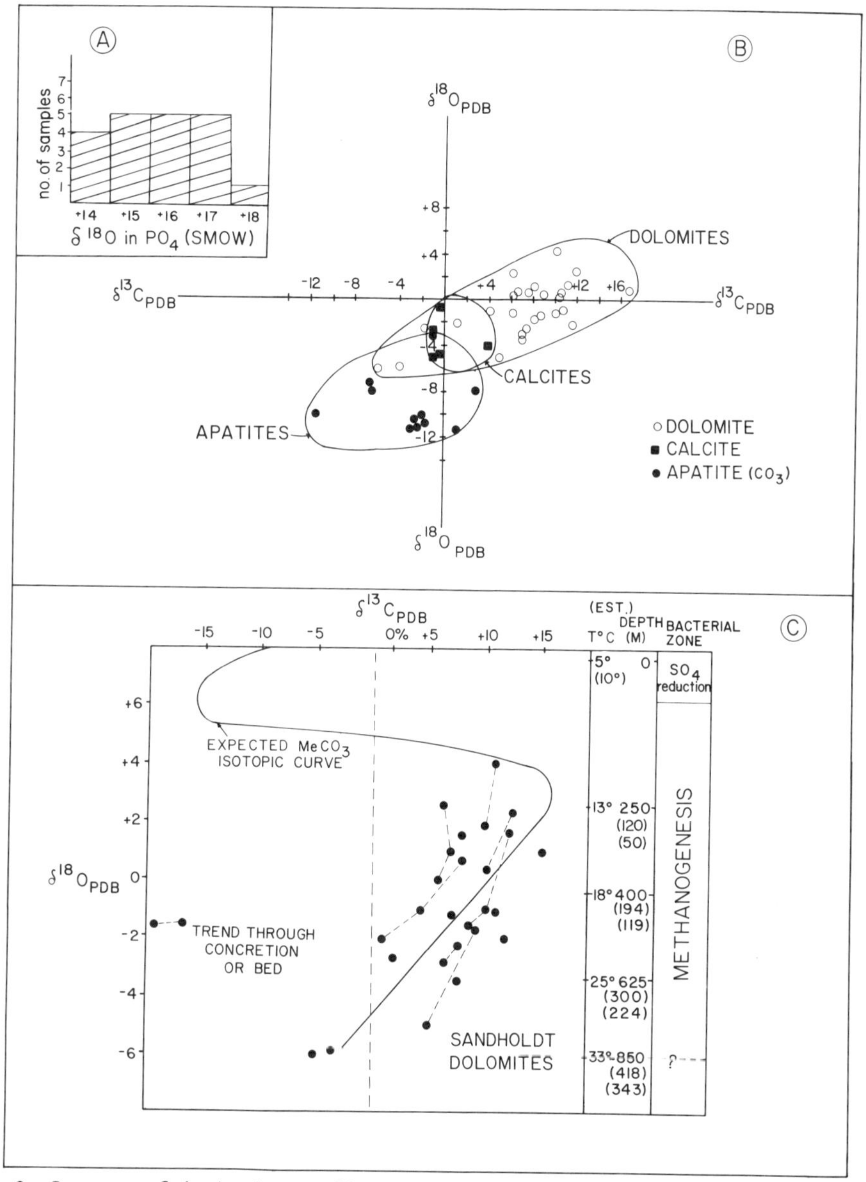

Fig. 8: Summary of isotopic results on Sandholdt apatites and dolomites. (A) Oxygen-18 values from PO_4^{3-} in apatite concretions and pelletal phosphorites; (B) Summary plot of $\delta^{18}O$ versus $\delta^{13}C$ from CO_3^{2-} in dolomites, apatite concretions and pelletal phosphorites, and whole-rock, calcite-rich samples; (C) Plot of dolomite isotopic values (from 8b) superimposed on the expected isotopic curve for carbonate forming in various zones of microbial activity during burial (Pisciotto and Mahoney, 1981; Irwin et al., 1977). Burial temperatures and depths of dolomite formation are estimates based on $\delta^{18}O$ data. Bottom water temperatures are assumed to be 5°C, but may range as high as 10°C which would result in shallower depths of formation. Temperature and depth values outside of parentheses are based on: (1) an estimated average linear geothermal gradient of 32°-33°C/Km; (2) an <u>average</u> depletion in pore water $\delta^{18}O$ of -0.3 ^{o}/oo per 100 meters of estimated burial depth (Pisciotto, 1981b); and (3) a bottom water temperature of 5°C. However, since the geothermal gradient may increase by a factor of 2 or more in unconsolidated sediment (Isaacs, 1984; see discussion) compared to compacted shales, a much higher geotherm (approx. 67°C/Km) may have prevailed during early burial when dolomites were forming. Depth of formation estimates in <u>parentheses</u> reflect these effects. The first figure takes an increased geotherm into account, and the second number incorporates <u>both</u> an increased initial geothermal gradient and a bottom water temperature of 10°C. Both depths in parentheses assume a high estimate of pore water $\delta^{18}O$ depletion (approx. 0.6 ^{o}/oo/100 meters burial--see discussion). This higher value is used to estimate the <u>shallowest</u> probable depths of dolomite formation given a combination of warmer bottom water temperatures and a higher initial geothermal gradient. Dashes connect samples taken on a straight line through concretions or lensoid beds, always perpendicular to bedding. Trends suggest continued accretion and growth of these dolomite bodies with burial.

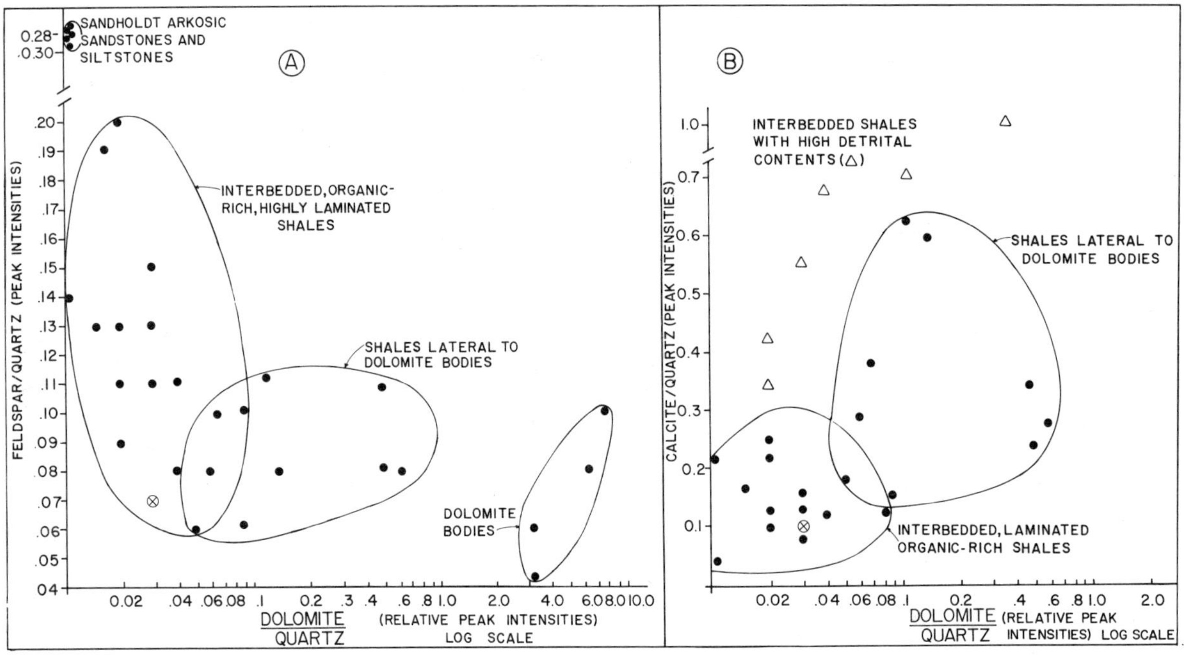

Fig. 9: (A) Plot of dolomite and feldspar XRD peak intensities, both relative to quartz, in Arroyo Seco organic-rich, laminated shales and dolomitized beds. Quartz is used as it is the largest peak, and is relatively constant in all samples (rarely changing more than ±25%). Feldspar is used as an approx. indicator of the relative amount of detrital material in the samples. Dolomite/quartz and feldspar/quartz show a strong inverse relationship, suggesting that less detrital-rich rocks may be more dolomitic. Normalizing the data to quartz avoids possible problems with dilation and volume expansion during dolomitization, and effects this would have on apparent detrital-richness. (B) The same samples are plotted for calcite versus dolomite, both relative to quartz. Shales lateral to dolomites tend to be more calcite-rich compared to interbedded shales. The triangles show samples of interbedded shales that are calcite-rich, but dolomite-poor. These rocks, however, are very detrital-rich which may affect dolomitization. Results of Fig. 9 are also suggested by chemical (XRF) data showing that shales lateral to dolomites contain less weight % aluminosilicates (and more dolomite) than interbedded shales.

Si (Hoffman and Hower, 1979; Ardito, 1983). Most kaolinite appears to form later than apatite (Fig. 2C; 6A,B), and pyrite (Fig. 6D,E). It is contemporaneous with (Fig. 7B), or post-dates dolomite precipitation (Fig. 7D). In the latter case, euhedral dolomite rhombs commonly have kaolinite books filling interstices between rhombs. Kaolinite-quartz relationships are unclear, but some kaolinite may have formed prior to diagenetic quartz (Fig. 7F), this being consistent with concepts that most kaolinite forms during relatively shallow burial (Singer and Müller, 1983). It can form, however, at greater depths (Kisch, 1983). Other clay minerals, identified by XRD and SEM (Table I), include mixed-layer illite/smectites, small amounts of illite, and glauconite; the latter being locally important in pelletal phosphorites and arkosic sandstones.

Quartz and Calcite

Diagenetic quartz is confirmed by XRD and XRF (Isaacs, 1980) as the main silica phase in almost all Sandholdt samples. Opal-CT occurs throughout only one section (Graves Creek - Fig. 1), and at the Sandholdt-Hames contact at Bear Trap Road (Fig. 1). Diagenetic quartz was differentiated from detrital quartz by methods outlined in Isaacs (1980), the latter rarely exceeding 15 weight percent. Diagenetic quartz post-dates formation of apatite, pyrite and dolomite and usually occurs as euhedral, hexagonal, pyramids (2-10 microns) in SEM (Fig. 7C,F). In thin-section, diagenetic quartz appears as a low-birefringent, microcrystalline groundmass. Silica transformations were not studied in detail, but the common presence of diagenetic quartz suggests that these rocks have experienced burial temperatures in excess of approximately 80°C (Pisciotto, 1981a; Isaacs et al., 1983).

Calcite constitutes approximately 4 to 35 weight % of Sandholdt shales, although it is difficult to gauge original amounts due to phosphatization, dolomitization, and silification of calcareous components. Isotopic data (Figure 8B), suggest that Sandholdt calcite reflects a "normal marine" $\delta^{13}C$ signature (Milliman, 1974). Analyses were run on calcite-rich, whole-rock samples which contained insignificant amounts of dolomite and/or apatite.

DISCUSSION

Mineralogy, Organic matter and Burial History

Organic geochemistry and mineralogic data put constraints on Sandholdt burial history, and allow clarification of the sequence and timing of authigenic phase formation. Additionally, knowledge of diagenesis yields important clues about the original depositional environment. The approximate timing of authigenic mineral formation during burial is schematically portrayed in Figure 10. No time scale is implied. The diagram is drawn to show the Sandholdt at its point of maximum, pre-uplift burial. Authigenic minerals are shown where they were thought to have formed when the unit was at shallower burial depths. The temperature scale is based on isotopic data from dolomites, silica phases (Isaacs et al., 1983), and organic geochemical data. A bottom water temperature of approximately 5°C is assumed based on values measured at depths of 1000-1500 meters in California Borderland basins (Douglas, 1981), depths similar to those hypothesized for much of the Sandholdt (Graham, 1976). Sub-bottom depths are based on section thicknesses from measured surface outcrops (Durham, 1974), and well core data (courtesy of Texaco Inc., Los Angeles). It is important to note that Sandholdt thickness, as well as total section thickness, varies appreciably from locality to locality. Figure 10 shows two Sandholdt sections at their point of maximum burial: (1) a thin (150 m) section (Bear Trap Road - Fig. 1) that is buried a maximum of 2400 meters; and (2) a thicker section (Arroyo Seco area - Fig. 1) where the Sandholdt is approximately 460+ meters thick. The entire section, measured from the base of the Sandholdt, experienced maximum burial depths of approximately 3300+ meters. The temperature scale is based on an average linear geothermal gradient of 32-33°/Km, although this may have been initially higher (Fig. 10). The geotherm is based on measured values of section thickness, inferred maximum burial temperatures based on organic geochemistry and mineralogy (see discussion below), and on heat flow data from the Monterey Formation in several southern California basins (Isaacs, 1984).

Apatite

Petrographic and isotopic results suggest that *in situ* apatite formed during earliest burial (Fig. 10), most likely within the zone of bacterial sulphate reduction. Early precipitation of apatite is well documented for modern environments off Peru (Burnett, 1977; Burnett et al., 1980, 1982); Mexico (Jahnke et al., 1983); southwest Africa (Baturin and Bezrukov, 1979; Birch, 1979; Baturin, 1981); the Gulf of California (H. Schrader, written comm. 1982); and off east Australia (O'Brien et al., 1981). Precipitation may occur in the top 10-30 cm of sediment (Jahnke et al., 1983; Burnett, 1977; Burnett et al., 1982) as evidenced by sediment pore water profiles showing rapid depletion of PO_4^{3-} and F^- with depth, and simultaneous precipitation of apatite (Froelich et al., 1982). Early diagenetic formation of apatite is also reported for several ancient examples including: the Permian Phosphoria in the western U.S. (Sheldon, 1981); Cretaceous phosphorites in Israel (Soudry and Champetier, 1983; Kolodny, 1980a; Nathan and Soudry, 1980); the Monterey Formation in California (Pisciotto and Garrison, 1981); and Pennsylvania Shales in Kansas and Oklahoma (Kidder, 1983).

"Light" carbon-13 isotopic values in apatite imply early authigenic formation, possibly associated with bacterial SO_4^{2-} reduction (Kolodny, 1980a; Benmore et al., 1983; McArthur et al., 1980). Oxygen-18 values in PO_4^{3-} are consistently "light" (Fig. 8A) suggesting diagenetic temperatures as high as 27-40°C (Shemesh et al., in press; Longinelli and Nuti, 1973). This apparent contradiction with earliest burial formation is enigmatic. $\delta^{13}C$ data from CO_3^{2-} in the apatite lattice (see below and Fig. 8B) suggest formation during earliest burial, however calculated temperatures are anomalously "warm." Higher diagenetic temperatures *may* reflect the "aging" process of apatite whereby an early amorphous, hydroxyl-apatite gel is converted to highly crystalline carbonate-fluorapatite, or francolite (Mahmoud, 1976). However, diagenetic recrystallizations at depth should alter the isotopic signature of carbon in apatite CO_3^{2-} as well as the $\delta^{18}O/\delta^{16}O$ ratio in PO_4^{3-}. Possible explanations may include a much higher initial geothermal gradient (see below), combined with higher bottom water temperatures; or pore waters that are highly depleted in $\delta^{18}O$, even at shallow burial depths.

Pyrite

Pyrite forms during earliest burial and is commonly associated with apatite in Sandholdt shales (see Observations and Data). If apatite does indeed form under SO_4^{-2} reducing conditions, then it follows that reduced sulphur compounds may be associated. In fact, pyrite (FeS_2) is a very common byproduct of bacterial sulphate reduction (Claypool and Kaplan, 1974; Kaplan, 1983; Altschuler et al., 1983; Leventhal, 1983). These reactions (see below) result in release of organic PO_4^{3-}, and H_2S gas which, if not oxidized, can react with iron to produce pyrite or an iron-sulphide precursor.

Dolomite

Sandholdt dolomites form consistently later than apatite and pyrite. Isotopic data (Fig. 8B) are distinct from apatite values and suggest that these dolomites formed within the zone of bacterial methanogenesis over a range of burial temperatures and depths (approx. 12-33°C and 100-450 meters, respectively). These values are consistent with those reported for similar dolomites in the Monterey Formation (Pisciotto, 1981c; Hennessy and Knauth, 1983; Murata et al., 1969). Results (Fig. 8C) closely conform to the expected metal-CO_3^{2-} isotopic curve derived by progressive fractionation between methane and CO_2 by action of methanogenic bacteria during burial and breakdown of organic matter (Claypool and Kaplan, 1974; Irwin et al., 1977; Pisciotto and Mahoney, 1981; Kelts and McKenzie, 1982).

Estimated depths and temperatures of formation (Fig. 8C) should be viewed with caution due to a combination of variables, each of which can profoundly effect calculated values. Temperatures, based on $\delta^{18}O$ data, can fluctuate depending on the $\delta^{18}O$ value of equilibrating pore waters (Pisciotto, 1981b; Lawrence et al., 1975). Based on D.S.D.P. cores, pore waters reflect an average depletion in $\delta^{18}O$ of -0.3‰ per 100 meters of burial depth (assuming a value of 0.0 °/oo SMOW near the sediment-water interface). However, this can range up to -0.6 °/oo/100 m or greater. Estimated depths

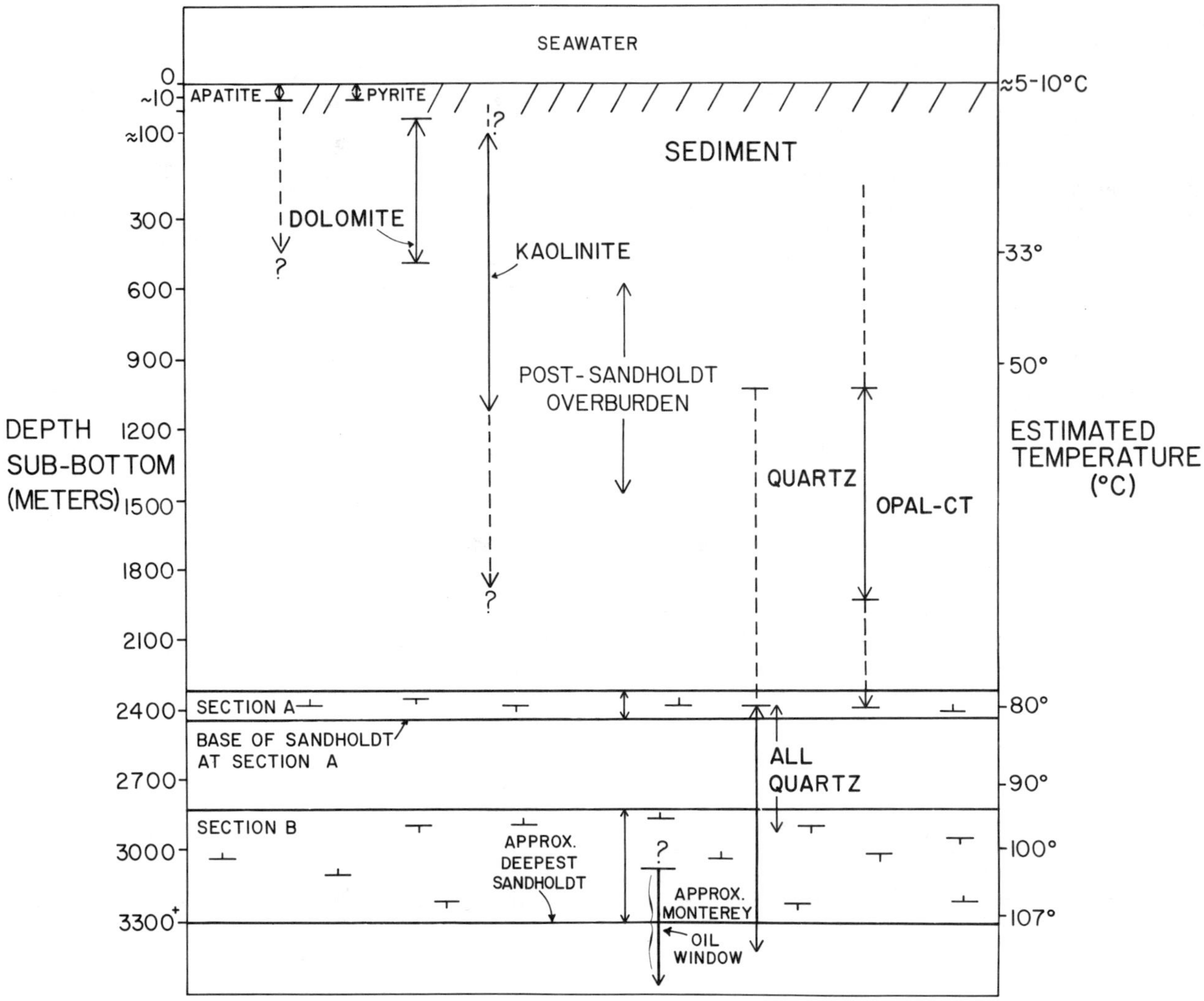

Fig. 10: Schematic plot depicting maximum (pre-uplift) burial for two representative sections of the Sandholdt (see text for elaboration on the sub-bottom depth and temperature scales). Case A, Bear Trap Road (Fig. 1), depicts a thin Sandholdt section that experienced burial depths no greater than 2400 m. Case B depicts the Sandholdt in the Arroyo Seco-Sycamore Flat region (Durham, 1974). Here the unit has been buried in excess of 3300 m. Authigenic phases are portrayed at temperatures and depths at which they were thought to have formed during earlier stages of burial. The temperature scale reflects the probability that the geothermal gradient was initially twice as high as the average, linear gradient of 32-33°C/Km due to low heat conductivity in unconsolidated, decompacted Monterey shales (Isaacs, 1984). Independent estimates of maximum burial temperatures based on paleomagnetic blocking temperatures (R. Coe, written comm., 1984) indicate a temperature of 85°C ± 13°, in close agreement with that reached using authigenic minerals and organic geochemical data.

of formation are also a function of bottom water temperature and the geothermal gradient. Bottom water temperatures are assumed to be near 5°C, however, this may have been higher (approx. 10°C) at shallower sites, or because bottom water temperatures were thought to be somewhat higher than present during the early Middle Miocene based on benthonic foraminifera (Savin et al., 1981). In addition, the geothermal gradient (see below) may have been initially much higher because conductivity values in unconsolidated, decompacted Monterey sediments are lower than in compacted shale (Isaacs, 1984). This would "shallow" the depth of dolomitization for any given isotopic value (Fig. 8C). In spite of these considerations, the systematic variation of $\delta^{13}C$ and $\delta^{18}O$ along the theoretical curve is strongly suggestive that these dolomites did form during relatively early burial (Fig. 8C, 10) in response to bacterial methanogenesis, and continued to accrete with progressive burial (Kelts and McKenzie, 1982; Pisciotto and Mahoney, 1981).

Kaolinite

Authigenic kaolinite ($Al_2Si_2O_5(OH)_4$) places additional constraints on burial history. It is generally agreed that kaolinite is limited to lower temperature assemblages, rarely exceeding 150°C (Hoffman and Hower, 1979; Kisch, 1983; Boles and Franks, 1979). It most commonly forms during early diagenesis (or "eodiagenesis"--Foscolos and Powell, 1979), but can persist into deeper burial stages (Singer and Müller, 1983; Kisch, 1983) if: (1) the permeability and porosity are low enough, and (2) if the cation supply (Fe, Mg, K) is so low as to prevent conversion to illite or chlorite. However, kaolinite may be unstable at temperatures as low as 80-100°C in permeable sediments when appropriate cations are available. Kaolinite is common in shales containing "immature" organic matter (vitrinite reflectances values < 0.5), and generally disappears during catagenesis, and significant generation of hydrocarbons (Foscolos and Powell, 1979; Heroux et al., 1978).

The presence of mixed layer illite/smectites in Sandholdt shales suggests that some kaolinite may form through breakdown of smectites to form mixed layer clays (Hoffman and Hower, 1979), perhaps by the following reaction, one that commences well before the upper temperature limits of kaolinite are reached (Singer and Müller, 1983).

$$(1)\ K^{+} + \text{X-SMECTITE} \rightarrow \text{I/S} + \text{KAOLINITE} + \text{QUARTZ} + X^{+}$$

Alternatively, kaolinite can form by breakdown of K-feldspars (Ardito, 1983), a process that may have played a role in detrital-rich Sandholdt shales.

By itself, the presence of kaolinite suggests a maximum Sandholdt burial temperature of approximately 80-150°C. Permeability and porosity values following 2000-3000+ meters of burial are probably quite low (Isaacs, 1981b; Isaacs, 1984), which would favor the upper temperature limit. Kisch (1983) however, suggests that cation availability may be the more important variable in judging when kaolinite will convert to illite or chlorite. Since Fe, Mg and K appear to be in sufficient supply (Fe-rich dolomites, illite/smectite, K-feldspar) the upper temperature-stability limit may be somewhat lower than 150° (perhaps 110°-130°C). If so, maximum Sandholdt burial temperatures were certainly lower than these values.

Silica Phases, Organic Matter, and the Onset of Petroleum Formation

Figure 10 suggests that the onset of significant hydrocarbon generation (i.e., the Monterey "oil window") occurs near 100°C. This is based on data from other Monterey sections in various Neogene basins in California (Isaacs et al., 1983). The position of this window is controversial, however, and may range from as low as 80°C (Surdam and Stanley, 1981) or as high as 120-140°C as suggested by Kruge (in prep.) for the Lost Hills Field in the San Joaquin Valley. Since most estimates suggest a lower temperature for the oil window, a value of 95°-100°C is considered representative.

T-MAX (from ROCK-EVAL pyrolosis) and vitrinite reflectance data (Table I) from Sandholdt shales (Mertz et al., 1983; Mertz, in prep.) show that most samples are sub-mature, while a few have crossed the threshold into the oil window. This suggests maximum burial temperatures of approximately 80°-105°C assuming that most Monterey rocks do indeed reach the main phase of petroleum generation at temperatures of 95-100°C. Additionally, almost all Sandholdt shales contain diagenetic quartz as the main silica phase. This, and the absence of opal-CT, suggests that these rocks have been heated at least to 80°C (Isaacs et al., 1983; Pisciotto, 1981a; Pisciotto, 1978). It is suggested, then, that most of the Sandholdt sections west of the Reliz-Rinconada fault zone have experienced maximum burial temperatures between approximately 80-105°C (Fig. 10). Maximum burial depths are highly variable, but can be in excess of 3300 meters (Durham, 1974). A geothermal gradient is calculated based on these inferred temperatures, and measured section and well core thicknesses. Values range from 28-38°/Km (ave. = 32-33°C/KM), and are assumed relatively constant through time. These data are similar to those reported from several California Neogene basins (Pisciotto, 1981a; Bostick et al., 1978). Decompaction of the Sandholdt, however, results in an early burial unconsolidated sediment package at least three times as thick as compacted shales (Isaacs, 1981b). Since pre-compacted Monterey sediments have lower heat conductivity values (Isaacs, 1984) than compacted shales, the geothermal gradient will rise accordingly, due to insulation effects. Since conductivity values in non-consolidated shale are approximately 2.1x lower than compacted shale, the geothermal gradient may have been, initially, over twice as high (approx. 67°C/Km) as the average, linear gradient (32°C/Km). An increased geothermal gradient during early burial would have profound effects on the timing of dolomitization, resulting in formation at shallower sub-bottom depths (Fig. 8C).

Apatite, Pyrite, Dolomite: Burial Environments

Authigenic apatites and dolomites commonly occur together along continental margins dominated by deposition of organic-rich, laminated, hemipelagic sediments. The formation of these phases may, in part, be interrelated (Kastner et al., 1983). The geochemical environment conducive to precipitation of apatite and pyrite may play an important role in "setting the stage" for subsequent formation of dolomite. Recent studies by MacArthur (1980) and Benmore et al. (1983) suggest that carbon-13 and sulphur-34 isotopic data (from CO_3^{2-} and SO_4^{2-} in the apatite lattice) can clarify mechanisms of formation and burial environments of apatite. Sulphur isotopic values, if heavier than average seawater $\delta^{34}S$, would suggest precipitation within the zone of bacterial sulphate reduction. This results from the use of "light" SO_4^{2-} (S^{32}-enriched) by bacteria, leaving pore waters enriched in S-34 (Nathan and Neilson, 1980; Kaplan, 1983). Similarly, $\delta^{13}C$ values would reflect the isotopic signature of the CO_2 reservoir in which the apatite formed.

Most authors concur that sedimentary apatite derives, in large part, from the interstitial breakdown of available organic matter (Burnett, 1977; Baturin, 1981; Sheldon, 1981; Kolodny, 1980b), which may include a significant contribution from the intra-sediment bacterial biomass (O'Brien et al., 1981; Soudry and Champetier, 1983). At present, seawater concentrations of PO_4^{3-} and F^- are too low to allow significant apatite precipitation in the water column or even at the sediment-water interface (Froelich et al., 1982; Sheldon, 1981), although this may have been different in the past.

Three important sources of PO_4^{3-} are suggested for Sandholdt shales: (1) Marine plankton, and other water column organic material, much of which may have reached the sediment through fecal pellet transport (Dunbar and Berger, 1981); (2) Bacterial biomass - bacterial-algal mats, reported in organic-rich sediments off Peru (Reimers, 1982), and in the Monterey formation (Williams and Reimers, 1983), are common in Sandholdt shales. These structures, thought to be cell chains and sheaths of non-photosynthetic, sulphur oxidizing bacteria, form binding mats at the sediment-water interface. Following burial and compaction, these form organic-rich "mats" that parallel lamination in many Sandholdt samples, and may contribute significantly to the biomass of the rock. They may also be an important source, following bacterial breakdown, of PO_4^{3-} for apatite (Soudry and Champetier, 1983; O'Brien et al., 1981); and (3) Fish bones - although these are an important source of apatite off Peru (Suess, 1981) and in the Cretaceous of Israel (Nathan and Soudry, 1980), they are not common in Sandholdt rocks, although this may reflect a preservational effect.

Release of PO_4^{3-} is thought to occur within the zone of sulphate reduction by action of SO_4^{2-}-utilizing bacteria including Desulfovibrio desulfuricans (Anderson and Arthur, 1983; Claypool and Kaplan, 1974; Berner, 1982), possibly by the following simplified reaction:

(2) $R(CHO) + SO_4^{2-} \longrightarrow HS^- + HCO_3^- + PO_4^{3-}$

(organic matter) — Sulphate Reducing Bacteria

$\searrow S^o + FeS \rightarrow FeS_2$ (pyrite)

Phosphate concentrations of 40-160 µM/Kg (up to 50 times that of seawater) accumulate in interstitial pore waters (Froelich et al., 1982; Burnett, 1977; Jahnke et al., 1983), vastly exceeding the solubility product of apatite. Additionally, slightly lowered pH values and high F^- concentrations may favor apatite precipitation over calcite (Nathan and Sass, 1981). Nucleation proceeds by (1) direct authigenic precipitation (Burnett, 1977; Burnett et al., 1982; Baturin and Bezrukov, 1979; Jahnke et al., 1983; Benmore et al., 1983), primarily as small *in situ*, displacive concretions of nearly pure carbonate fluorapatite, commonly occurring among organic-rich mat structures; or (2) by replacement of precursor carbonate material, and hydroxyl-apatite fish debris (Manheim et al., 1975; Starinsky et al., 1982; Birch et al., 1983; Shemesh et al., in press). In Sandholdt shales, the first process appears to be dominant. SO_4^{2-} reduction processes are assumed important due to the ubiquitous presence of pyrite, a common by-product of these reactions. Early apatite formation is also suggested by the common incorporation of sulphate into the apatite lattice (confirmed by X-ray analysis on SEM images; Table II). The range of carbon-13 isotopic values (Fig. 8B) in apatite CO_3^{2-} reflects the isotopic signature of available HCO_3^-. "Light" carbon-13 values (-4 ^{o}/oo to -12 ^{o}/oo PDB) reflect input of organic CO_2 derived from breakdown of organic matter via sulphate reduction (Anderson and Arthur, 1983). Values between +2 ^{o}/oo and -4 ^{o}/oo PDB suggest replacement of precursor carbonate, or a combination of the two processes. Skeletal carbonate includes foraminifera and nannofossils, the latter known to precipitate carbonate tests with $\delta^{13}C$ values ranging from approximately +2 to -3 ^{o}/oo PDB (Anderson and Arthur, 1983).

SO_4^{2-} reduction in organic-rich sediments is often limited to the upper few meters of sediment depending on degree of oxygenation of overlying seawater, rates of diffusion of sulphate into the sediment, and sedimentation rates (Claypool and Kaplan, 1974; Demaison and Moore, 1980; Anderson and Arthur, 1983). Depletion of SO_4^{2-} by bacterial activity can be rapid, especially under anoxic water bodies where sedimentation rates are high, leading to rapid burial and preservation of organic matter (Pisciotto and Mahoney, 1981; Claypool and Kaplan, 1974). Concurrent increase in alkalinity (as HCO_3^-), and removal of SO_4^{2-} serve to facilitate dolomite formation. In fact, sulphate has been shown to significantly inhibit the rate of dolomite nucleation and growth (Baker and Kastner, 1981; Kastner et al., 1983; Kushnir, 1983). Its removal may be critical to subsequent formation of methane zone dolomites.

Dolomitization in response to bacterial methanogenesis (Fig. 8C) is supported by consistently "heavy" carbon isotopic values in dolomite CO_3^{2-} (see Pisciotto and Mahoney, 1981; Kelts and McKenzie, 1982; Irwin et al., 1977). Systematic variation in $\delta^{13}C$ and $\delta^{18}O$ isotopic values suggest nucleation and growth of concretions over a range of temperatures and burial depths (Fig. 8C). Dolomites apparently form in response to bacterial breakdown of organic matter through reduction of CO_2. Preferential uptake of isotopically light HCO_3^- (C-12 rich) by methane-producing bacteria results in isotopically light methane, leaving the residual supply of HCO_3^- enriched in C-13. This signature is carried by newly formed authigenic carbonates, perhaps as expressed by Claypool and Kaplan (1974):

$$(3)\ Me^{++} + 2HCO_3^- + 8H^+ \rightarrow CH_4 + MeCO_3 + 3H_2O$$

(Dolomite)

The trend towards lighter $\delta^{13}C$ values with progressive burial (Fig. 8C) is either a result of increased input of isotopically light CO_2 due to thermocatalytic decarboxylation reactions at depth, or by depletion of heavy CO_2 through precipitation of abundant dolomite (Kelts and McKenzie, 1982).

Episodicity of dolomitized beds in Sandholdt sections remains enigmatic. Occurring regularly every 3 to 5 meters in organic-rich sections, dolomite seems to favor zones that are less detrital-rich, and more calcite-rich than interbedded shales. The presence of increased

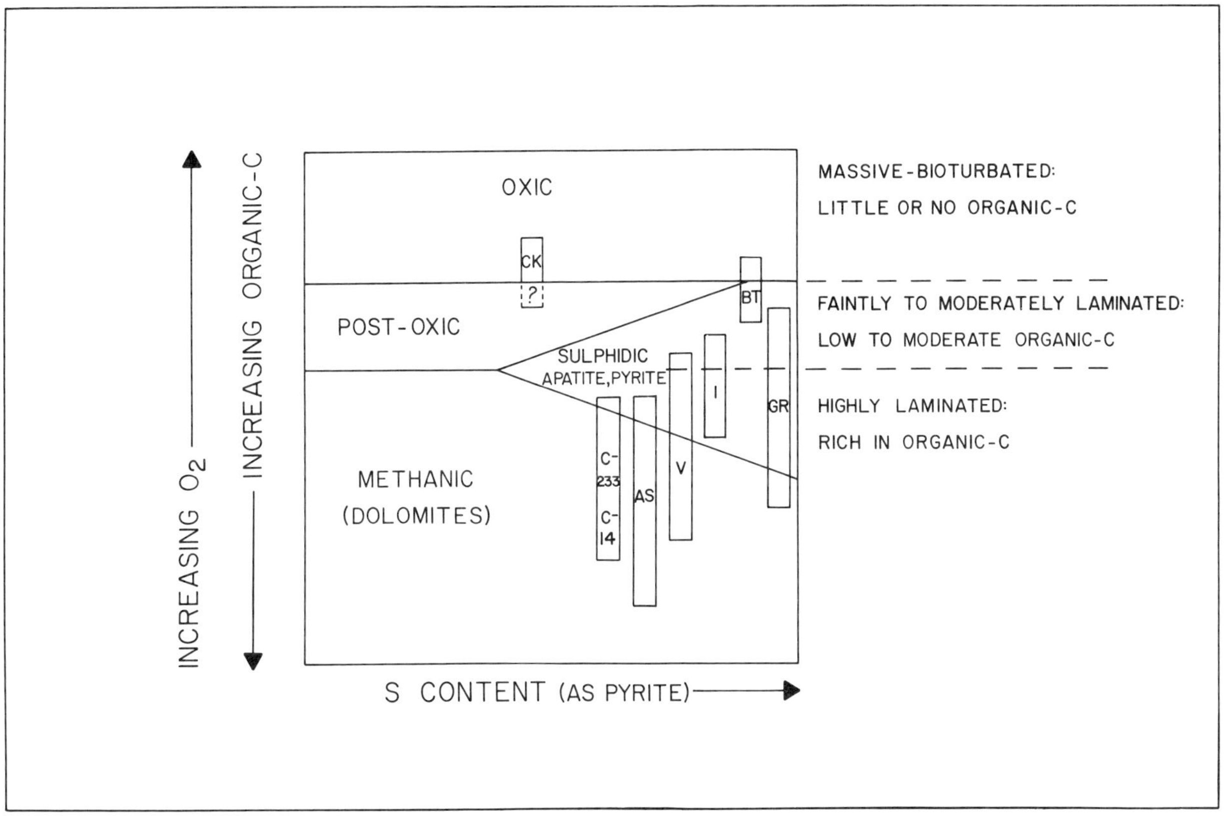

Fig. 11: Geochemical classification of Sandholdt sedimentary environments (modified from Berner, 1981; Maynard, 1982). See text for explanation of environments. Localities (Fig. 1) include AS = Arroyo Seco; V = Vaqueros Canyon; I = Indians Ranch; GR = Graves Creek (Atascadero); BT = Bear Trap Road (near Lockwood); CK = Cosio Knob; C-233 = well core near Reliz Canyon. Columns on diagram show only the range of lithologies and environments present in a given section and the approximate proportion of a section influenced by a given environment. Thickness is not implied.

precursor calcite would, perhaps, facilitate subsequent dolomite nucleation, and is consistent with observations from the Gulf of California (Kelts and McKenzie, 1982) suggesting that beds with increased precursor calcite may favor dolomitization. Additionally, their studies suggest that these zones may correlate to large-scale climate changes (glacial-interglacial), with dolomite exhibiting a preference for carbonate-enriched beds deposited during interglacial times.

In sum, apatite-dolomite associations in laminated, organic-rich Sandholdt shales suggest that SO_4^{2-} reduction reactions, instrumental in formation of apatite and pyrite, play an important role in the dolomite process through depletion of SO_4^{2-} and increased alkalinity. Under increasing burial temperatures, dolomitization can proceed in pore waters containing only "adequate" Mg (Mg/Ca < 5:1: Pisciotto and Mahoney, 1981). Fe-rich dolomites are common, and may reflect: (1) relatively low Mg^{2+} concentrations and increased incorporation of Fe^{2+} into the dolomite lattice during progressive burial (Pisciotto, 1981c); or (2) increased availability of Fe^{2+}.

Although apatite co-exists with dolomite in laminated, organic-rich shales, P_2O_5 values (Fig. 5A) rarely exceed 0.5%. Alternatively, apatite is more abundant in moderately to faintly laminated and massive-bioturbated samples where dolomite is often negligible or absent. Several Sandholdt sections, while rich throughout in apatite, contain only a few isolated dolomite bodies. This suggests that dolomitization may not necessarily be a prerequisite for apatite precipitation, despite the fact that removal of Mg^{2+} is known to facilitate crystallization of carbonate-fluorapatite (Martens and Harriss, 1970). This is consistent with studies by Jahnke et al.. (1983) suggesting that apatite can form in interstitial environments in the presence of Mg^{2+}.

Authigenic Phases and Depositional Environments

Sedimentary environments can be classified on the basis of geochemical data (Berner, 1981; Maynard, 1982). Figure 11 portrays Berner's geochemical classification of environments based on three principal variables: (1) O_2 supply; (2) organic-carbon supply; and (3) sulphur content (as H_2S in recent sediments, or sulphide minerals in the rock record). Environments are classified as either oxic or anoxic depending on oxygen supply. Anoxic environments are further subdivided into sulphidic and non-sulphidic depending on sulphur content. The non-sulphidic environment can be divided into: (1) post-oxic, characterized by low sulphur content, minor amounts of organic-C, and presence of glauconite and other Fe-silicates; and (2) the methane zone characterized by authigenic carbonates (e.g., Fe-dolomites, siderite), sulphides previously formed by sulphate reduction, reducing conditions, and abundant organic matter that survived previous oxidation reactions. Fe-rich carbonates are common due to buildup of reduced Fe^{2+} that is no longer being removed by precipitation of sulphides. Sulphidic environments can be organic-rich, and are characterized by abundant H_2S gas and/or sulphide phases (pyrite, marcasite). This zone is dominated by bacterial sulphate reduction, whereas the post-oxic zone is dominated by oxidation reactions that precede SO_4^{2-} reduction (i.e., nitrate and iron reduction, once O_2 is depleted).

Maynard (1982) applies this scheme to Paleozoic shelf-slope-basin sediments in the Appalachian basin. All of Berner's environments are represented on the basis of color, organic-carbon contents, authigenic minerals, and degree of bioturbation (reflecting O_2 supply). Sandholdt lithologies are classified on a similar basis (Fig. 11). The columns represent sections or well cores, and are positioned to portray the variety of environments represented in any given section. As an example, Arroyo Seco (AS) shales are generally well laminated, and organic-rich. The presence of apatite and pyrite suggests that these rocks experienced SO_4^{2-} reduction prior to burial through the methane zone and formation of Fe-rich dolomites. Other sections are placed depending on organic-richness, preservation of lamination, and the presence or absence of critical minerals. Figure 11 summarizes the observations that: (1) most Sandholdt rocks are organic-rich, deposited under low oxygen to anoxic conditions. Anaerobic conditions clearly prevailed within the sediment, although two localities (BT and CK) show evidence of macrobioturbation suggestive of temporary oxic interstitial conditions; (2) that the occurrence of pyrite and apatite suggests that most sections experienced SO_4^{2-} reducing conditions (sulphidic environment) prior to burial within the methane zone; and (3) that there is considerable variability within some sections (i.e., GR and BT), often on a bed-to-bed basis, in degree of lamination and organic-richness.

Sandholdt deposition occurred under the influence of a generally strong, but variable, oxygen minimum zone. Evidence includes good preservation of lamination and abundant, hydrogen-rich organic matter (Ingle, 1981; Demaison and Moore, 1980); low-oxygen, "stressed" planktonic foraminiferal assemblages (G. Keller, written comm., 1983); and low oxygen-tolerant benthonic foraminifera (Blake, 1981; Pisciotto and Garrison, 1981; Graham, 1976). Sections deposited primarily under the influence of the core of the O_2-minimum (i.e., Arroyo Seco, Vaqueros Canyon) contain well preserved lamination and abundant, hydrogen-rich organic matter (Fig. 11). Shales from these localities are dolomite-rich, but contain only small quantities of apatite (Fig. 5). This contrasts markedly with sections (i.e., Indians Ranch, Graves Creek, Bear Trap Rd. - Fig. 11) thought to be deposited nearer the edges of the O_2-minimum (or outside of it) where apatite is consistently more abundant, and dolomite less so. That apatite seems to form more readily under sub-oxic conditions (at the edges of the O_2-minimum zone) has been reported off Peru and Chile (Burnett et al., 1980). Other authors (Sandstrom, 1982; Jahnke et al., 1983; Bein and Amit, 1982; Kidder, 1983) suggest that apatite, although associated with laminated, organic-rich hemipelagic sediments, tends to form more readily in beds or zones where organic matter has undergone more extensive oxidation and breakdown than in surrounding organic-rich shales. In the Sandholdt, apatite is consistently more abundant in rocks with lower TOC values, more oxygen-rich and less hydrogen-rich organic matter, and less dolomite (Figures 4, 5). That apatite does, however, occur in small amounts in laminated shales is consistent with observations that it forms in sediments throughout the O_2 minimum zone off Peru. It is suggested, however, that increased bioturbation, increased diffusion of oxidants (i.e., SO_4^{-2}), and more extensive microbial breakdown of organics near the edges of the O_2-minimum may be instrumental in release of abundant PO_4^{3-} and increased apatite precipitation.

Lithologic variability both vertically in individual sections, and laterally within the basin (Fig. 2B, 11), is thought to result from two main factors. First and most important are fluctuations in the position and intensity of the oxygen minimum zone. This commonly occurs in modern environments in response to many oceanographic factors including variable productivity, changes in intermediate level circulation, sea level changes, and variability in upwelling intensity linked to climatic effects and oceanic circulation patterns (Ingle, 1981). Examples include the Gulf of California (Schrader, 1980), off Peru and Chile (Burnett, 1980), and the southern California Borderland basins (Douglas, 1981). Secondly, tectonic position of the sea floor relative to the O_2-minimum (mainly through regional uplift and subsidence) plays a role in large-scale lithologic variability (Garrison et al., 1979), but probably cannot explain bed-to-bed fluctuations in organic-richness, degree of lamination, and mineralogy.

Fluctuations in oxygenation and degree of organic preservation directly affect interstitial microbial behavior, which in turn, plays an important role in authigenic mineral genesis, resulting in the wide range of lithologies observed in Sandholdt sections. This kind of lithotypic variability is common along continental margins, or in restricted basins, where hemipelagic shales are being deposited today (Soutar et al., 1981; Donegan and Schrader, 1981; Schrader, 1980; Sandstrom, 1982). Bed-by-bed variability in degree of preservation of organic matter and lamination is also recorded in the record. Examples include the Monterey Formation in California (Govean and Garrison, 1981; Pisciotto and Garrison, 1981), and Cretaceous and Devonian black shales (Pratt and Claypool, 1983).

Among modern continental margin environments that are characterized by seasonal upwelling, high organic productivity, and well developed oxygen-minima, precipitation of apatite and dolomite is commonly reported (Peru-Chile: Burnett et al., 1982; Kulm et al., 1982; Gulf of California: Kelts and McKenzie, 1982; H. Schrader, written comm.,

1982; S. W. Africa: Baturin, 1981; Mexico: Jahnke et al., 1983). A similar setting is favored to explain formation of in situ apatite in Sandholdt shales, and subsequent authigenic dolomite. Caution is advised, however, by numerous authors (O'Brien et al., 1981; Birch et al., 1983; Riggs, 1979) who suggest that apatite can form on non-upwelling, low-nutrient margins as well, often by bacterial or replacement mechanisms.

SUMMARY

(1) Sandholdt shales range from highly laminated and organic-rich, to moderately and faintly laminated, to massive (bioturbated) and organic-poor. Common authigenic constituents, in approximate order of formation, include apatite and pyrite, dolomite and kaolinite, and diagenetic quartz. Laminated rocks contain abundant, hydrogen rich organic matter, whereas more poorly laminated samples have lower TOC values and contain less hydrogen-rich, and more oxygen-rich organic matter. Organic carbon in dolomite bodies is sparse and oxygen-rich.

(2) Apatite and dolomite commonly occur together, however, dolomite is the dominant phase in laminated shales, whereas apatite becomes increasingly abundant in rocks with lower TOC values and more poorly preserved lamination. In faintly laminated to massive rocks dolomite is insignificant.

(3) Apatite and pyrite formed during earliest diagenesis through action of bacterial SO_4^{2-} reduction. Decreased SO_4^{2-} and increased HCO_3^-, as a result of these reactions, probably facilitated dolomite formation at depth (approx. 100-450 m or 12-33°C) under the influence of bacterial methanogenesis. Dolomitization, however, does not appear to be a prerequisite for apatite crystallization.

(4) Apatite occurs both as a direct precipitate (mainly microconcretions) and replacement product (calcareous plankton, and benthonic foraminifera), in response to elevated interstitial PO_4^{3-} concentrations. Sources include breakdown of phytoplankton organic matter, bacterial mat structures, and fish bones. Apatite concretions are commonly associated with bacterial-algal mat laminations. Dolomite appears to form by replacement of host shales in zones that contain: (1) organic matter with high oxygen indices (H-poor); (2) less detrital material; and (3) more calcite than interbedded organic-rich shales.

(5) Authigenic kaolinite is common in many shales, usually filling foraminiferal chambers or along porous, organic-rich, "mat" lamination zones. It is thought to derive from breakdown of K-feldspars and smectites during early (low-temperature) burial.

(6) Authigenic mineralogy and organic geochemistry constrain maximum burial temperatures to approximately 80-105°C (depths = 2400-3300+ meters). Most rocks are organically "immature," although a few samples have entered the oil window. Mineral phases, organic richness and quality, and degree of lamination suggest that Sandholdt shales were deposited under the influence of a strong, but variable, O_2-minimum zone. Conditions varied from anoxic (laminated, organic-rich, dolomitic rocks), to weakly oxic (apatite-rich, moderately to faintly laminated samples) and, occasionally, more oxic conditions (massive, organic-poor samples).

ACKNOWLEDGEMENTS

Funding in support of this project was provided by the donors to American Chemical Society/Petroleum Research Fund grant #ACS/PRF 13953-AC2, and Geological Society of America Student research grant #3030-82. Organic geochemical analyses were kindly provided by Gerard Demaison, Chevron Overseas Petroleum Co., and Kenneth Pisciotto, SOHIO Petroleum, both in San Francisco. Isotopic analyses on dolomites and apatites were run courtesy of Miriam Kastner's lab at Scripps Institute of Oceanography, and through Michael Arthur, University of Rhode Island. Isotopic analyses of oxygen in phosphate was kindly provided by Y. Kolodny, Hebrew University in Jerusalem. Scanning electron microscopy and energy dispersive X-ray analyses were conducted at the SEM facility at the University of California at Santa Cruz under the direction of Jan Nowell. Planktonic foraminifera were identified by Gerta Keller at the U.S.G.S. Nannofossils were analyzed courtesy of C. H. Ellis, SOHIO Petroleum in San Francisco. Conversations with M. Kastner, Y. Kolodny, M. Arthur, B. Garrison, K. Pisciotto and G. Demaison helped clarify ideas in this paper. Special thanks go to David Hollander at U.C.S.C. for assistance in laboratory analyses, and Sandra Cherk, presently with SOHIO, for thin-section preparation. Well core data and samples were provided courtesy of Texaco, Inc., Los Angeles. The manuscript was reviewed by Donald Zenger, Pomona College; Miriam Kastner, Scripps Institute of Oceanography; and Robert Garrison, University of California, Santa Cruz.

REFERENCES

Altschuler, Z. S., Schnepfe, M. M., Silber, C. C., and Simon, F. O. 1983, Sulphur diagenesis in Everglades peat and origin of pyrite in coal: Science, v. 221, pp. 221-227.

Anderson, T. F and Arthur, M. A., 1983, Stable isotopes of oxygen and carbon and their application to sedimentologic and paleoenvironmental problems: S.E.P.M. Short Course No. 10, Dallas, 1983, p. 1-1 to 1-151.

Ardito, P. A., 1983, Mineral-groundwater interactions and formation of authigenic kaolinite within the southeastern intake beds of the Great Australian (Artesian) basin, New South Wales, Australia: Sedim. Geol., v. 35, p. 249-261.

Atwater, T., 1970, Implications of plate tectonics for the Cenozoic tectonic evolution for western North America: G.S.A. Bull., v. 81, p. 3513-3566.

Baker, P. A. and Kastner, M., 1981, Constraints on the formation of sedimentary dolomite: Science, v. 213, p. 214-216.

Baturin, G. N., 1981, Phosphorites in the Sea Floor--Origin, Composition and Distribution: Developments in Sedimentology 33, Elsevier Publishing Co., 343 p.

Baturin, G. N. and Bezrukov, P. L., 1979, Phosphorites on the sea floor and their origin: Marine Geology, v. 31, p. 317-327.

Bein, A. and Amit, O., 1982, Depositional environments of the Senonian chert, phosphorite and oil shale sequence in Israel as deduced from their organic matter composition: Sedimentology, v. 29, p. 81-90.

Bein, A. and Sandler, A., 1983, Early diagenetic oxidation and maturation trends in organic mater extracted from Eocene chalks and cherts: Chem. Geol., v. 38, p. 213-224.

Benmore, R. A., Coleman, M. L., and McArthur, D. M., 1983, Origin of sedimentary francolite from its sulphur and carbon isotope composition: Nature, v. 302, p. 516-518.

Berner, R. A., 1981, A new geochemical classification of sedimentary environments: Jour. Sed. Petrol., v. 51, p. 359-366.

Berner, R. A., 1982, Burial of organic carbon and pyrite sulfur in the modern ocean: its geochemical and environmental significance: Amer. Jour. Sci., v. 282, p. 451-473.

Birch, G. F., 1979, Phosphatic rocks on the western margin of South Africa: Jour. Sed. Petrology, v. 49, p. 93-110.

Birch, G. F., Thompson, J., McArthur, D. M., and Burnett, W. C., 1983, Pleistocene phosphorites off the west coast of South Africa: Nature, v. 302, p. 601-603.

Blake, G. H., 1981, Biostratigraphic relationship of Neogene benthic foraminifera from the southern California outer continental borderland to the Monterey Formation. In: R. E. Garrison et al. (Editors), The Monterey Formation and Related Siliceous Rocks of California. Pacific Sect. S.E.P.M. Spec. Publ., pp. 1-14.

Blake, M. C., Jr., Cashman, S. M., McCulloh, T. H., and Waddell, C. T., 1978, Neogene basin formation in relation to plate tectonic evolution of the San Andreas Fault system, California: AAPG Bull., v. 62, No. 3, p. 344-372.

Boles, J. R. and Franks, S. G., 1979, Clay diagenesis in Wilcox sandstones of southwest Texas: Implication of smectite diagenesis on sandstone cementation: Jour. Sed. Petrol., v. 49, p. 55-70.

Bostick, N. H., Cashman, S. M., McCulloh, T. H. and Waddell, C. T., 1978, Gradients of vitrinite reflectance and present temperature in the Los Angeles and Ventura Basins, California, in Oltz, D. F., ed., Symposium in geochemistry: Low temperature metamorphism of kerogen and clay minerals: Pacific Section S.E.P.M., p. 65-96.

Burnett, W. C., 1977, Geochemistry and origin of phosphorite deposits off Peru and Chile: G.S.A. Bull., v. 88, p. 813-823.

Burnett, W. C., 1980, Apatite-glauconite associations off Peru and Chile: palaeo-oceanographic implications: Journal Geol. Soc. London, v. 137, p. 757-764.

Burnett, W. C., Veeh, H. H. and Soutar, A., 1980, U-series, oceanographic and sedimentary evidence in support of recent formation of phosphate nodules off Peru, in Bentor, Y. K., ed., Marine Phosphorites--Geochemistry, Occurrence, Genesis: S.E.P.M. Spec. Publication No. 29, p. 61-72.

Burnett, W. C., Beers, M. J., and Roe, K. K., 1982, Growth rates of phosphate nodules from the continental margin off Peru: Science, v. 215, p. 1616-1618.

Claypool, G. E. and Kaplan, I. R., 1974, The origin and distribution of methane in marine sediments. In: I. R. Kaplan (Editor), Natural Gases in Marine Sediments. Plenum Press, New York, N.Y., pp. 99-139.

Craig, H., 1957, Isotopic standards for carbon and oxygen correction factors for mass spectrometric analysis of carbon dioxide: Geochim. Cosmochim. Acta, v. 12, p. 133-149.

Crowell, J. C., 1974, Origin of Late Cenozoic basins in southern California, in Dickinson, W. R., ed., Tectonics and Sedimentation: S.E.P.M. Spec. Pub. 22, p. 190-204.

Demaison, G. J. and Moore, G. T., 1980, Anoxic environments and oil source bed genesis: AAPG Bull., v. 64, p. 1179-1209.

Dickinson, W. R. and Synder, W. S., 1979, Geometry of triple junctions related to San Andreas transform: EOS, v. 84, no. B2, p. 561-572.

Donegan, D. and Schrader, H., 1981, Modern analogues of the Miocene diatomaceous Monterey Shale of California: evidence from sedimentologic and micropaleontologic study, in Garrison, R. E., ed., The Monterey Formation and Related Siliceous Rocks of California: Spec. Publications of the Pacific Section S.E.P.M., May 29, 1981, p. 149-158.

Douglas, R. G., 1981, Paleoecology of continental margin basins: a modern case history from the borderland of southern California, in Douglas, R. G. et al., eds., Depositional systems of active continental margin basins: S.E.P.M. Pacific Section, short course notes, San Francisco, May, 1981, p. 121-156.

Dunbar, R. B. and Berger, W. H., 1981, Fecal pellet flux to modern bottom sediment of the Santa Barbara Basin (California) based on sediment trapping: G.S.A. Bull., v. 92, p. 212-218.

Durham, D., 1974, Geology of the southern Salinas Valley area, California: U.S.G.S. Prof. Paper 819, 111 p.

Espitalie, J., Madec, M. and Tissot, B., 1977, Source rock characterization method for petroleum exploration: paper presented at 9th Offshore Technology Conference, Houston, Texas, May 2-5, 1977, p. 439-444.

Foscolos, A. E. and Powell, T. G., 1979, Mineralogy and geochemical transformation of clays during burial-diagenesis (catagenesis): relation to oil generation, in Mortland, M. M. and Farmer, V. C., eds., Developments in sedimentology No. 27, International clay conference, 1978, p. 261-270.

Fritz, P. and Smith, D. C. W., 1970, The isotope composition of secondary dolomites: Geochim. Cosmochim. Acta, v. 34, p. 1161-1163.

Froelich, P. N., Bender, M. L., Luedtke, N. A., Heath, G. R., and De Vries, T., 1982, The marine phosphorous cycle: American Journal of Science, v. 282, p. 474-511.

Garrison, R. E., Stanley, R. G., and Horan, L. J., 1979, Middle Miocene sedimentation on the southwestern edge of the Lockwood High, Monterey County, California, in Graham, S. A., ed., Tertiary and Quaternary geology of the Salinas Valley and Santa Lucia Range, Monterey County, California: Pacific Section S.E.P.M., Field Guide No. 4, p. 51-65.

Goldstein, J. I. et al., 1981, Scanning electron microscopy and X-ray microanalysis: Plenum Press, New York.

Govean, F. M. and Garrison, R. E., 1981, Significance of laminated and massive diatomites in the upper part of the Monterey Formation, California. in R. E. Garrison et al. (Editors), The Monterey Formation and Related Siliceous Rocks of California, Pacific Sect. S.E.P.M. Spec. Publ. pp. 181-198.

Graham, S. A., 1976, Tertiary sedimentary tectonics of the central Salinian block of California: Ph.D. Thesis, Stanford University, Stanford, California, 510 p.

Graham, S. A., 1979, Tertiary paleotectonics and paleogeography of the Salinian Block, in Armentrout, J. M. et al.., eds., Cenozoic paleogeography of the western United States: S.E.P.M. Pacific Section, Pacific Coast Paleogeography Symposium 3, March 1979, p. 45-52.

Hennessy, J. W. and Knauth, L. P., 1983, Isotopic variation in dolomite concretions from the Monterey Formation of California: G.S.A. Abstracts with Programs, Annual Meeting, Indianapolis, v. 15, No. 6, p. 595.

Heroux, Y., Chagnon, A., and Bertrand, R., 1979, Compilation and correlation of major thermal maturation indicators: Bull., AAPG, vol. 63, no. 17, p. 2178-2144.

Hoffman, J. and Hower, J., 1979, Clay mineral assemblages as low grade metamorphic geothermometers: application to the thrust faulted disturbed belt of Montana, U.S.A., in Scholle, P. A. and Schluger, P. R., eds., Aspects of diagenesis: S.E.P.M. Spec. Pub. No. 26, p. 55-80.

Ingersoll, R. V. and Graham, S. A., 1983, Recognition of the shelf-slope break along ancient, tectonically active continental margins, in, Stanley, D. J. and Moore, G. T., eds., The shelfbreak: critical interface on continental margins: S.E.P.M. Spec. Pub. No. 33, p. 107-120.

Ingle, J. C., Jr., 1981, Origin of Neogene diatomites around the north Pacific rim, in Garrison, Ed., The Monterey Formation and related siliceous rocks of California: S.E.P.M. Pacific Section Research Symposium, San Francisco, May 29, 1981, p. 159-180.

Irwin, H., Curtis, C. D., and Coleman, M., 1977, Isotopic evidence for source of diagenetic carbonates formed during burial of organic-rich sediments: Nature, v. 269, p. 209-213.

Isaacs, C. M., 1980, Diagenesis in the Monterey Formation examined laterally along the coast near Santa Barbara, California: Ph.D. thesis, Stanford University, Stanford, California, 329 p.

Isaacs, C., ed., 1981a, Guide to the Monterey Formation in the California coastal area, Ventura to San Luis Obispo: AAPG Pacific Section, v. 52, 91 p.

Isaacs, C. M., 1981b, Porosity reduction during diagenesis of the Monterey Formation, Santa Barbara coastal area, California, in, R. E. Garrison et al. (Editors), The Monterey Formation and Related Siliceous Rocks of California, Pacific Sect. S.E.P.M. Spec. Publ., pp. 257-271.

Isaacs, C. M., 1984, The Monterey--key to offshore California boom: Oil and Gas Journal, v. 82, no. 2, p. 75-81.

Isaacs, C. M., Pisciotto, K. A., and Garrison, R. E., 1983, Facies and diagenesis of the Miocene Monterey Formation, California: A summary, in, Iijima, A., et al., eds., Silicious Deposits in the Pacific Region: Developments in Sedimentology, v. 36, Elsevier Publishing Co., Amsterdam, p. 247-282.

Jahnke, R. A. Emerson, S. R., Roe, K. K., and Burnett, W. C., 1983, The present day formation of apatite in Mexican continental margin sediments: Geochim. et Cosmochim. Acta, v. 47, p. 259-266.

Kaplan, I. R., 1983, Stable isotopes of sulfur, nitrogen and deuterium in recent marine environments: S.E.P.M. Short Course No. 1, Dallas, 1983, p. 2-1 to 2-108.

Kastner, M., Kushnir, J., and Anderson, G. E., 1983, Diagenetic relationships of dolostone-phosphorite-chert: a geochemical interpretation: G.S.A. Abstracts with Programs, Annual Meeting, Indianapolis, v. 15, no. 6, p. 607.

Keller, G. and Barron, J., 1983, Paleoceanographic implications of Miocene deep sea hiatuses: G.S.A. Bull., v. 94, p. 590-613.

Kelts, K., and McKenzie, J., 1982, Diagenetic dolomite formation in Quaternary anoxic diatomaceous muds of D.S.D.P. Leg 64, Gulf of California, in, Curray, J. R., et al., eds., Init. Reports D.S.D.P., v. 64, part 2, Washington, U.S. Government Printing Office, p. 553-569.

Kidder, D. L., 1983, Petrology and origin of midcontinent Pennsylvanian phosphate nodules: G.S.A. Abstracts with Programs, Annual Meeting, Indianapolis, v. 15, No. 6, p. 612.

Kisch, H. J., 1983, Mineralogy and petrology of burial diagenesis (burial metamorphism) and incipient metamorphism in clastic rocks in, Larsen, G. and Chilingar, G. V., eds., Diagenesis in sediments and sedimentary rocks 2: Developments in sedimentology No. 25B, p. 289-494.

Kolodny, Y., 1980a, Carbon isotopes and depositional environment of a high productivity sedimentary sequence--the case of the Mishash-Ghareb Formations, Israel: Israel Jour. Earth Sciences, v. 29, p. 147-156.

Kolodny, Y., 1980b, The origin of phosphorite deposits in light of occurrences of recent sea-floor phosphorites (extended abstract), in Bentor, ed., Marine Phosphorites--Geochemistry, Occurrence, Genesis: S.E.P.M. Spec. Publication No. 29, p. 249.

Kruge, M. A., 1983, Diagenesis of Miocene biogenic sediments in Lost Hills Oil Field, San Joaquin Basin, California: in prep.

Kulm, L. D. et al., 1981, Late Cenozoic carbonates on the Peru continental margin: Lithostratigraphy, biostratigraphy, and tectonic history, in Kulm, L. D., Dymond, J., Dasch, E. J. and Hussong, D. M., eds., Nazca Plate: Crustal Formation and Andean Convergence: Geol. Soc. Am. Mem. 154, p. 469-508.

Kushnir, J., 1983, The rate of dissolution of carbonate minerals in the presence of sulphate ions: G.S.A. Abstracts with Programs, Annual Meeting, Indianapolis, v. 15, No. 6, p. 620.

Lawrence, J. R., Gieskes, J., and Broecker, W. S., 1975, Oxygen isotope and cation composition of DSDP pore waters and the alternation of Layer II basalts: Earth Planet. Sci. Lett. v. 27, p. 1-10.

Leventhal, J. S., 1983, An interpretation of carbon and sulfur relationships in Black Sea sediments as indicators of environment of deposition: Geochim. Cosmochim. Acta, v. 47, p. 133-137.

Longinelli, A. and Nuti, S., 1968, Oxygen isotopic composition of phosphorites from marine formations: Earth Planet. Sci. Lett., v. 5, p. 13-16.

Longinelli, A. and Nuti, S., 1973, Revised phosphate-water isotopic temperature scale: Earth Planet. Sci. Lett., v. 19, p. 373-376.

Mahmoud, F. J., 1976, Geochemistry and genesis of sedimentary marine carbonate fluorapatite: Ph.D. Dissertation, University of California, Santa Barbara, 167 p.

Manheim, F., Rowe, G. T., and Jipa, D., 1975, Marine phosphorite off Peru: Jour. Sed. Petrology, v. 45, p. 243-251.

Mann, P., et al.., 1983, Development of pull-apart basins: Jour. of Geology, v. 91; p. 529-544.

Martens, C. S. and Harriss, R. C., 1970, Inhibition of apatite precipitation in the marine environment by magnesium ions: Geochim. Cosmochim. Acta, v. 34, p. 621-625.

Maynard, J. B., 1982, Extension of Berner's "New geochemical classification of sedimentary environments" to ancient sediments: Jour. Sed. Petrol., v. 52, p. 1325-1331.

McArthur, J. M., Coleman, M. L., and Bremner, J. M., 1980, Carbon and oxygen isotopic composition of structural carbonate in sedimentary francolite: Jour. Geol. Soc. London, v. 137, p. 669-673.

McCrea, J. M., 1950, On the isotopic chemistry of carbonates and a paleotemperature scale: J. Chem. Phys., v. 18, p. 849-857.

McWilliams, M. O. and Howell, D. G., 1982, Exotic terranes of western California: Nature, v. 297, p. 215-217.

Mertz, K. A., Jr., Demaison, G., and Garrison, R. E., 1983, Lithofacies, organic matter, and source rock potential: Sandholdt Member (Middle Miocene), Monterey Formation, Northern Santa Lucia Mountains, California: A.A.P.G.-S.E.P.M. Pacific Section meeting, Sacramento, California, May, 1983, Programs and Abstracts, p. 113.

Mertz, K. A., Jr., in prep., Organic matter, petrology and depositional environments: Sandholdt Member, Miocene Monterey Formation, Santa Lucia Range, California: Abstract submitted to SEPM, San Jose, August 1984.

Milliman, J. D., 1974, Marine Carbonates: Heidelberg (Springer-Verlag), pp. 311-313.

Milner, C. W. D., 1982, Geochemical analyses of sedimentary organic matter and interpretation of maturation and source potential: S.E.P.M. Short Course No. 7, p. 217-252.

Murata, K. J., Friedman, I., and Madsen, B. H., 1969, Isotopic composition of diagenetic carbonates in marine Miocene formations of California and Oregon: U.S.G.S. Prof. Paper 614-B, 24 p.

Nathan, Y. and Nielsen, H., 1980, Sulphur isotopes in phosphorites, in Bentor, ed., Marine Phosphorites--Geochemistry, Occurrence, Genesis: S.E.P.M. Spec. Publications No. 29, p. 73-78.

Nathan, Y. and Sass, E., 1981, Stability relations of apatites and calcium carbonates: Chem. Geol., v. 34, p. 103-111.

Nathan, Y. and Soudry, D., 1980, Phosphate peloids from the Negev phosphorites: Jour. Geol. Soc. London, v. 137, p. 1-7.

O'Brien, G. W., Harris, J. R., Milnes, A. R., and Veeh, H. H., 1981, Bacterial origin of East Australian continental margin phosphorites: Nature, v. 294, p. 442-444.

Page, B. M., 1981, The southern coast ranges, in, Ernst, W. G., ed., The geotectonic development of California, v. 1: Prentice-Hall, New Jersey, p. 330-417.

Page, B. M., 1982, Migration of Salinian composite block, California, and disappearance of fragments: Amer. Jour. Sci., v. 282, p. 1694-1734.

Pisciotto, K. A., 1978, Basinal sedimentary facies and diagenetic aspects of the Monterey Shale, California: Ph.D. Thesis, University of California, Santa Cruz, California, 450 p.

Pisciotto, K. A., 1981a, Diagenetic trends in the siliceous facies of the Monterey Shale in the Santa Maria region, California: Sedimentology, v. 28, p. 547-571.

Pisciotto, K. A., 1981b, Distribution, thermal histories, isotopic compositions and reflection characteristics of siliceous rocks recovered by the Deep Sea Drilling Project, in Warme, J. E. et al., eds., The Deep Sea Drilling Project: A decade of progress: S.E.P.M. Spec. Publ. No. 32, p. 129-149.

Pisciotto, K. A., 1981c, Review of secondary carbonates in the Monterey Formation, California, in Garrison, ed., The Monterey Formation and related siliceous rocks of California: S.E.P.M. Pacific Section Research Symposium, San Francisco, May 29, 1981, p. 273-284.

Pisciotto, K. A. and Garrison, R. E., 1981, Lithofacies and depositional environments of the Monterey Formation, California, in Garrison, ed., The Monterey Formation and related siliceous rocks of California: S.E.P.M. Pacific Section Research Symposium, San Francisco, May 29, 1981, p. 97-122.

Pisciotto, K. A. and Mahoney, J. L., 1981, Isotopic survey of diagenetic carbonates, Deep Sea Drilling Project Leg 63: Initial Reports of the Deep Sea Drilling Project, v. 63, p. 595-609.

Pratt, L. and Claypool, G. E., 1983, Geochemical imprint of depositional and diagenetic environment on organic matter in Cretaceous and Devonian black shales and interbedded siltstones or limestones: G.S.A. Abstracts with Programs, Annual Meeting, Indianapolis, v. 15, No. 6, p. 664.

Redfield, A. C., Ketchum, B. H., and Richards, F. A., 1963 in Hill, M. N., ed., The Sea, V. 2, p. 26, Interscience, N.Y., 1963.

Reimers, C. E., 1982, Organic matter in anoxic sediments off central Peru: relations of porosity, microbial decomposition and deformation properties: Marine Geology, v. 46, p. 175-197.

Riggs, S. R., 1979, Phosphorite sedimentation in Florida--A model phosphogenic system: Econ. Geol., v. 74, p. 285-314.

Roehl, P. O., 1981, Dilation brecciation--a proposed mechanism of fracturing, petroleum expulsion and dolomitization in the Monterey Formation, California, in Garrison, ed., The

Monterey Formation and related siliceous rocks of California: S.E.P.M. Pacific Section Research Symposium, San Francisco, May 29, 1981, p. 285-314.

Sandstrom, M. W., 1982, Organic geochemistry of phosphorites and associated sediments: Ph.D. dissertation, Research School of Earth Sciences, Australian National University, 346 p.

Savin, S. M., et al., 1981, Miocene benthic foraminiferal isotope records: A synthesis: Marine Micropaleontology, v. 6, p. 423-450.

Schrader, H., 1980, Laminated diatomaceous sediments from the Guaymas Basin Slope (central Gulf of California): 250,000--year climate record: Science, v. 207, p. 1207-1211.

Sheldon, R. P., 1981, Ancient marine phosphorites: Ann. Rev. Earth. Planet. Sci., v. 9, p. 251-284.

Shemesh, A., Kolodny, Y. and Luz, B., in press, Oxygen isotope variation in phosphate of biogenic apatites: II. Phosphorite rocks: Earth Planet. Sci. Letters.

Silvermann, S. R., Fugat, R. K., and Weiser, J. D., 1952, Quantitative determination of calcite associated with carbonate-bearing apatites: Amer. Mineralogist, v. 37, p. 211-222.

Singer, A. and Muller, G., 1983, Diagenesis in argillaceous sediments, in Larsen, G. and Chilingar, G. V., eds., Diagenesis in sediments and sedimentary rocks, 2: Developments in Sedimentology No. 25B, p. 115-212.

Soudry, D. and Champetier, Y., 1983, Microbial processes in the Negev phosphorites (southern Israel): Sedimentology, v. 30, p. 411-423.

Soutar, A., Johnson, S. R., and Baumgartner, T. R., 1981, In search of modern depositional analogs to the Monterey Formation, in Garrison, R. E. et al., eds., The Monterey Formation and related siliceous rocks of California: S.E.P.M. Pacific Section Research Symposium, San Francisco, May, 1981, p. 123-148.

Starinsky, A., Katz, A. and Kolodny, Y., 1982, The incorporation of uranium into diagenetic phosphorite: Geochim. Cosmochim. Acta, v. 46, p. 1365-1374.

Suess, E., 1981, Phosphate regeneration from sediments of the Peru continental margin by dissolution of fish debris: Geochim. Cosmochim. Acta, v. 45, p. 577-588.

Surdam, R. C. and Stanley, K. O., 1981, Diagenesis and migration of hydrocarbons in the Monterey Formation, Pismo syncline, California, in R. E. Garrison et al., eds., The Monterey Formation and related siliceous rocks of California: Pacific Sect. S.E.P.M. Spec. Publ., p. 317-327.

Thorup, R. R., 1941, Vaqueros Formation at its type locality, Junipero Serra Quad., Monterey County, California: G.S.A. Bull., v. 52, p. 1957-1958.

Tissot, B., Durand, B., Espitalie, J. and Combaz, A., 1974, Influence of the nature and diagenesis of organic matter on the formation of petroleum: Am. Assoc. Petr. Geol. Bull., v. 58, no. 3, p. 499-506.

Tissot, B. P. and Welte, D. H., 1978, Petroleum Formation and Occurrence, Springer-Verlag, New York, N.Y., 536 p.

Tudge, A. P., 1960, A method of analysis of oxygen isotopes in orthophosphates--its use in measurements of paleotemperatures: Geochim. Cosmochim. Acta, v. 18, p. 81.

Williams, L. A. and Reimers, C., 1983, Role of bacterial mats in oxygen-deficient marine basins and coastal upwelling regimes: Preliminary report: Geology, v. 11, no. 5, p. 267-270.

Woodruff, F., Savin, S. M., and Douglas, R. G., 1981, Miocene stable isotope record: a detailed deep Pacific Ocean study and its paleoclimatic implications: Science, v. 212, p. 665-668.

THE ASSOCIATION OF DOLOMITITE-PHOSPHORITE-CHERT: CAUSES AND POSSIBLE DIAGENETIC SEQUENCES

M. Kastner
Scripps Institution of Oceanography
La Jolla, CA 92093

K. Mertz, D. Hollander
and R. Garrison
Earth Science Board
University of California
Santa Cruz, CA 95064

ABSTRACT

Dolomitites and phosphorites are two common early diagenetic sedimentary rocks which form in organic-rich shelf and continental margin sediments by the replacement of precursor $CaCO_3$ particles as well as by direct precipitation. Their association varies from the superposition of lithological units, through discrete layers or concretion in a given section, to intimate interbedding and lateral transition.

In the Monterey Formation, California, $\delta^{13}C$ values of dolomite vary from about -30 to 20 $^o/oo$ (PDB). This large range and the $\delta^{18}O$ values indicate formation at different burial depths and in various diagenetic zones: in the anaerobic sulfate and carbonate reduction zones, and most probably also in the zone of organic matter thermocatalytic decarboxylation. The majority of the few available $\delta^{13}C$ values of the CO_3^{-2} in apatite from the Monterey Formation range from 3 to -12 $^o/oo$ (PDB), indicating formation by replacement of "normal" marine $CaCO_3$ or by direct precipitation in the aerobic respiration, sulfate reduction, and possibly decarboxylation zones. Oxygen isotopes from PO_4^{-3} suggest formation at temperatures of approximately 30° to 50°C, and from CO_3^{-3} of approximately 30° to 80°C.

On the basis of the $\delta^{13}C$ data, dolomite forms continuously from early to late diagenesis, but apatite seems to be an early to medial diagenetic phase.

Dolomitization of $CaCO_3$ is inhibited by the high concentration of dissolved sulfate in sea water. Apatite formation is kinetically inhibited by the high Mg^{+2}/Ca^{+2} ratio of sea water. Aging, conversion of amorphous Ca-phosphate to crystalline apatite is enhanced by low Mg^{+2}/Ca^{+2} ratios and high phosphate concentrations. Thus, in and below the sulfate reduction zone, dolomite formation is promoted by both the removal of dissolved sulfate and by enrichment in dissolved carbonate in the pore water. Apatite formation is promoted by pore water enrichment in dissolved phosphate and by depletion of the Mg^{+2}/Ca^{+2} ratio through $CaCO_3$ dolomitization. These dolomites and apatites will have either light or heavy $\delta^{13}C$ values. Alternatively, although as yet not commonly observed in nature, amorphous Ca-phosphate may already form prior to dolomitization. Subsequent dolomite formation will enhance its conversion into crystalline apatite by depressing the Mg^{+2}/Ca^{+2} ratio in the pore water. This dolomite and the recrystallized apatite may also have either light or heavy $\delta^{13}C$ values.

The biogenic sediments phosphorite and chert form in high-productivity regions, hence their frequent association.

Dolomite formation may retard the opal-A to opal-CT transformation. During medial to late diagenesis, and/or in areas of high geothermal gradients, the rate of opal-A to opal-CT transformation may exceed the rate of $CaCO_3$ dolomitization, even during early diagenesis, when temperatures reach higher than 40° to 50°C, in the presence of Mg^{+2} and OH^-. The dolomitization reaction could then be retarded. During late diagenesis, when opal-CT transforms to quartz, the Mg^{+2} and OH^--bearing nuclei (most likely sepiolite) will be released and some dolomitization promoted by the following reaction: opal-CT + sepiolite + calcite → quartz + dolomite + H_2O.

INTRODUCTION

In organic-rich shelf and continental slope sediments, dolomitites and phosphorites are two common early diagenetic sedimentary rocks. The dolomite mineral $CaMg(CO_3)_2$ is usually (but not necessarily) Ca-rich and often also Fe-rich and somewhat disordered. The main phosphate-bearing mineral in the phosphorites is carbonate fluorapatite (francolite) $(Ca, Na, Mg, Sr)_{10}(PO_4, CO_3, SO_4)_6F \geq 2$.

The characteristic triple association of phosphorite, porcelanite, or chert and organic matter has been recently reviewed by Bentor (1980), Kolodny (1981), Baturin (1982), and Calvert and Price (1983). The porcelanite and chert are composed of opal-CT and/or quartz which were diagenetically derived from diatom and Radiolaria oozes in Cretaceous and younger sediments, as already recognized by Bramlette (1946), or from Radiolaria and spongiae in Paleozoic and early Mesozoic rocks. This triple association signifies upwelling environments.

in Garrison, Robert E.; Kastner, Miriam and Zenger, Donald H., eds., 1984, **Dolomites of The Monterey Formation and Other Organic-Rich Units:** Pacific Section S.E.P.M., Vol. 41, p. 75-86.

Apatite concretions and lenses, millimeters to centimeters long, forming in Recent upwelling environments within anoxic diatomaceous sediments, were described from the Namibian and Peruvian shelves (for example, Baturin, 1969; Baturin et al., 1982; Calvert and Price, 1983; and references therein).

Holocene diagenetic dolomite is, as well, widespread in the shelf and continental margin off Peru (e.g., Kulm et al., this volume). Are these upwelling environments Recent analogs of the Miocene Monterey Formation, where dolomitites, phosphorites, porcelanites, and cherts are widespread diagenetic rocks?

Similar to the problem of dolomite formation, two mechanisms for apatite formation were invoked: phosphatization of a carbonate precursor and direct precipitation from the pore fluids (e.g., Gulbrandsen, 1969; Baturin, 1969, 1982; Kennedy and Garrison, 1975; Manheim et al., 1975; and references therein).

Oxygen and carbon isotopes of the carbonate ion in apatite and dolomite and oxygen isotopes of the phosphate ion in apatite provide important information about the temperatures and diagenetic zones of their formation (for example, Kolodny and Kaplan, 1970; Longinelli and Nuti, 1973; McArthur et al., 1980; Shemesh et al., 1983). Such information is essential for recognizing various possible sequences of formation of these phases and the relationships of each sequence with oceanographic, sedimentologic, and diagenetic factors.

The magnesium ion plays an important role, affecting rates of diagenetic reactions in each of these three systems, as follows. In the pore fluid, a $Mg^{+2}/Ca^{+2} \geq 1$ is necessary for dolomite formation. Apatite formation is kinetically inhibited by the high Mg^{+2}/Ca^{+2} ratio of sea water (Martens and Harris, 1970; Nathan and Lucas, 1976; Gulbrandsen et al., 1984). As apatite is a diagenetic phase, depletion of Mg^{+2} of pore fluids by diagenetic reactions and formation of Mg-bearing clay minerals sepiolite, palygorskite, smectites or glauconite, or by dolomite formation were invoked (Gulbrandsen, 1960; Burnett, 1974; Baturin, 1982). Jahnke et al. (1983), however, suggest Recent apatite formation in pore waters undepleted in Mg^{+2} below its sea water value, in western Mexican continental margin sediments. The rate of opal-A to opal-CT transformation is enhanced in the presence of a magnesium hydroxide compound (MHC), most likely sepiolite, which serves as a nucleus for the crystallization of opal-CT (Kastner et al., 1977; Kastner and Gieskes, 1983).

A preliminary attempt to unravel the causes and possible diagenetic sequences responsible for a few occurrences of dolomite, phosphorite, chert, and porcelanite in the Monterey Formation, California is hereby presented.

METHODS OF STUDY

Thin sections were stained for calcite with Alizarin Red-S and for Fe-rich dolomite with potassium ferricyanide (Friedman, 1959).

Mineralogical analyses of bulk samples were carried out by X-ray diffraction (XRD) using a Philips APD 3600 instrument, at a scanning rate of $1°2\theta$/min.

Oxygen and carbon isotopic ratios were measured on a 6 cm VG Micromass 602 C mass spectrometer by analysis of CO_2 gas released at 50°C from the dolomites and apatites by reaction with 100% phosphoric acid (McCrea, 1950). The isotopic analyses were corrected according to Craig (1957), and the results are reported as per mil ($^o/oo$) relative to the PDB standard for carbonate in apatites and for the dolomites and relative to the SMOW standard (Craig, 1961) for the phosphate in apatites. The standard deviation is ±0.15 for $\delta^{18}O$ and ±0.20 for $\delta^{13}C$ values, because of some organic matter interference; 0.8 $^o/oo$ were subtracted from all $\delta^{18}O$ values of dolomite (Sharma and Clayton, 1965).

Prior to isotopic analyses of the dolomites and apatites, the small amounts of calcite present were dissolved with either an acetic acid-Na acetate buffer (pH = 4.7) or by treatment with Silverman's solution (Silverman et al., 1952). Complete removal of calcite was certified by XRD. The oxygen and carbon isotopic composition of the dolomites and apatites were not affected by this mild dissociation treatments, as verified by subjecting samples of pure dolomite and apatite to the same chemical treatments. The isotopic compositions of the untreated and treated samples were the same.

Temperatures of formation were calculated using the experimentally-determined fractionation factors for protodolomite by Fritz and Smith (1970) and using a constant fractionation of 3.1 $^o/oo$ between dolomite and calcite, given by Friedman and O'Neil (1977), and not the experimentally-determined fractionation factors for dolomite by Northrop and Clayton (1966) or Matthews and Katz (1977) for the following main reasons: (1) the experiments by Fritz and Smith (1970) were conducted between 25° and 79°C, the temperature range of formation of most dolomites in the Monterey Formation; and (2) most dolomites in the Monterey Formation are protodolomites or Fe-rich protodolomites. Also, temperatures of dolomite formation calculated from the experimental protodolomite-water fractionation expression are intermediate, between higher temperatures calculated according to Northrop and Clayton (1966) and somewhat lower temperatures calculated according to Matthews and Katz (1977).

The isotopic fractionation of the reaction between the carbonate in apatite and 100% phosphoric acid is unknown and assumed to be the same as that for the liberation of CO_2 from calcite (Kolodny and Kaplan, 1970). The $\delta^{18}O_{CO_3}$ of the apatites should thus be considered approximate, as suggested by Kolodny and Kaplan (1970). The $\delta^{13}C_{CO_3}$ of the apatites, however, are more reliable. Temperatures of formation were calculated from $\delta^{18}O_{CO_3}$ in apatite using the carbonate-water temperature scale of Epstein et al. (1953). The oxygen isotopic composition of the phosphate in apatite was determined by Y. Kolodny using the method described by Longinelli and Nuti

(1973) and Kolodny et al. (1983). Temperatures of formation from $\delta^{18}O_{PO_4}$ were calculated using the revised temperature equation by Kolodny et al. (1983). Certainly, whenever available these latter temperatures of formation will be used in the discussion instead of those calculated from $\delta^{18}O_{CO_3}$.

For all the equations $\alpha_{CO_2-H_2O}$ at 25°C of 1.0412 was used (Friedman and O'Neil, 19777). It was also assumed that the dolomites and apatites equilibrated with a fluid phase, having $\delta^{18}O$ = 0.0 °/oo relative to SMOW. The results are listed in Tables 1 and 2 and shown in Figure 6.

In order to investigate the influence of opal-A (diatomite) on the calcite dolomitization reaction, hydrothermal experiments with mixtures of opal-A and calcite were conducted, following the procedures described in Kastner et al. (1977), Baker and Kastner (1981), and Kastner and Gieskes (1983). The experiments were carried out at 200°C in a dolomitizing solution of Mg^{+2}/Ca^{+2} of ~1.3 (0.0765 M $MgCl_2$, 0.0575 M $CaCl_2$) at sea water ionic strength (I) of 0.7. The results are summarized in Table 3 and shown in Figure 7.

FIELD OCCURRENCE

Except for two samples of dolomite and apatite from the Point Sal Formation at Point Sal in the western Santa Maria Basin (Woodring and Bramlette, 1950), the samples studied are from the Sandholdt Member, Monterey Formation, late early Miocene (Durham, 1974) in the northern part of the Santa Lucia Range, Salinas Basin, Monterey County. Specific localities for the latter area are near Arroyo Center along the Arroyo Seco, at Vaqueros Canyon south of the Arroyo Seco, and near the Indians Ranch near the San Antonio River. Fresh and unweathered-appearing dolomite and apatite samples were collected from in situ dolomitite and phosphorite concretions and from dolomitite layers in laminated calcareous-siliceous shales.

In the Sandholdt Member, dolomite and apatite occur in a range of types of shaly rocks (see Mertz, this volume). In a given section, dolomite and apatite either form separate bodies (as discrete concretions and/or layers; Figures 1, 2, and 3), or are intimately interbedded (Figure 4), or are mutually associated with each other in any of these rock bodies, as for example in some of the light-colored concretions and layers shown in Figure 4.

According to Graham (1976), in this area the Sandholdt Member was deposited on a slope at 500 to 2500 m water depth; and according to Mertz and Garrison (1983), sedimentation rates were at least as high as 400 m/10^6 yr.

MINERALOGY AND PETROLOGY

On the basis of petrographic analyses, using stained thin sections and X-ray diffraction analyses, in addition to the insoluble residue, most samples studied here were composed of either chiefly of dolomite or apatite (± calcite) or of

Figure 1. A typical lensoid body of dolomitite, with its long axis parallel to bedding, in shale, Sandholdt Member, Monterey Formation, near Arroyo Center, along the Arroyo Seco, Monterey County.

Figure 2. Typical habit of in situ apatite concretions in Sandholdt shale. Same locality as Figure 1.

Figure 3. Dolomite layer (in lower third of photograph) and apatite concretion above it, in Sandholdt shale. Same locality as Figure 1.

Figure 4. Intimately interbedded dolomite (in dark host rock) and apatite (in light colored thin layers and concretions). In several of the light colored zones apatite is associated with dolomite (± calcite).

various mixtures of dolomite and apatite (± calcite). Detrital quartz and some feldspars, clay minerals--mainly mixed-layer smectite/illite--and ubiquitous diagenetic pyrite constitute the insoluble residues. Calcite, although present in many of the samples, was not an abundant phase. On the basis of textural relations, a small fraction of the calcite seems to have a diagenetic origin. The dolomite is Ca- and Fe-rich; many of the thin sections stained with potassium ferricyanide (Friedman, 1959) show more than one generation of dolomite formation, each having a different concentration of Fe in the dolomite and different oxygen and carbon isotopic values (see Table 2). This type of information is usually lost by routine X-ray diffraction analysis, which only provides approximate chemical averages of such dolomite mixtures.

Using the binocular microscope and a tungsten needle, several light-colored (yellowish-brown to white) thin layers and concretions, supposed to be composed of apatite, were separated from the rock shown in Figure 4 and from the samples from Point Sal, and were analyzed by XRD. A few contain apatite only, most contain various mixtures of apatite, dolomite, and calcite, and others contain only dolomite and calcite. Representative XRD results are shown in Figure 5. Several dolomite samples were independently separated from the dark-colored host dolomitite (Figure 4). These dolomites, which are from immediately above or below these light-colored concretions and layers, are more Fe-rich than the dolomite within the light sediment (for example, Samples AS-1, 3, and 5 in Table 2).

On the basis of textural relationships and mass balance calculations, the concretions and layers studied are composed of various mixtures of replacement and void-filling dolomite and apatite. Some primary biogenic apatite must be present as well.

STABLE ISOTOPES OF THE ASSOCIATION OF DOLOMITITE-PHOSPHORITE

Dolomite and Apatite in Discrete Concretions and Layers

The $\delta^{18}O$ values of PO_4^{-3} of apatites and the $\delta^{18}O$ and $\delta^{13}C$ of CO_3^{-2} of apatites and dolomites are listed in Table 1. Calculated temperatures of formation of both the dolomites and apatites in concretions and layers in the Sandholdt Member, of the Monterey Formation, are distinctly higher (between approximately 30° and 80°C) than about 6° to 20°C temperatures of formation calculated for Recent apatites from the Namibian and Peruvian shelves (Kolodny and Kaplan, 1970; McArthur et al., 190; Shemesh et al., 1983). Such medium to high temperatures of formation do not support apatite formation at or close to the sediment/water interface. On the basis of $\delta^{18}O_{PO_4}$ values for apatite and $\delta^{18}O_{CO_3}$ values for dolomites, several of the dolomites formed at lower temperatures, therefore earlier, than the neighboring apatites (Samples AS-25, V-8, and V-15), and dolomite AS-16 and the adjacent apatite AS-15 formed at about the same temperature (~45°C) at about the same time. None of these dolomites formed at higher temperatures, later than the apatites. Depending on the oxygen isotopic composition of the pore fluids the temperatures of formation of both dolomite and apatite could be shifted to either lower or to higher temperatures. In these calculations it was assumed that $\delta^{13}O_{pore\ fluid} = 0\ ^{o}/oo$ (SMOW). In this temperature range, a $1\ ^{o}/oo$ (SMOW) depletion in the $\delta^{18}O$ of the pore fluid will decrease the calculated formation temperature of each of these minerals by about 5°C. Similarly, a 1% enrichment in the pore fluid $\delta^{18}O$ value will increase calculated temperatures by about 5°C.

Carbon isotopes of diagenetic carbonates and of apatites reflect the $\delta^{13}C$ of the pore fluid $\sum CO_2$ and are sensitive indicators of organic matter degradation reactions (for example, Rosenfeld and Silverman, 1959; Claypool, 1974; Claypool and Kaplan, 1974; Nissenbaum et al., 1977; McArthur et al., 1980). The $\delta^{13}C$ values of all these dolomites are highly positive, between +7.39 and +16.66 $^{o}/oo$ (PDB); except for Sample V-9A, with a $\delta^{13}C$ value of 3.05 $^{o}/oo$, the apatites have negative $\delta^{13}C$ values between −0.85 and −3.00 $^{o}/oo$ (PDB), as seen also in Figure 6. The light carbon isotope values of apatites in the Monterey Formation had been assumed to indicate formation in the sulfate reduction zone or even at shallower subsurface depths, close to the sediment/water interface, in the aerobic respiration zone. The calculated high temperatures of formation in Table 1 do not support this assumption.

Subsequent isotopic exchange is not supported by (1) the evidence by Shemesh et al. (1983) that oxygen in apatite phosphate is much less exchangeable than the oxygen in carbonates and possibly even in cherts, and (2) by the evidence in Kushnir and Kastner (this volume) that no chemical and isotopic exchange could have occurred in a dolomite concretion from this same area. It is therefore suggested that these light $\delta^{13}C$ values may reflect instead isotopically lighter carbonate

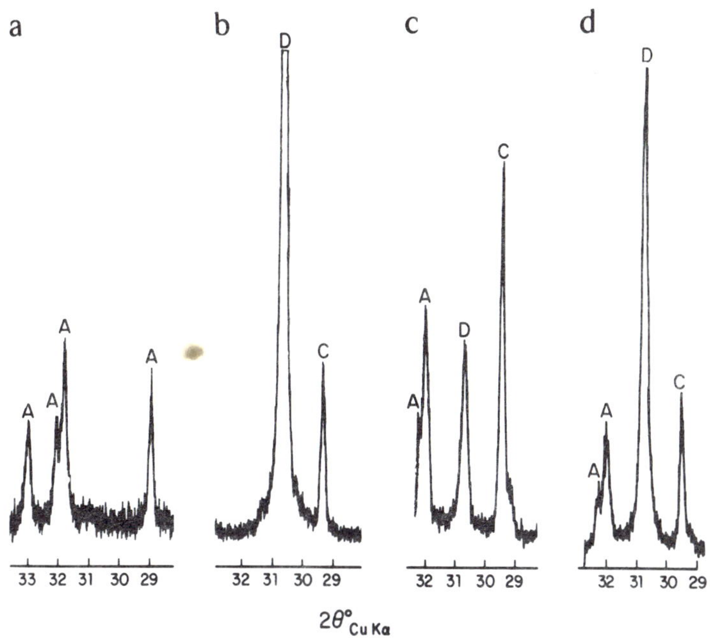

Figure 5. X-ray diffraction diagrams of variable light colored zones; separated from the dolomitite shown in Figure 4.

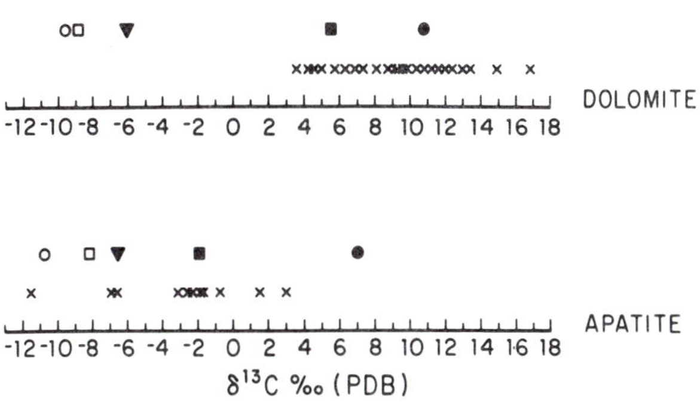

Figure 6. $\delta^{13}C$ values of dolomite and apatite in discrete concretions and layers, Sandholdt Member, at Arroyo Seco (X). The other symbols designate mutually associated dolomite and apatite in light colored concretions and layers. ●, ■, ▲ represent three dolomite-apatite pairs, separated from diferent parts of the dolomitite, shown in Figure 4. □ and O designate two such pairs from the Point Sal Formation, Point Sal. These sample pairs are described in Table 2.

ions diffused and/or advected from below from the zone of thermocatalytic decarboxylation. This suggestion is supported by the calculated high temperatures of apatite formation. The dolomites, however, formed already in the anaerobic carbonate reduction zone.

Mutually Associated and Intimately Interbedded Dolomite and Apatite

The apatite samples listed in Table 2 were not analyzed for their phosphate oxygen isotopes. Their calculated temperatures of formation on the basis of $\delta^{18}O_{CO_3}$ should therefore be considered approximate. The temperatures of formation of most dolomites and apatites in this table lie between 30° and 60°C. At a geothermal gradient of 60° to 70°C/km (Isaacs, 1984) and bottom temperatures of about 7°C, these dolomites and apatites formed between about 350 and 800 m burial depth, and not close to the sediment/water interface.

Samples AS-1, AS-3, and AS-5 are subsamples of the rock shown in Figure 4: AS-1 from the left third, AS-5 from the center, and AS-3 from the right third. The dark-colored host rock is a dolomitite. Most of the light-colored thin layers and concretions consist of apatite, dolomite, and

Table 1. Isotopic Composition of Phosphate in Apatites and of Carbonate in Dolomites in the Monterey Formation, California

Sample #		Type of Sample	Locality	Monterey Formation	$\delta^{18}O_{PO_4}$ † °/∘∘ (SMOW)	Temperature* (°C)	$\delta^{18}O_{CO_3}$ °/∘∘ (SMOW)	Temperature** (°C)	$\delta^{13}C$ °/∘∘ (PDB)
AS-26	} ∿ 8 m	Apatite concretion in shale	Arroyo Seco	Sandholdt Member	---	---	19.80	79	-1.83
AS-25	} ∿ 20 m	Dolomite	" "	"	---	---	28.84	39	10.42
AS-22		Apatite concretion in shale	" "	"	15.60	45	20.26	76	-2.61
AS-15	} ∿ 1 m	Apatite concretion in calcareous shale	" "	"	15.50	45	19.92	78	-3.00
AS-16		Dolomite	" "	"	---	---	27.28	47	7.39
V-9a	} ∿ 0.8 m	Apatite concretion in shale	Vaqueros Canyon	Top of Sandholdt Member	16.32	42	22.73	59	3.05
V-8		Dolomite	"	"	---	---	30.52	30	16.66
V-14	} ∿ 0.2 m	Apatite concretion in shale	"	Base of Sandholdt Member	14.47	50	27.50	32	-0.85
V-15		Dolomite	"	"	---	---	27.81	44	11.34

*Calculated according to Kolodny et al. (1983)

**Calculated according to Epstein et al. (1953)

†Analyzed by Dr. Y. Kolodny, Hebrew University, Jerusalem, Israel

Table 2. Isotopic Composition of Carbonate in Apatites and of Associated Dolomites

Sample		Type of Sample	Locality	Formation	$\delta^{13}C$ ‰ (PDB)	$\delta^{18}O_{CO_3}$ ‰ (PDB)	Temperature* (°C)
AS-1a	(●)	Apatite in concretion	Arroyo Seco	Monterey Fm., Sandholdt Member	6.93	-3.43	33
AS-1b	(●)	Dolomite in same concretion	" "	"	10.62	-0.39	30
AS-1c		Dolomite host rock	" "	"	8.96	-2.70	43
AS-3a		Apatite in concretion	Arroyo Seco	"	-2.86	-7.82	59
AS-3b		Dolomite host rock	" "	"	10.14	-1.47	36
AS-5a	(■)	Apatite in concretion	Arroyo Seco	"	-2.05	-6.60	52
AS-5b	(■)	Dolomite in same concretion	" "	"	5.45	-4.16	52
AS-5c		Dolomite host rock	" "	"	7.87	-3.34	47
I-7a	(▼)	Apatite in concretion	Indian Ranch	"	-6.51	-7.34	56
I-7b	(▼)	Dolomite host rock	" "	"	-6.16	-7.21	74
PS-30Ia	(□)	Apatite in concretion	Point Sal	Point Sal Formation	-8.15	-5.28	43
PS-30Ib	(□)	Dolomite in same concretion	" "	"	-8.78	-6.60	68
PS-30IIa	(○)	Apatite in layer	Point Sal	Point Sal Formation	-10.80	-4.41	38
PS-30IIb	(○)	Dolomite in same layer	" "	"	-9.58	-4.94	59

*Calculated according to Kolodny _et al_. (1983)

■ ● ▼ □ ○ -- See Figure 6.

calcite. The host rock dolomite is Fe-rich; the dolomites within the light zones contain significantly less Fe. The $\delta^{18}C$ values of the mutually-associated dolomites and apatites are also plotted in Figure 6. The $\delta^{13}C$ values of some of the dolomites associated with apatite are much lighter than of the dolomites in the host rock and of most of the dolomites analyzed at and in the vicinity of Arroyo Seco. Except for Sample AS-1a, the apatites have negative $\delta^{13}C$ values.

These results indicate that not all apatites formed before and at lower temperatures than the adjacent dolomites; or alternatively, that both formed at about the same time, depth, and temperatures. Clearly, some dolomites formed earlier and at lower temperatures than the associated apatites.

The rather wide range of apatite formation temperatures and the distinct differences in the isotopic composition and temperatures of formation of the dolomites within the light-colored concretions and layers versus those in the host rock, strongly support a multigeneration origin of the dolomitites and phosphorites in the Monterey Formation.

THE ASSOCIATION OF DOLOMITITE-CHERT

Biogenic silica (opal-A), once deposited in the sediment, generally undergoes evolutionary diagenetic changes with time, burial depth, and temperature, to porcelanite or chert composed of opal-CT and/or quartz. Experimental work between

Table 3. Effect of Amorphous Silica on Dolomite Formation from Calcite, 200°C, 2 Weeks, in 0.080 M $MgCl_2$ + 0.060 M $CaCl_2$ + 0.280 M NaCl Solution, I = 0.7

Starting Solids	Calcite/ Ludox Si (Weight % Ratio)	Molar Ca^{+2} Final Solution	Molar Mg^{+2} Final Solution	Molar Mg^{+2}/Ca^{+2} Final Solution	H_4SiO_4 (μM) Final Solution	Carbonates XRD Results
100 mg Calcite 50 mg Ludox Si	2	0.11769	0.0204	0.17	5328	Calcite
100 mg Calcite 25 mg Ludox Si	4	0.09055	0.0483	0.53	1703	Calcite
50 mg Calcite 20 mg Ludox Si	2.5	0.08791	0.0532	0.61	1431	Calcite
50 mg Calcite 10 mg Ludox Si	5	0.07876	0.0663*	0.84	824	Calcite
20 mg Calcite 10 mg Ludox Si	2	0.07610	0.0643	0.85	769	Calcite
20 mg Calcite 5 mg Ludox Si	4	0.07325	0.0690	0.94	549	Dolomite > Calcite
20 mg Calcite 2 mg Ludox Si	10	0.07223	0.0697	0.97	344	Dolomite plus traces of Calcite
Starting Solution	--	0.05753	0.0765	1.33	--	--

*Value from 9 days' experiment.

25° and 150°C by Kastner et al. (1977) and Kastner and Gieskes (1983) has demonstrated that a magnesium hydroxide compound (HMC), most probably sepiolite, may serve as a nucleus which is chiefly responsible for the observed enhanced transformation rate of opal-A to opal-CT, in environments where dissolved Mg^{+2} and $(OH)^-$ are available. Experimental work by Baker and Kastner (1981) and Kastner and Baker (1982) on dolomitization of calcite and aragonite showed that at 200°C, and in a dolomitizing solution of Mg^{+2}/Ca^{+2} ratio of 1.3, calcite completely transforms to dolomite in 12 to 14 days.

The main objective of the present experimental study was to determine the effect of opal-A on the rate of calcite dolomitization in such dolomitizing solutions. The results are listed in Table 3 and shown in Figure 7.

Various weight ratios (between 10 and 2) of calcite to Ludox Silica, a pure synthetic opal-A, were reacted at 200°C for two weeks in a solution of sea water ionic strength and a molar Mg^{+2}/Ca^{+2} of 1.3. The Mg^{+2}/Ca^{+2} ratios of the final solutions are significantly lower than of the starting solutions, and do not correlate with the weight ratios of calcite/Ludox silica; see, for example, the significantly different Mg^{+2}/Ca^{+2} ratios of the final solutions in two experiments, each with calcite/Ludox silica ratios of 2 and 4, in Table 3 and the upper portion of Figure 7. When, however, the Mg^{+2}/Ca^{+2} ratios of the final solutions are plotted against the amount of Ludox silica, a straight line with a negative slope is obtained, shown in the lower part of Figure 7.

These results indicate that at this elevated temperature, the rate of opal-A to opal-CT transformation reaction is greater than the rate of calcite transformation to dolomite. In the presence of large amounts of opal-A, its transformation to opal-CT rapidly consumed most of the dissolved Mg^{+2}, shifting the solution Mg^{+2}/Ca^{+2} ratio below about 1, the minimum ratio for dolomite formation. Dolomite formed only in the experiments with just 2 and 5 mg Ludox silica; in these experiments, the opal-A to opal-CT reaction reduced the magnesium concentrations in the solution slowly and to a lesser extent. This allowed dolomite formation. Not all the calcite transformed to dolomite, even in the experiment with just 2 mg opal-A; there was still enough opal-A present to reduce the Mg^{+2}/Ca^{+2} ratio of the final solution to slightly below 1 (Table 3, Fig. 7).

CAUSES AND POSSIBLE DIAGENETIC SEQUENCES

Dolomite is the thermodynamically stable carbonate phase in sea water. Dolomitization at $CaCO_3$ is, however, inhibited by the high concentration of dissolved sulfate in sea water (Baker and Kastner, 1981; Kastner and Baker, 1982). Martens and Harris (1970) showed experimentally that apatite precipitation is strongly dependent on the Mg^{+2}/Ca^{+2} ratio in the solution, that its precipitation is inhibited by the high Mg^{+2}/Ca^{+2} ratio of sea water. From sea water they obtained only an amorphous Ca-phosphate precipitate. This precipitate had a lower molar Ca/P (1.35) ratio than the ratio in apatite 1.67). During eight

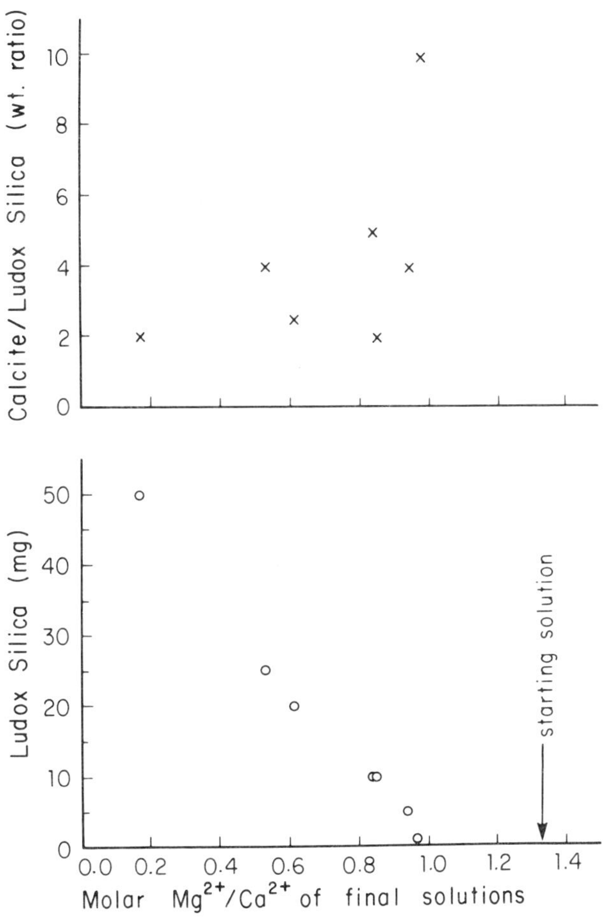

Figure 7. Changes in the molar Mg^{+2}/Ca^{+2} ratio of a starting solution as a function of: (a) various calcite/Ludox silica weight ratios, and (b) the amount of Ludox silica present. These hydrothermal experiments were conducted at 200°C, for 2 weeks, in a solution of 0.0765 M $MgCl_2$, 0.0575 M $CaCl_2$, and NaCl, at sea water ionic strength of 0.7.

months the amorphous precipitate did not transform to crystalline apatite. When, however, immersed in Mg-free artificial sea water it quickly transformed to apatite. In these experiments apatite did not crystallize above a Mg^{+2}/Ca^{+2} ratio of 0.2, which is an extremely low ratio relative to the ratio of 5.3 in sea water. Nathan and Lucas (1976) have shown that the limiting Mg^{+2}/Ca^{+2} ratio for apatite crystallization is pH dependent; the ratio increases with decreasing pH. Gulbrandsen et al. (1984) described the synthesis of carbonate fluorapatite in sea water. Several types of seed crystals and fluorine (7.6 mg F/l) were added to sea water (sea water has ~1.3 F/l). During the first five months phosphate was also slowly added. At concentrations between 90 to 98 mg PO_4/l (total phosphate) the precipitation of an amorphous phosphate phase began in each of the experiments. After nearly ten years, in the experiments with calcite seeds, all the former precipitates disappeared and only apatite was detected by XRD. In the experiments with apatite seeds, an unidentified phase formed in addition to apatite. In the presence of quartz and fluorite seeds, however, only the very beginning of apatite crystallization was observed after nearly ten years.

Thus, as stated before, the Mg^{+2}/Ca^{+2} ratio of sea water kinetically inhibits apatite crystallization. An amorphous phosphate phase precipitated first. Time is a most important factor in the heterogeneous nucleation and growth of apatite at sea water Mg^{+2}/Ca^{+2} ratio.

It thus seems reasonable that diagenetic dolomite formation, which reduces the Mg^{+2}/Ca^{+2} ratio of sea water should promote apatite formation, as already suggested, for example by Gulbrandsen (1960) and Burnett (1974). No direct evidence for this paragenesis has been proposed by them. Unit recently it was also assumed that dolomitization requires a Mg^{+2}/Ca^{+2} molar ratio higher than that of sea water (5.3).

We suggest the following dolomite-apatite paragenesis: In and below the sulfate reduction zone of organic-rich sediments, dolomite formation is promoted by both the removal of dissolved sulfate and by enrichment in dissolved carbonate in the pore fluids*. Apatite formation, which is promoted by pore water enrichment in dissolved phosphate from organic matter degradation, and by depletion of Mg^{+2}/Ca^{+2} ratio through dolomite formation, could immediately follow early dolomite formation. Depending on sedimentation rates, organic matter concentrations in the sediments and geothermal gradients, these dolomites and apatites may have either light or heavy $\delta^{13}C$ values. Dolomite formation before and at lower temperatures than apatite is supported by several of the results in Tables 1 and 2. This suggested sequence of diagenetic events is presented schematically in Figure 8.

Alternatively, amorphous Ca-phosphate may already form prior to dolomite formation. Subsequent dolomitization and increased phosphate concentration from organic matter decomposition will enhance its conversion into the crystalline phase, apatite, by depressing the Mg^{+2}/Ca^{+2} ratio in the pore fluids. Deposition of amorphous Ca-phosphate was described by Baturin (1982) in the Namibian shelf phosphorites, by Kulm et al. (this volume) in the Trujillo Basin off Peru (there more than 4% of the rock is composed of amorphous Ca-phosphate) and by Mahmoud (1976) in the Monterey Formation, California.

The isotopic values of these apatites may record the diagenetic environment of conversion to apatite and not their original environment of deposition. Thus, the observed medium to high temperatures of formation of apatite, in the Monterey Formation, may reflect this sequence of diagenetic reactions, shown schematically in Figure 9.

*Extensive dolomitization is, however, possible only if precursor $CaCO_3$ phases are present.

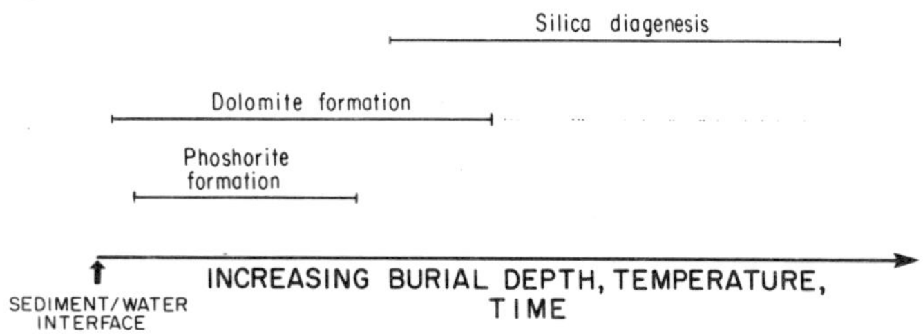

Figure 8. A sequence of diagenetic events: (a) dolomite formation, (b) apatite and dolomite formation, (c) dolomite formation, (d) silica phase transformations and some dolomite formation.

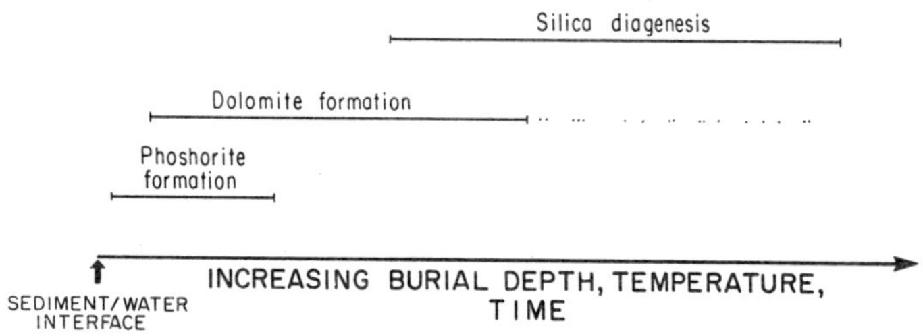

Figure 9. A sequence of diagenetic events: (a) amorphous Ca-phosphate formation, (b) dolomite and apatite formation, and crystallization of amorphous Ca-phosphate to apatite, (c) dolomite formation, (d) silica phase transformations and some dolomite formation.

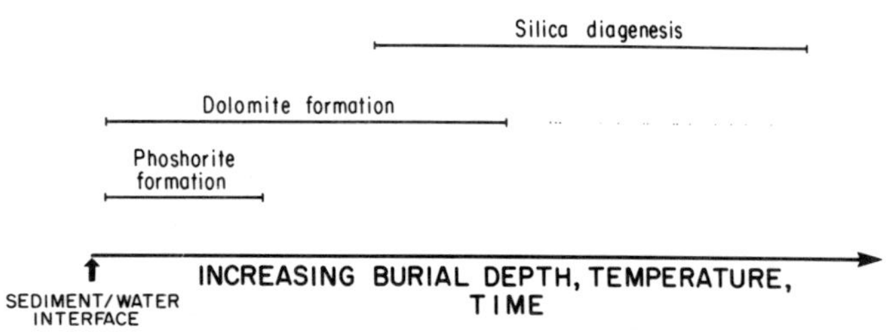

Figure 10. A sequence of diagenetic events: (a) amorphous Ca-phosphate and dolomite formation, (b) dolomite and apatite formation and crystallization of amorphous Ca-phosphate to apatite, (c) dolomite formation, (d) silica phase transformation and some dolomite formation.

An early origin of some phosphate nodules in Monterey rocks is suggested by the occurrence of intraformational conglomerates composed of exhumed and redeposited phosphate nodules (e.g., at Lions Head, in the Santa Maria Basin, Pisciotto, 1981; near Naples Beach along the Santa Barbara coast, Isaacs, 1981). They may have formed initially as other phosphate phases which subsequently crystallized to apatite during burial.

It is also possible that amorphous Ca-phosphate and dolomite form simultaneously, as suggested in Figure 10.* The early forming amorphous Ca-phosphate would then crystallize into apatite. Hence, at different times, depths and temperatures, different phosphate phases probably form: amorphous Ca-phosphate, carbonate fluorapatite with and without SO_4^{-1}.

Even in diagenetic environments with little or no dolomite formation amorphous Ca-phosphate will crystallize to apatite. This is supported by the experimental results of Gulbrandsen et al. (1984), and by the Recent formation of apatite in western Mexican continental margin sediments in the presence of pore water with a Mg^{+2}/Ca^{+2} ratio similar to the sea water ratio (Jahnke et al., 1983). It also is not surprising that amorphous phosphate material was described from the two Recent upwelling systems in which phosphorites form (Namibian and Peruvian shelves) and not from geologically old phosphorites.

We suggest that in the Monterey Formation, California, not all apatites formed close to the sediment/sea water interface, as previously suggested. Some amorphous Ca-phosphates most probably have formed at or close to this interface and subsequently transformed to crystalline apatite at greater depths and higher temperatures; others seem to have formed later, directly as apatite. Isotopically it may be difficult to distinguish between apatite formation at moderate to elevated temperatures and amorphous Ca-phosphate crystallization to apatite under similar diagenetic conditions. Apatite or amorphous Ca-phosphate which form above and within the sulfate reduction zone may contain SO_4^{-2}. Those formed below the sulfate reduction zone, in the anaerobic carbonate reduction zone, for example apatite sample AS-1a (Table 2) and some apatite concretions from the Namibian shelf, sample 3424 in McArthur et al. (1980), or in the thermocatalytic decarboxylation zone, would not contain SO_4^{-2}. Additional, as yet undescribed geochemical differences may exist as well.

As for the association of dolomitite and porcelanite or chert, dolomite formation may retard the opal-A to opal-CT transformation. During medial to late diagenesis, and/or in areas of high geothermal gradients even during early diagenesis, the rate of opal-A to opal-CT transformation reaction may exceed the rate of dolomite formation. This occurs at temperatures higher than 40° to 50°C in environments of deposition where Mg^{+2} and

*In Figures 8, 9, 10 silica diagenesis designates the transformation of diatomite to porcelanite or chert. Diagenetic reactions within diatomite are not included.

$(OH)^-$ are available, e.g., in organic-rich and/or carbonate-rich sediments. The dolomitization reaction could then be retarded.

During late diagenesis, when opal-CT transforms to quartz, the Mg^{+2} and $(OH)^-$ bearing nuclei will be released and some late dolomitization promoted, especially in organic-rich sediments in which the carbonate ion may be available from thermocatalytic decarboxylation reactions. The following reaction of late dolomite formation is suggested: opal-CT + sepiolite + calcite → quartz + dolomite + H_2O.

In organic-rich sediments dolomite formation may start in the sulfate reduction zone, and continue to form in the carbonate reduction and thermocatalytic decarboxylation zones, as shown in Figures 8, 9 and 10.

ACKNOWLEDGEMENTS

We are grateful to Dr. D. H. Zenger, for his suggestions on improving the manuscript. Special thanks to Dr. Y. Kolodny for the analyses of the oxygen isotopes of phosphate in apatites. We thank Ms. G. Anderson and Ms. L. Henderson for their invaluable help in the laboratory. This research was supported by ONR contract No. USN N00014-80-C-0440 (M.K.), by NSF grant #OOE80-24630 (M.K.), and by Exxon Production Research Co., Houston (M.K.). Partial support for this research was derived through grants from the Petroleum Research Fund (PRF #13953-AC2 and #12432-AC2)(R.E.G.) and NSF Grant #EAR80-09128 (R.E.G.).

REFERENCES CITED

Baker, P. A. and Kastner, M., 1981, Constraints on the formation of sedimentary dolomite: Science, v. 213, p. 214-216.

Baturin, G. N., 1982, Phosphorites on the sea floor, origin, composition and distribution: Developments in Sedimentology, v. 33, Elsevier Sci. Public., Amsterdam, 343 p.

Baturin, G. N, 1969, Authigenic phosphate concretions in recent sediments of the southwest Africa shelf: Doklady Akademii Nauk S.S.S.R., v. 189, p. 227-230.

Baturin, G. N, Merkulova, K. I. and Chalov, P. I., 1972, Radiometric evidence for recent formation of phosphatic nodules in marine self sediments: Mar. Geol., v. 3, p. 37-41.

Bentor, Y. K., 1980, Phosphorites--the unsolved problems: SEPM Spec. Publ. 29, p. 3-18.

Bramlette, M. N., 1946, The Monterey Formation of California and the origin of its siliceous rocks: U.S. Geol. Surv. Prof. Paper 212, 57 p.

Burnett, W. C., 1974, Phosphorite deposits from the sea floor off Peru and Chile. Radiochemical and geochemical investigations concerning their origin: Ph.D. Thesis, Hawaii Inst. of Geophysics, 164 p.

Burnett, W. C., Beers, M. S. and Roe, K. K., 1982, Growth rates of phosphate nodules from the continental margin off Peru: Science, v. 215, p. 1616-1617.

Calvert, S. E. and Price, N. B., 1983, Geochemistry of Namibian shelf sediments: In: Coastal Upwelling, pt. A, Suess, E. and Thiede, J., eds, Plenum Pub. Corp., p. 337-375.

Claypool, G. E., 1974, Anoxic diagenesis and bacterial methane production in deep sea sediments: Ph.D. Thesis, Univ. of Calif., Los Angeles.

Claypool, G. E. and Kaplan, I. R., 1974, The origin and distribution of methane in marine sediments, In: Natural Gases in Marine Sediments, Kaplan, I. R., ed., p. 99-139, Plenum, New York.

Craig, H., 1957, Isotopic standards for carbon and oxygen and correction factors for mass spectrometer analyses of carbon dioxide: Geochim. Cosmochim. Acta, v. 12, p. 133-149.

Craig, H., 1961, Standard for reporting concentrations of deuterium and oxygen-18 in natural waters: Science, v. 133, p. 1833.

Durham, D. C., 1974, Geology of the southern Salinas Valley: U.S. Geol. Surv. Prof. Paper 814, 111 p.

Epstein, S. Buchsbaum, R., Lowenstein, H. A. and Urey, H. C. 1953, Revised carbonate-water temperature scale: Geol. Soc. Amer. Bull. v. 62, p. 417-426.

Friedman, G. M., 1959, Identification of carbonate mineral by staining methods: J. Sed. Petrol., v. 29, p. 87-97.

Friedman, I. and O'Neil, J. R., 1977, Compilation of stable isotope fractionation factors of geochemical interest: U.S. Geol. Survey Prof. Paper 440-kk.

Fritz, P. and Smith, D. G. W., 1970, The isotopic composition of secondary dolomites: Geochim. Cosmochim. Acta, v. 34, p. 1161-1173.

Graham, S. A., 1976, Tertiary sedimentary tectonics of the central Salinian Block of California: Ph.D. Thesis, Stanford Univ., Stanford, Ca., 510 p.

Gulbrandsen, R. A., 1960, Petrology of the Mead Peak phosphatic shale member of the Phosphoria Formation at Coal Canyon, Wyoming: U.S.G.S. Bull. 111C, 71-146.

Gulbrandsen, R. A., 1969, Physical and chemical factors in the formation of marine apatite: Econ. Geol., v. 64, p. 365-682.

Gulbrandsen, R. A., Roberson, C. E. and Neil, S. T., 1984, Time and the crystallization of apatite in sea water: Geochim. Cosmochim. Acta, v. 48, p. 213-218.

Isaacs,, C. M., 1981, Field trip guide for the Monterey Formation, Santa Barbara Coast, California, In, Guide to the Monterey Formation in the California Coastal area, Ventura to San Luis Obispo, Isaacs, C. M., ed., Pacific Sec. AAPG, p. 55-71.

Isaacs, C. M., 1984, The Monterey--key to the offshore California boom: Oil and Gas Journal, Jan. 9, 1984, p. 75-81.

Jahnke, R. A., Emerson, S. R., Roe, K. K. and Burnett, W. C., 1983, The present day formation of apatite in Mexican continental margin sediments: Geochim. Cosmochim. Acta, v. 47, 259-266.

Kastner, M. and Baker, P. A., 1982, A new insight into the origin of sedimentary dolomite: In, 1982/193 Yearbook of Science and Technology, 406-408, McGraw Hill.

Kastner, M. and Gieskes, J. M., 1983, Opal-A to opal-CT transformation: a kinetic study, In, Siliceous deposits in the Pacific region, Iijima, A., Hein, J. R. and Siever, R., eds., Developments in Sedimentology, v. 36, Elsevier, p. 211-227.

Kastner, M., Keene, J. B. and Gieskes, J. M., 1977, Diagenesis of siliceous oozes. I. Chemical controls on the rate of opal-A to opal-CT transformation--an experimental study: Geochim. Cosmochim. Acta, v. 41, p. 1041-1059.

Kennedy, W. J. and Garrison, R. W., 1975, Morphology and genesis of nodular phosphates in the Cenomanian Glauconitic Marl of southeast England: Lethaia, v. 18, p. 339-360.

Kolodny, Y., 1981, Phosphorites: The Sea, v. 7, Emiliani, C., ed., Wiley, p. 981-1023.

Kolodny, Y. and Kaplan, I. R., 1970, Carbon and oxygen isotopes in apatite-CO_2, and co-existing calcite from sedimentary phosphorite: J. Sed. Petrol., v. 40, p. 954-959.

Kolodny, Y., Luz, B. and Navon, O., 1983, Oxygen isotope variations in phosphate of biogenic apatites, I. Fish bone apatite--rechecking the rules of the game: Earth Planet. Sci. Let., v. 64, p. 398-404.

Longinelli, A. and Nuti, S., 1973, Oxygen isotope measurements of phosphate from fish teeth bones: Earth Planet. Sci. Let., v. 20, p. 337-340.

Mahmoud, F. J., 1976, Geochemistry and genesis of sedimentary marine carbonate fluorapatite: Ph.D. Thesis, Univ. California Santa Barbara, 167 p.

Manheim, F. T., Rowe, G. T. and Jipa, D., 1975, Marine phosphorite formation off Peru: J. Sed. Petrol., v. 45, p. 243-251.

Martens, C. S. and Harris, R. C., 1970, Inhibition of apatite precipitation in the marine environment by Mg-ions: Geochim. Cosmochim. Acta, v. 34, p. 621-625.

Matthews, A. and Katz, A., 1977, Oxygen isotope fractionation during the dolomitization of calcium carbonate: Geochim. Cosmochim. Acta, v. 41, p. 1431-1438.

McArthur, J. M., Coleman, M. L. and Bremner, J. M., 1980, Carbon and oxygen isotopic composition of structural carbonate in sedimentary francolite: J. Geol. Soc. London, v. 137, p. 669-673.

McCrea, J. M., 1950, On the isotopic chemistry of carbonate and a paleotemperature scale: J. Chem. Phys., v. 18, p. 849-857.

Mertz, K. A. and Garrison, R. E., 1983, Notes on the Sandholdt Member of the Monterey Formation in the Arroyo Seco area, Monterey County, California, In, Fieldguide to the Monterey Formation, northern Santa Lucia Mountains and Monterey Area, Monterey County, California, Soc. Econ. Pelontol. Mineral. Pac. Sec. p. 89-96.

Nathan, Y. and Lucas, J., 1976, Experiences sur la precipitation discrete de l'apatite dans l'eau de mer; implications dans la genesis des phosphorites: Chem. Geol., v. 18, p. 181-186.

Nissenbaum, A., Presley, B. J. and Kaplan, I. R., 1972, Early diagenesis in a reducing fjord, Saanich Inlet, British Columbia, I. Chemical and isotopic changes in major components of interstitial water: Geochim. Cosmochim. Acta, v. 36, p. 1007-1027.

Northrop, D. A. and Clayton, R. N., 19666, Oxygen-isotope fractionation in systems containing dolomite: J. Geol., v. 74, p. 174-196.

Pisciotto, K. A., 1981, Notes on Monterey rocks near Santa Maria, California, In, Guide to the Monterey Formation in the California coastal area, Ventura to San Luis Obispo, Isaacs, D. M., ed., Pacific Sec. AAPG, p. 73-81.

Rosenfeld, W. D. and Silverman, S. R., 1959, Carbon isotope fractionation in bacterial production of methane: Science, v. 130, p. 1658-1659.

Sharma, T. and Clayton, R. N., 1965, Measurement of $^{18}O/^{16}O$ ratios of total oxygen of carbonates: Geochim. Cosmochim. Acta, v. 29, 1347-1353.

Shemesh, A., Kolodny, Y. and Luz, B., 1983, Oxygen isotope variations in phosphate of biogenic apatites, II. Phosphorite rocks: Earth Planet. Sci. Lett., v. 64, p. 405-416.

Silverman, S. R., Fugat, R. K. and Weiser, J. D., 1952, Quantitative determination of calcite associated with carbonate-bearing apatites: Amer. Mineral., v. 37, p. 211-222.

Veeh, H. H., Burnett, W. C. and Soutar, A., 1973, Contemporary phosphorites on the continental margin of Peru: Science, v. 181, p. 844-845.

Woodring, W. P. and Bramlette, M. N., 1950, Geology and paleontology of Santa Maria district, California: U.S. Geol. Surv. Prof. Paper 222, 185 p.

EARLY DIAGENETIC DOLOMITES AND THE ORIGIN Of DOLOMITE-BEARING BRECCIAS, LOWER MONTEREY FORMATION, ARROYO SECO, MONTEREY COUNTY, CALIFORNIA

Robert E. Garrison
Earth Sciences Board
University of California
Santa Cruz, California 95064

Stephen A. Graham
Dept. of Applied Earth Sciences
Stanford University
Stanford, California 94305

ABSTRACT

Debris and gravity flow deposits interbedded with lower Monterey hemipelagic slope rocks in the Arroyo Seco area contain reworked dolomite concretions and fragments of these concretions. These dolomite clasts are identically petrologically and isotopically to *in situ* dolomite concretions which occur in the interbedded hemipelagic slope rocks, and they are apparently products of submarine erosion and resedimentation of early formed concretions. Limited isotopic data suggests the dolomites formed in the zone of carbonate reduction at temperatures which ranged from about 22° to 35°C. Employing a spectrum of assumed geothermal gradients and estimated sedimentation rates, we suggest these concretions formed over time spans of hundreds of thousands of years and at sub-bottom depths in excess of 100 meters. Therefore, simple erosional scour by individual gravity flow events cannot by itself explain the seafloor erosion and resedimentation of these concretions. A more likely explanation is that multiple episodes of gravity flow erosion, perhaps acting in tandem with other events such as channeling and slumping on the upper slope, acted to exhume the concretions. Two types of early diagenetic fracturing are evident in these concretions. This fracturing predated the erosion and reworking of the concretions by gravity flows, and formed the breakage planes along which the concretions became fragmented during erosion and transportation.

INTRODUCTION

The Monterey Formation consists largely of basinal hemipelagic sediments. Most of the unit is siliceous, having been originally a diatomaceous ooze that commonly became converted to porcelanite or chert during burial diagenesis. Dolomite concretions are not uncommon in these siliceous rocks but are volumetrically minor, constituting perhaps only one percent of the thickness of a typical section (Murata et al., 1969). In several parts of the California Coast Ranges, the lower part of the Monterey is highly calcareous and phosphatic, as in the section herein described, and lenticular beds and concretions of dolomite are volumetrically more significant than in the siliceous facies.

Bramlette (1946), in his classic study of Monterey rocks, observed that most carbonate concretions (both calcite and dolomite) in the unit have the shape of flattened or oblate spheroids with long axes parallel to bedding. He noted they apparently formed by cementation and replacement of the enclosing sediment; because laminae in the enclosing host rock can be traced into concretions where they are three to four times thicker, he believed many concretions formed relatively early during diagenesis. Subsequent isotopic geochemical work (see following discussion) has suggested complicated and varied histories for these concretions, a conclusion strengthened by the papers in this volume.

Field evidence from a locality in the central California Coast Ranges (Fig. 1) shows clearly that Monterey dolomite concretions there formed relatively early during diagenesis, but were uncovered, eroded and redeposited by slumping and by submarine debris and grain flows. Field, petrographic and geochemical information presented in the following sections sheds light on the processes which formed the concretions as well as the physical properties of Monterey dolomites during the earlier stages of diagenesis. In addition, these data yield insights on the sedimentology of Monterey slope deposits.

PREVIOUS WORK ON DOLOMITES IN THE MONTEREY FORMATION AND SIMILAR ORGANIC-RICH UNITS

Dolomites in the Monterey Formation and related units have been investigated by Spotts and Silverman (1966); by K. J. Murata, I. Friedman and their co-workers in the U.S. Geological Survey (Murata et al., 1967, 1969; Murata et al., 1972; Friedman and Murata, 1979); and by Pisciotto (1978, 1981a). Most Monterey dolomite is non-stoichiometric, calcium-rich dolomite with 49 to 56 mole % $CaCo_3$ and iron contents up to 16 mole % $FeCo_3$. Much interest has focused on the carbon isotopic compositions of these dolomites as indicators of the carbon reservoirs and diagenetic zones where they formed. Murata et al. (1972) subdivided Monterey dolomites into three groups based on chemical and isotopic characteristics, and these were subsequently interpreted by Pisciotto (1981a) as follows:

1. Light carbon-low iron dolomites, with carbon-13 values of -26 to -4 $^o/oo$ (PDB) and <2 mole % $FeCo_3$, interpreted by Pisciotto to have formed either: (a) at shallow depths in the

in Garrison, Robert E.; Kastner, Miriam and Zenger, Donald H., eds., 1984, **Dolomites of The Monterey Formation and Other Organic-Rich Units:** Pacific Section S.E.P.M., Vol. 41, p. 87-101.

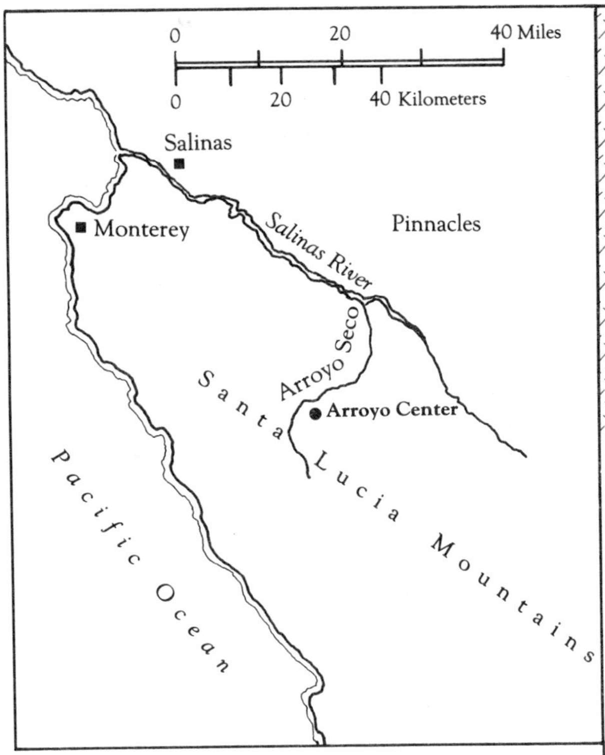

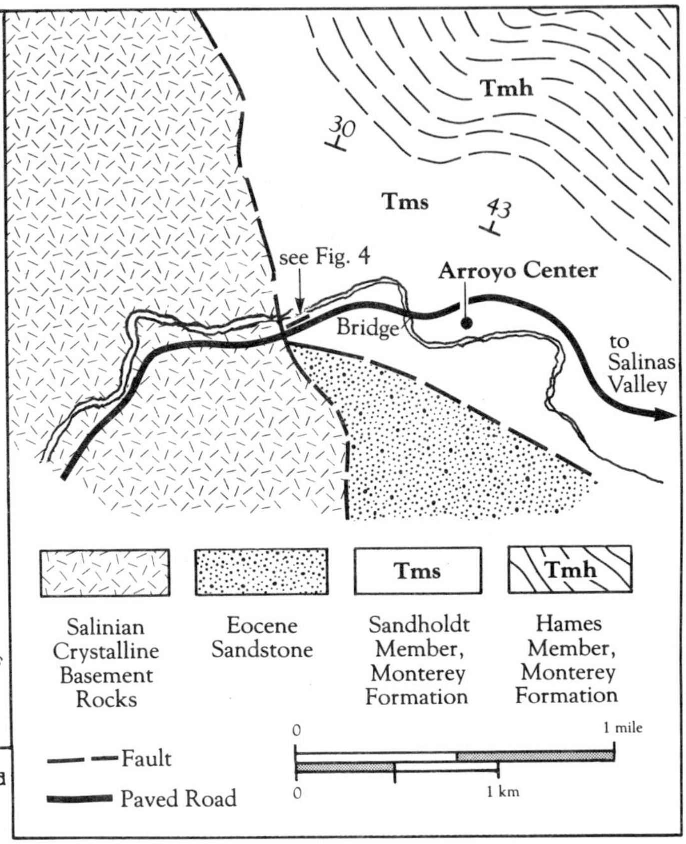

Figure 1. Location and geologic sketch maps of field area along the Arroyo Seco near Arroyo Center. Note location of measured section shown in Figure 4.

anaerobic zone of sulfate reduction (low iron content due to depletion of iron from a finite iron reservoir during formation of pyrite), or (b) as a result of aerobic respiration.

2. Heavy carbon-low iron dolomites, with carbon-13 values of +5 to +21 $^{o}/oo$ (PDB) and <2 mole % $FeCo_3$, formed according to Pisciotto in the anaerobic zone of active methane production; the low iron content results from its earlier depletion in the zone of sulfate reduction.

3. Heavy carbon-high iron dolomite, carbon-13 values of -2 to +10 $^{o}/oo$ (PDB) and > 2 mole % $FeCo_3$. Pisciotto suggested these dolomites formed in deeper zones of methanogenesis and thermocatalytic decarboxylation; the higher iron content reflects gradual depletion of Mg^{++} in pore waters and progressive substitution of Fe^{++} for Mg^{++} in later formed dolomites.

Studies of dolomites recovered from organic-rich sediments in the present ocean basins have helped to establish limits on the timing and conditions of this kind of dolomitization. Among the most important of these studies have been those of Hein et al. (1979) on Neogene dolomite- and calcite-cemented diatom oozes in the Bering Sea; by Pisciotto and Mahoney (1981) on Neogene dolomites from offshore southern and Baja California; by Kelts and McKenzie (1982; see also this volume) on dolomites in laminated Quaternary diatom oozes in the Gulf of California; and by Kulm et al. (1981; see also this volume) on dolomites and dolomitic limestones associated with organic-rich sediments dredged from marginal basins off Peru.

Because it focused on the depth and timing of dolomitization, the study by Pisciotto and Mahoney (1981) has particular relevance to the investigation reported herein. Expanding on earlier work by Irwin et al. (1977) Pisciotto and Mahoney viewed the wide variation in carbon isotopes of Monterey and Monterey-like dolomites as a consequence of precipitation in different diagenetic zones having differing carbon isotope reservoirs. They documented the fact that within any sedimentary section dolomite may form simultaneously at several different depths or at different times within the same interval. The critical factor in their view is the sedimentation rate which controls both the preservation of organic matter and the diffusion of oxygen and sulfate from the overlying sea water into the sediment column. Slow sedimentation rates allow more complete oxidation of organic matter near the sea floor by aerobic oxidation or anaerobic sulfate reduction, yielding light carbon carbonates. Faster rates inhibit oxygen and sulfate diffusion, thus favoring methanogenesis and precipitation of heavy carbon carbonates at shallow depths. This trend toward heavy carbon would be reversed at greater depth by thermal decarboxylation.

Baker and Kastner (1981; see also Kastner and Baker, 1982) have provided additional insight on the origin of dolomites in organic-rich sediments by showing experimentally that dolomitization can be significantly enhanced when the concentrations of inhibiting sulfate ions are low, as for example within and below the zone of bacterial sulfate reduction.

LOCATION, STRATIGRAPHY AND SEDIMENTARY-TECTONIC SETTING

The redeposited dolomite concretions occur in the Sandholdt Member of the Monterey Formation along the Arroyo Seco, a large E-W trending

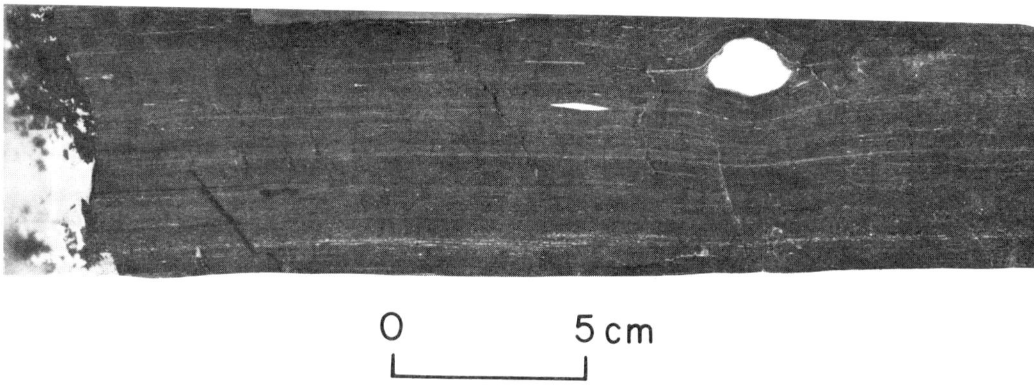

Figure 2. Polished surface of organic-rich laminated foraminiferal-coccolith shale with phosphate nodules (white). Sample from the Sandholdt Member of the Monterey Formation, near Arroyo Center.

drainage which cuts into the northern part of the Santa Lucia Range in northern Monterey County (Fig. 1). The specific outcrops occur along the banks of the Arroyo Seco for several hundred meters upstream from the bridge at Arroyo Center.

Compton (1966), Durham (1974) and Graham (1976) have described the geology of this area, and Mertz and Garrison (1983) and Mertz (this volume) have provided petrological and geochemical data on Sandholdt rocks. The Sandholdt is the lower member of the Monterey Formation over most of the Santa Lucia Range. Although originally assigned formation status by Thorup (1943), we follow Durham (1974) and Graham (1976) in treating the Sandholdt as a member within the Monterey Formation. The Sandholdt is of early to middle Miocene age; more specifically, benthic foraminiferal assemblages indicate assignment in the general Santa Lucia Mountain area to the upper Saucesian through Luisian stages of the California Miocene stages (Graham, 1976).

More recent nannofossil dating of the Sandholdt along the Arroyo Seco near Arroyo Center (C. H. Ellis, written comm., 1983; Mertz and Garrison, 1983 and Mertz, this volume) shows that the unit here can be assigned largely to the upper subzone of the *Helicosphaera ampliaperta* and *Sphenolithus heteromorphus* zones of Bukry (1981); the uppermost part of the Sandholdt near Arroyo Center belongs to the *Coccolithus miopelagicus* subzone of the *Discoaster exilis* zone, corresponding approximately to the Luisian-Mohnian stage boundary. Thus the Sandholdt Member near Arroyo Center can be rather precisely assigned to part of the Relizian and all of the Luisian stages, and it encompasses a time span between approximately 17.0 to 14.0 mybp at this locality.

The Sandholdt Member in the Arroyo Seco area is a transitional unit lying between shallow water neritic sandstones of the underlying Vaqueros Formation and bathyal siliceous deposits of the overlying Hames Member of the Monterey Formation. Benthic foraminiferal assemblages suggest deposition at middle bathyal depths of about 500 to 2500 meters (Graham, 1976). Paleogeographic reconstructions for middle Miocene time indicate the Arroyo Seco area was part of a deeply submerged E-W trending trough that lay between a shallow shelfal region to the north (the Gabilan Mesa Shelf) and a submarine ridge to the south (the Lockwood High). Evidence discussed later suggests that deposition of the Sandholdt occurred on a slope bordering the trough. The overall geographic-tectonic setting resembled the present borderland region off southern California and was similarly the product of wrench tectonism.

LITHOLOGIES AND SEDIMENTATION OF THE SANDHOLDT MEMBER

Hemipelagic Rocks

The predominant lithology in the Sandholdt Member is dark gray foraminiferal rock which varies from fissile calcareous shale and siltstone to platy or somewhat blocky calcareous mudstone and silty, fine-grained limestone. These rocks are commonly laminated and contain scattered phosphatic nodules and pellets (Fig. 2) as well as scattered small (<1 cm) pyrite nodules. Microscopically they comprise silt to sand size grains (foraminiferal tests, siliciclastic grains, fish debris, phosphate pellets) in a matrix of micritic calcite (containing coccoliths) and variable amounts of microcrystalline quartz, clay minerals, dolomite, pyrite and finely disseminated organic matter. The Sandholdt rocks near Arroyo Center lie within the quartz diagenetic zone. Total organic carbon values range between 2.5 and 7.0 percent (Mertz and Garrison, 1983).

The original sediments were hemipelagic muds composed mainly of calcareous tests (coccoliths, benthic and planktonic foraminifera) and subordinate siliceous skeletons (mainly diatoms, now converted to microcrystalline quartz) and probably deposited in dysaerobic to anaerobic environments which commonly excluded a burrowing infauna. The most likely environment was within or near the edges of the intersection of the oxygen minimum zone with a submarine slope, an interpretation compatible with the middle bathyal depths indicated by the benthic foraminifera; this is compatible also with the presence of phosphatic nodules somewhat similar to Quaternary nodules formed during early diagenesis in anoxic, organic-rich sediments associated with the oxygen minimum zone off Peru and Chile (Veeh et al., 1973; Burnett, 1977, 1980).

Figure 3. Hammer rests on large in situ dolomite concretion which resides in laminated phosphatic shales like that shown in Figure 2. Laminations pass from the laminated host shale onto the concretion where they are three to five times thicker than outside the concretion.

Dolomite Concretions and Lenticular Beds

Dolomite occurs in these laminated hemipelagic rocks as resistant lenticular beds and ellipsoidal concretions (Fig. 3) which are dark gray but weather to a conspicuous yellow brown or ochre. The concretions are up to three meters in maximum dimension, and lenses are hundreds of meters long and of varying thickness; both tend to be concentrated within specific stratigraphic intervals. Paralleling Bramlette's (1946) observations, we noted that packets of laminae commonly can be traced from adjacent hemipelagic rocks into individual concretions where they thicken by a factor of three to five times. The dolomites are thus of secondary origin and formed before the completion of compaction. Their petrology will be considered in a later section.

Sandstones and Breccias

Interbedded with the hemipelagic rocks in the lower part of the Sandholdt near Arroyo Center are resedimented beds of sandstone and breccia (Fig. 4). Three main types are present:

(1) Classical turbidite sandstones with partial Bouma divisions (e.g., unit 7, Fig. 4); these are relatively rare.

(2) Thick, massive sandstones, some of which contain scattered large clasts (e.g., units 6, 9, 14, and 18, Fig. 4); these occur in beds up to 20 meters thick and were probably deposited from highly concentrated grain flows (Walker, 1978).

(3) Breccias comprising large clasts in a sandstone matrix (e.g., units 10, 12A and 12C and 16, Figs. 4, 5 and 7), probably the products of submarine debris flows.

Redeposited dolomite concretions occur in the latter two types of beds. Several flute casts shown NE to SW paleocurrent directions and indicate a possible source of the Gabilan Mesa Shelf to the north and northeast. A few sandstone beds and one interval of hemipelagic mudstone have slump folds suggesting deposition on a slope.

Sandstones in all three kinds of beds are feldspathic, and their grains are coarse to medium and angular to subrounded. In addition to siliciclastic grains, some sandstones also contain reworked glauconite, phosphatic pellets, foraminiferal tests and other skeletal debris (e.g., oyster and pectinid fragments) suggestive of redeposition from an adjacent neritic zone such as the Gabilan Mesa Shelf.

One of the massive sandstone beds contains a coarse basal part that includes cobble to boulder size clasts of granitic rock, asphanitic volcanic rocks, sandstone, shale, dolomite, and abraded megafossils, chiefly oysters, pectinids and barnacles (unit 14, Figs. 6 and 7). The debris flow breccias have comparatively less exotic clasts; these are ungraded and contain chiefly

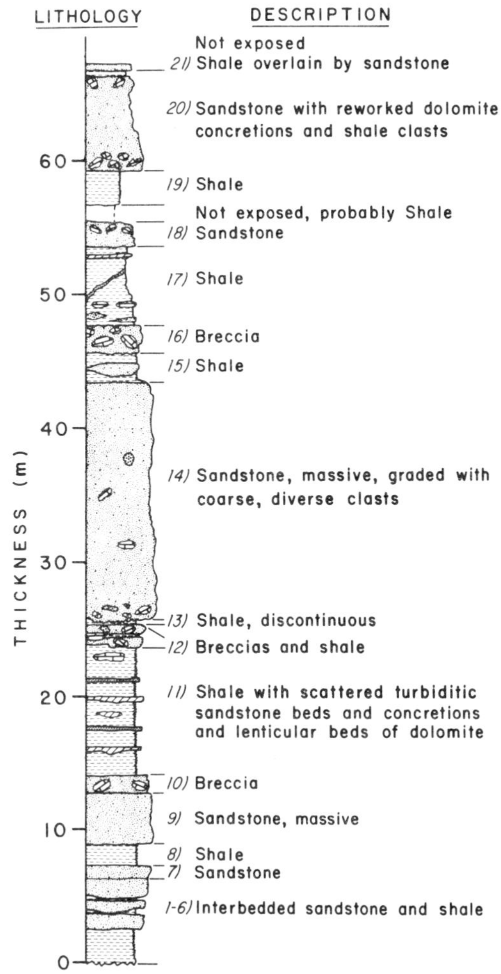

Figure 4. Measured section of interbedded hemipelagic slope shales and gravity flow beds. This section lies along the south bank of the Arroyo Seco (cf. Fig. 1) and is subdivided into 21 units.

reworked dolomite, shale and sandstone clasts which are dispersed chaotically in a sandstone matrix (Fig. 5).

Post-depositional mobilization of redeposited sand is indicated in unit 17 (Fig. 4) by sandstone dikes and small sandstone pods which appear to be sills.

Redeposited Dolomite Concretions

Redeposited dolomite in the massive sandstones and breccias occurs mainly as subangular to subrounded cobbles and boulders up to 50 centimeters in maximum dimension. These appear to be fragmented portions of concretions which were broken during resedimentation. A few redeposited dolomite clasts appear to be whole concretions that were eroded, transported perhaps a short distance and redeposited *in toto*; these attain maximum dimensions up to 2.5 meters and were noted only in debris flow breccias (Fig. 5).

Detailed field observations shed light on the processes which formed these dolomite-bearing gravity flow deposits. Figure 6 is a field photograph of parts of units 11 to 14 (Fig. 4), and Figure 7 is a schematic portrayal of some of the crucial sedimentologic relationships in these units. Unit 11 is a laminated hemipelagic shale which contains *in situ* concretions of dolomite (e.g., A in Fig. 7). Unit 12 actually comprises three beds; beds 12A and 12C are sandy debris flow deposits with clasts of mainly dolomite (e.g., D) and laminated shale, and bed 12B is a laminated shale with a very large *in situ* laminated dolomite concretion (C in Fig. 7). Unit 13 is a thin laminated hemipelagic shale, and unit 14 is a graded, massive grain flow sandstone with coarse basal clasts, which include dolomite and shale as well as more allochthonous components.

All of the gravity flow deposits shown in Figure 7 have pronounced scoured basal contacts, and all have suggestions of channeling (especially bed 12A). These observations indicate substantial seafloor erosion and suggest a ready source for the abundant shale ripup clasts. The basal contact of bed 12C rests partially on the top of large concretion C; sandstone from the matrix of bed 12C fills one large and many small fractures in the top part of concretion C (Fig. 7; see also Figs. 8E and 8F). These relationships indicate either that erosion during emplacement of bed 12 cut down to the level of concretion C, or that concretion C had become exposed on the sea floor by some process such as slumping prior to the deposition of bed 12C. We consider these possibilities in a later section. In either case, the field relations show that fracturing of concretion C had occurred prior to deposition of bed 12C; sandy matrix of this bed was injected into these fractures forcibly widening some of them. As discussed below, injections of this kind evidently were an important mechanism for the fragmentation of large concretions.

Another suggestion for the relatively early origin of the dolomite concretions is provided by slump folds in which dolomite concretions were rotated. The clearest example is on the east bank of the Arroyo Seco immediately upstream from Arroyo Center (Fig. 1) where a large slump fold occurs in a thick grain flow sandstone. The hemipelagic shales stratigraphically below the sandstone are also folded, and large dolomite concretions in the shales have been rotated about the fold axis, showing that they existed at the time of soft sediment folding.

Paleomagnetic measurements provide conformation of the displaced nature of the dolomite clasts. Whereas the *in situ* dolomites in the Arroyo Seco section contain a clear record of normal and reversed polarities, dolomite clasts in the gravity flow deposits show random orientations (Coe et al., abstracts in this volume).

PETROLOGY AND GEOCHEMISTRY OF DOLOMITE CONCRETIONS

Mineralogy and Petrography

Thin section and X-ray diffraction analyses indicate no substantial differences between the redeposited and *in situ* concretions. The dolomite in both is a non-stoichiometric, calcium-rich dolomite which has 42 to 48 mole % $MgCo_3$ (Table 1) according to XRD measurements (Goldsmith and Graf, 1958). Texturally the concretions vary from rocks

Figure 5. Debris flow deposit containing three large dolomite concretions (D), numerous smaller broken fragments of dolomite concretions, and abundant ripup clasts of hemipelagic shale. All of these components are dispersed through a sandstone matrix. Figure 9 is an enlarged view of the concretion at bottom center-left. This is part of unit 16 (Fig. 4).

Figure 6. Field photograph of units 12 to 14 (cf. Fig. 4). The numbers and letters correspond to those in Figure 7.

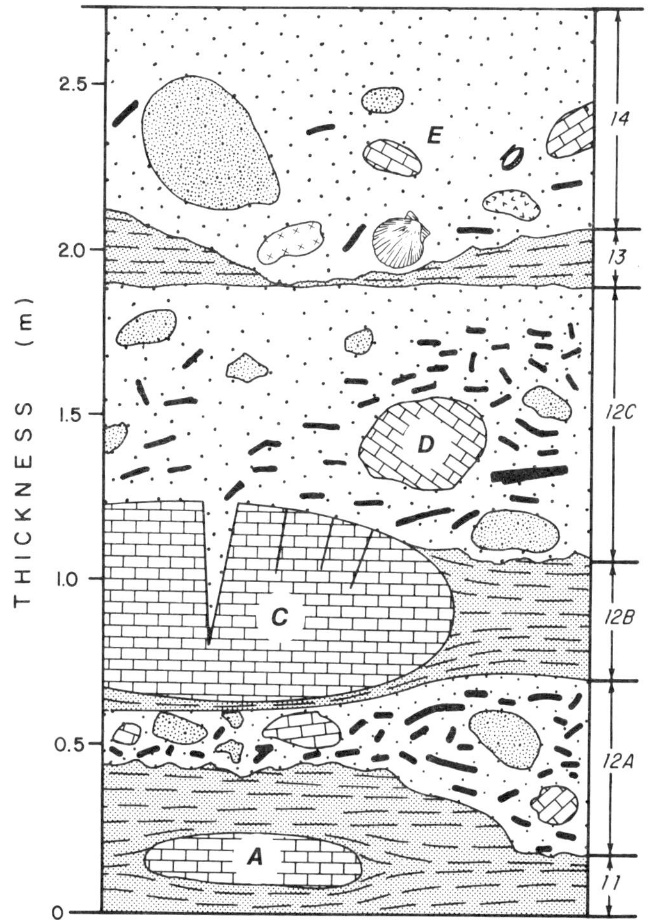

Figure 7. Schematic drawing of sedimentological relationships in beds 11 to 14. Vertical and horizontal scales are the same. See discussion in text.

that are predominantly dolomite (Figs. 8A and 8B) to dirtier rocks with abundant siliciclastic grains and only 20 to 30% dolomite as matrix or cement (Fig. 8D); the former appear to represent dolomitization of a foraminiferal ooze, the latter cementation of a silty mud. Most of the dolomites examined are intermediate between the above extremes and consist of 15-25% silt and sand size grains (foraminiferal tests--both calcitic and dolomitized, fish debris, pyrite, and siliciclastic and authigenic phosphate grains) in a matrix of dolomite. Dolomite forms rather uniform mosaics of euhedral to subhedral crystals mostly in the range of 5 to 15 microns (Fig. 8C). Poorly preserved microfossil tests occur as ghost-like outlines in a dolomite mosaic (Fig. 8B). Most of the siliciclastic grains show marginal replacement by dolomite (Fig. 8D). Several of the concretions have what appear to be patchy vugs that are 20 to 80 microns across, and dolomite crystals up to 50 microns across line these vugs as well as small fractures. Secondary kaolinite, apatite, pyrite and diagenetic quartz also fill some of the microfossil cavities and vugs (Mertz, this volume).

The petrographic evidence suggests the concretions formed by dolomitization of a range of host sediments, from relatively pure calcareous foraminiferal oozes to silty calcareous muds. One lenticular bed of dolomite was found to contain numerous impressions of small, thin-shelled pectinids, identified by Warren O. Addicott (personal communication, 1980) as belonging to the genus *Delectopectin*. Some of the dolomite evidently precipitated as void-filling cement, other parts of it formed by replacement of calcite as well as siliciclastic grains and probably clay minerals as well.

Thin section observations also suggest that the redeposited concretions may not have been fully coherent at the time they became cracked and injected by sand. Figure 8E and 8F show fracture-filling sand grains which include subangular to subrounded pieces of the host dolomite; apparently the dolomite concretion was quite friable and perhaps porous at the time of injection so that pieces of the wall rock were easily plucked and incorporated into the crack fill.

Light Stable Isotopes

Carbon and oxygen isotopic contents of six dolomite samples were measured by Geochron Laboratories of Cambridge, Massachusetts, and the results are tabulated in Table 1. Like the mineralogy and textures of these concretions, no significant differences exist between *in situ* and redeposited concretions. Carbon-13 values range from -7.21 °/oo to +7.89 °/oo (PDB), but most are strongly positive; the two *in situ* concretions measured have both negative (-7.21 °/oo) and positive (+7.89 °/oo) values whereas all but one of the redeposited concretions have positive values (+4.84 to +7.89 °/oo). Values of oxygen-18 range from -0.20 to +2.51 °/oo (PDB).

Assuming these dolomites equilibrated with a fluid having $\delta^{18}O$ = 0.0 (SMOW) during their precipitation (Pisciotto, 1981a; Pisciotto and Mahoney, 1981), we have attempted to calculate their temperature of formation (Table 2) using their oxygen isotope values with 0.8 °/oo subtracted to account for the acid fractionation relative to calcite. We have followed Pisciotto and Mahoney (1981) in using the protodolomite-water fractionation expression of Fritz and Smith (1970), employing the constant dolomite-calcite fractionation of 3.1 °/oo (SMOW) given in Friedman and O'Neil (1977). The temperatures thus derived range from 22° to 35°C (Table 1).

These data are too few to allow definitive conclusions, though the temperatures shown in Table 1 are in general agreement with those calculated by Mertz (this volume). On the basis of many more isotopic analyses of Arroyo Seco dolomites, he computed temperatures of dolomite formation of from 19° to 68°C (see also Kastner et al., this volume). Another limitation to our isotopic data is that they represent single specimens from individual concretions (generally collected from the outer edges), whereas serial sampling of carbonate concretions from the center to the edges commonly reveals substantial differences in both carbon and oxygen isotopes. As mentioned previously, authigenic dolomites of this kind may begin to form at one burial depth and continue to grow outward during progressive burial into other diagenetic zones, with higher temperatures and differing

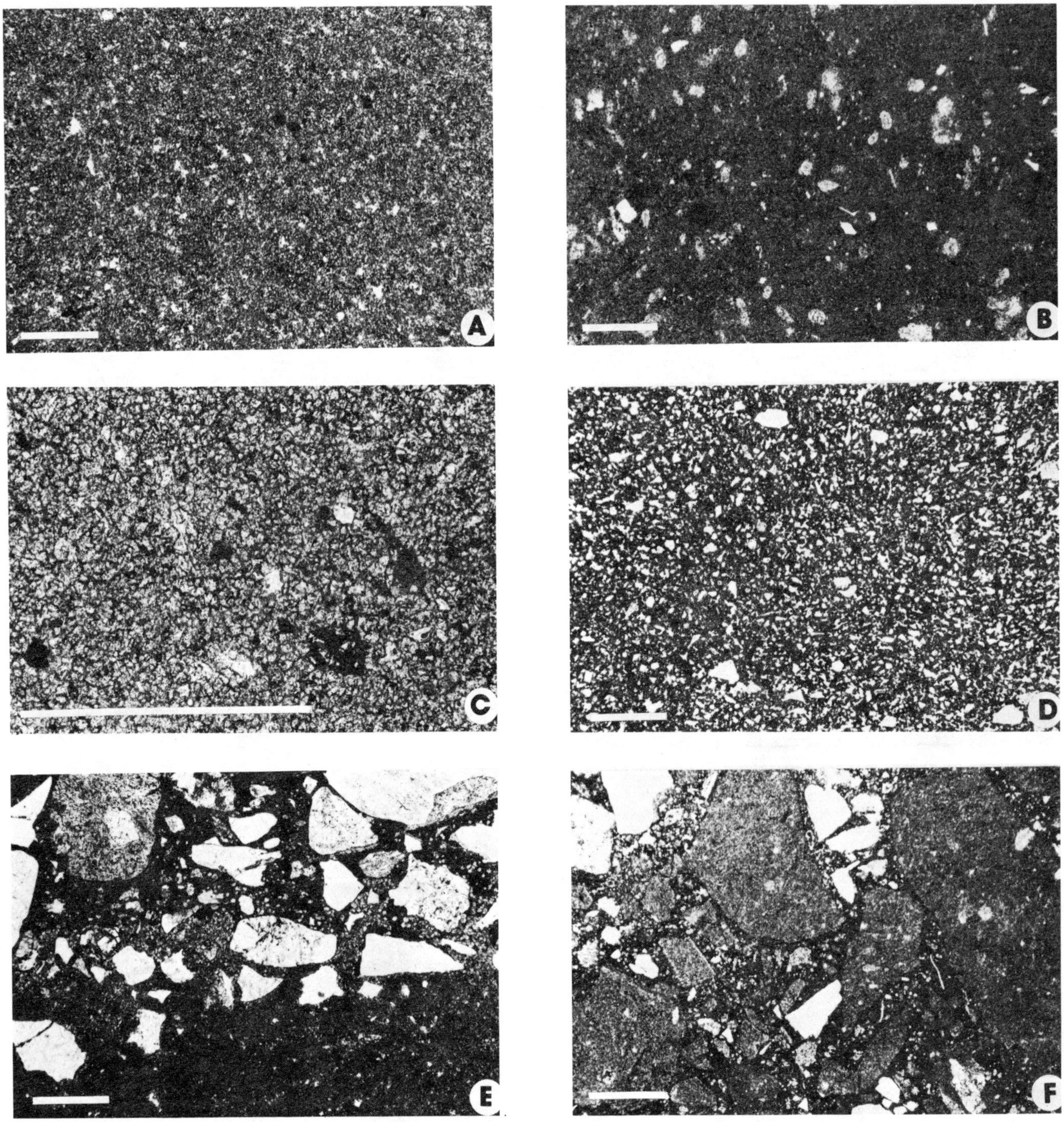

Figure 8. Photomicrographs of thin sections of rocks from the Sandholdt Member near Arroyo Center. All are in plane polarized light, and the scale bars indicate 500 microns.

A. Silty dolomite from redeposited concretion, bed 12C.

B. Relatively pure dolomite from concretion showing ghost-like remnants of dolomitized foraminiferal tests; from in situ concretion C, bed 12B (cf. Fig. 7).

C. Higher magnification view of texture in redeposited dolomite concretion, bed 12C.

D. Sandy and silty dolomite from redeposited concretion, bed 12A.

E. Sandstone-filled fracture in concretion C, bed 12B (cf. Fig. 7). Part of the dolomite concretion is at the bottom of the photograph.

F. Dolomite clasts and sand grains in fracture-filling within dolomite concretion C, bed 12B.

TABLE 1.
COMPOSITIONAL AND ISOTOPIC DATA AND CALCULATED TEMPERATURES OF FORMATION FOR DOLOMITES IN THE SANDHOLDT MEMBER, MONTEREY FORMATION, ARROYO SECO

A	B	C	D	E	F
Sample Number	Description	$MgCO_3$ in Dolomite (Mol %)	$\delta^{13}C$ (^{o}/oo)	$\delta^{18}O$ (^{o}/oo)	Calculated Temperatures of Formation (°C)**
AS-1	Redeposited concretion from Unit 12 (Bed 12C, Fig. 7)	46	+4.84	+1.88	25
AS-4	Redeposited concretion from Unit 10	42	-7.07	-0.20	35
AS-5	From outer edge of *in situ* concretion in Unit 12 (C in Bed 12B, Fig. 7)	48	-7.21	+1.54	26
AS-6	Redeposited concretion from Unit 12 (Bed 12C, Fig. 7)	45	+7.04* +7.08*	+1.58* +1.50*	26 27
AS-7	Redeposited concretion from Unit 12 (Bed 12A,Fig. 7)	48	+7.48	+1.91	25
AS-8	From edge of *in situ* concretion in Bed 11	42	+7.89	2.51	22

*Duplicate analyses.
**See text for method of calculation.

isotopic reservoirs (cf. Curtis et al., 1972; Irwin et al., 1977; Hudson, 1978; Kelts and McKenzie, 1982; Kushner and Kastner, this volume).

It is important also to emphasize that the temperatures in Table 1 are probably maximum values for several reasons. First, as noted, many of the samples analyzed are from the outer edges of the concretions; this dolomite thus likely formed later at greater depths and temperatures than the dolomite in the center part of the concretions. Second, our temperature calculations have assumed that the dolomites equilibrated with pore waters having $\delta^{18}O$ of 0.0 (SMOW). Studies of DSDP cores, however, have shown depletion of pore waters of up to 0.6 δ ^{o}/oo per 100 meters of burial (see discussion in Pisciotto, 1981b). Utilizing a depletion factor of about 0.3 δ ^{o}/oo per 100 meters estimated burial, Mertz's (this volume) temperatures (cited above) are recalculated to be in the range 12° to 33°C.

DEPTH AND TIME OF FORMATION OF THE DOLOMITE CONCRETIONS, ARROYO SECO AREA

Graham (1976) compiled thickness-time data for Sandholdt sections in several parts of the Santa Lucia Mountains and calculated undecompacted sedimentation rates ranging from 30 to 152 m/my. Sandholdt sedimentation rates in the Arroyo Seco area, judging by recent nannofossil age dating, were somewhat higher; Mertz (this volume) suggests that undecompacted rates there were between 115 and 200 m/my.

Isaacs (1984) has shown that highly porous diatomaceous Monterey sediments which were buried into the quartz diagenetic zone (as is the case for the Sandholdt in the Arroyo Seco area) have compacted to between six and seven times their original thickness. Also, because diatomites have very low thermal conductivities, geothermal gradients in unaltered opal-A sediments will be as much as two times greater than those in quartzose or opal-CT rocks (Isaacs, 1984).

Because Sandholdt hemipelagic sediments were a mixture of diatoms, coccoliths, foraminiferal tests and siliciclastic grains, their compaction and thermal histories doubtless differed from that of the relatively pure diatomites discussed by Isaacs (1984). Mertz (pers. commun.) has suggested that Sandholdt sediments buried into the quartz diagenetic zone experienced total compaction of three to four times, yielding estimated decompacted sedimentation rates between 460 and 600 m/my for the Arroyo Seco section.

In addition, the thermal properties of Sandholdt sediments must have differed from those of more purely siliceous deposits. For offshore regions of southern California, Isaacs (1984)

TABLE 2.
ESTIMATED SUB-BOTTOM DEPTHS AND TIMES OF FORMATION (IN 10^6 YEARS AFTER DEPOSITION) OF DOLOMITES IN THE SANDHOLDT MEMBER, MONTEREY FORMATION, ARROYO SECO

A	B	C	D	E	
Assumed Geothermal Gradient	Est. T(°C) of Dolomite Formation (Table 1)	Difference in T (°C), Sea Floor to Depth of Dolomite Formation*	Estimated Sub-bottom Depth of Dolomite Formation (m)	Time in 10^6 years of Dolomite Formation, assuming Sed. Rate of: 460 m/my	600 m/my
60°C/km	22 (Min.)	14	233	0.506	0.388
	35 (Max.)	27	583	1.267	0.972
70°C/km	22 (Min.)	14	200	0.435	0.333
	35 (Max.)	27	500	1.087	0.833
100°C/km	22 (Min.)	14	140	0.304	0.233
	35 (Max.)	27	350	0.761	0.583

*Assuming average bottom water temperature of 8°C.

estimated thermal gradients of 75°C/km in opal-A sediments and 44°C/km in opal-CT rocks, assuming a heat flow of 1.5×10^{-6} cal/cm^2-sec. Sandholdt sediments, being mixtures of siliceous, calcareous, and siliciclastic components, probably had higher thermal conductivities and lower geothermal gradients than pure diatomaceous sediments. Assuming the same heat flow noted above, a gradient of around 60° to 70°C/km appears reasonable for uncompacted to slightly compacted Sandholdt sediments. Our calculations (Table 2) have also included, as a possible upper limit, a geothermal gradient of 100°C/km, a value approximating that of Quaternary diatomaceous sediments of the Guyamas Basin (Kelts and McKenzie, 1982).

The temperature range of dolomite formation suggested by oxygen isotopes is 22° to 35°C assuming pore water isotopic compositions were the same as sea water. Most Sandholdt sediments accumulated prior to the onset of Antarctic glaciation and the development of cold bottom waters (Woodruff et al., 1981). Bottom water temperatures in a marginal basin such as this one were probably in the range of 5 to 10°C, and we have used a value of 8°C in our estimates. Thus the estimated differences between the sea floor temperature and the temperature of dolomite formation were 14° to 27°C (Column C in Table 2).

Columns D and E in Table 2 list estimated sub-bottom depths and times of formation for the Sandholdt dolomites isotopically analyzed in this study. The range of estimated sub-bottom depths (140 to 583 m) and times of formation (0.233 to 1.267×10^6 years) might be regarded as toward the upper limits because, as previously stated, temperature calculations (Table 1) are maximum values. Mertz (this volume), for example, on the basis of more complete isotopic data from the Arroyo Seco area and correcting for a possible depletion of as much as 0.6 ^{o}/oo per 100 meters, estimated that some Sandholdt dolomites formed at sub-bottom depths of as little as 80 meters.

Despite the limitations of our data, it is interesting to note that the range of our estimated depths of dolomite formation are comparable with values compiled for Quaternary dolomites in the Gulf of California (greater than 60 m; Kelts and McKenzie, 1982; see also Kelts and McKenzie, this volume) and also for Neogene dolomites off southern and Baja California (87-658 m; Pisciotto and Mahoney, 1981). We conclude therefore that dolomite concretions of the size found in the Arroyo Seco formed, at a minimum, at burial depths of several tens to hundreds of meters and over time spans of hundreds of thousands of years.

The mixture of heavy and light carbon (Table 1) suggests the dolomites formed in either the zone of anaerobic carbonate reduction or partly in this zone, partly in the zone of sulfate reduction. More complete isotopic analyses across individual concretions might reveal systematic changes reflecting progressive growth in different diagenetic zones (cf. Curtis et al., 1972; Irwin et al., 1977; Kushner and Kastner, this volume). However, the temperature and sub-bottom depth estimates discussed above (Table 2) suggest that most of the individual concretions developed in the zone of carbonate reduction.

EARLY DIAGENETIC FRACTURING IN DOLOMITE CONCRETIONS AND THE SOURCE OF DOLOMITE CLASTS

Field and petrographic relationships observed in dolomite concretions from the Arroyo Seco yield insights regarding very early fracturing of Monterey dolomites. Such early fracturing has been postulated by Roehl (1981) and Redwine (1981) in the Santa Maria Basin and may be an important component of fractured Monterey reservoirs in that basin.

A number of the reworked dolomite concretions have fractures or networks of fractures which are now filled with sediment; the fracture thus existed

Figure 9. Enlarged view of dolomite concretion in a debris flow deposit (unit 16, cf. Fig. 5) showing network of sediment-filled fractures.

at the time the concretions were redeposited. The fracture-fillings of the concretion shown in Figure 9 are a combination of sandstone and hemipelagic mudrock (Fig. 8E); the former forms the matrix of the debris flow rock and the latter probably was soft mud in which the concretion originally formed. Thus we may surmise that the fractures existed prior to uncovering of the concretion; some of the hemipelagic mud was forced onto the fractures at that time, and the sand was injected during subsequent transport in the debris flow. That this sand injection exerted considerable stress is suggested by clasts of dolomite mixed with siliciclastic sand grains in some of the sediment-filled fractures (Fig. 8F); apparently these clasts were plucked from the fracture walls during sediment injection.

Another instance of sand injection into early formed fractures is shown in Figure 7; as discussed earlier, sand from bed 12C injected downward to fill and perhaps expand pre-existing fractures in concretion C. Figure 10 illustrates how sand injections of this kind may begin to pry apart the interfracture areas of the concretions; continuation of this process would lead to their disintegration into smaller, angular or subangular clasts of the kind common in the gravity flow deposits.

These early formed fractures have at least two origins. Some form polygonal networks resembling mudcracks (Fig. 9) and appear related to volume changes imposed during concretion growth; they resemble the septarian dislocations of tensional origin discussed by Raiswell (1971) and Hudson (1978). Other early fractures may have resulted from tectonic stresses. Figure 11 is a concretion showing antithetic step faults which die out upward. These closely resemble the "fault-graded" beds described by Seilacher (1969) in Monterey rocks near Santa Barbara and interpreted by him as tectonically induced fractures formed during very early diagenesis. Such early fracturing might also result from gravity-induced stresses on a slope, or from dewatering (Williams, 1982). In Figure 11 most of the fracturing probably preceded the main phase of dolomitization by which the concretion acquired coherence. The intensely deformed lamina at one edge of the concretion suggest, however, that this part of the sediment remained plastic and undolomitized; it thus became deformed against the coherent nucleus of the concretion due to tectonic and/or compaction stresses, and only later during burial became itself dolomitized.

Early-formed dolomites may be intensely fractured for several reasons: (1) the mineral dolomite, unlike calcite, does not readily twin in response to stress; instead it tends to fracture, therefore dolomitic rocks are much more brittle than calcite ones (Griggs and Handin, 1960; Turner and Weiss, 1963). Fractures in interbedded dolomitic and calcitic rocks, like the section in the Arroyo Seco, will thus be particularly abundant in the dolomites. (2) Because they formed relatively early, within a few hundred thousand years following deposition of their host sediment, these dolomites would be subjected to fracture-inducing stresses over much longer time spans than, for example, opal-A diatomaceous sediments. Opal-A

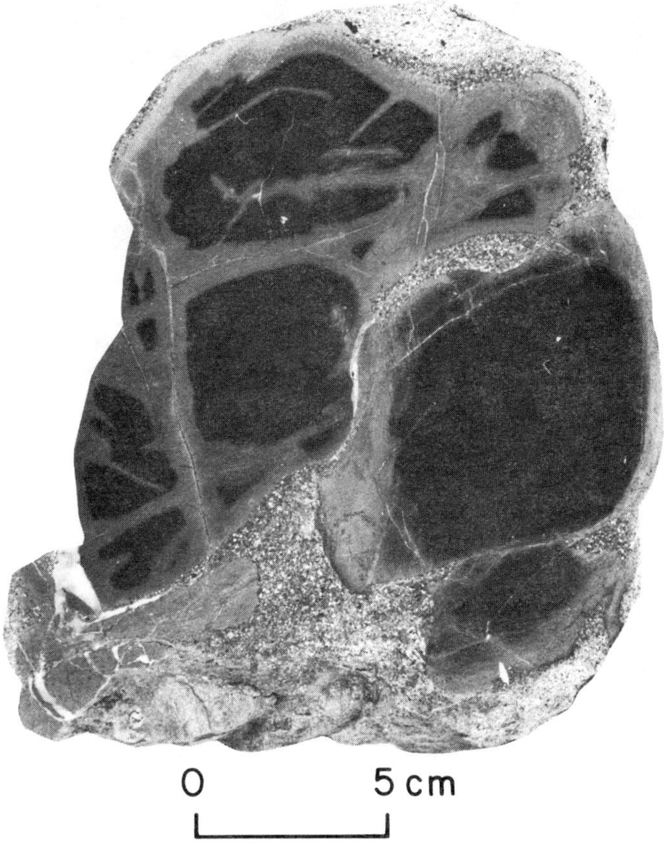

Figure 10. Slabbed and polished dolomite clast from debris flow deposit (bed 12C) showing sandstone injected along a network of fractures. Light gray areas are weathered dolomite (note how they are concentrated along fractures), dark areas are relatively unweathered dolomite.

Figure 11. Slabbed and polished laminated dolomite clast from debris flow deposit. The sequence of structures in this specimen resembles those in "seismites" described by Seilacher (1969). Note how antithetic faults in the center die out upward, and note also intensely deformed laminations along lower edge of the specimen. Calcite- and sediment-filled fractures in the upper part of the specimen (to the left and right of the letter A) are tensional structures of possible early diagenetic origin.

sediments may become very brittle and prone to intense fracturing by conversion to opal-CT or quartz only millions of years after they were deposited. Although opal-A rocks in the Monterey do show evidence of fracturing (Williams, 1982), they are much more ductile than either dolomite or siliceous rocks composed of opal-CT or quartz. (3) Fractures developed in early-formed dolomites will be natural conduits for fluids, which may include pore water expelled from compacting sediments, early-formed biogenic methane and thermally matured hydrocarbons. The dolomites may thus be prone to natural hydraulic fracturing processes (Redwine, 1981; Roehl, 1981) which act to amplify existing fractures and create some new ones.

ORIGIN OF REDEPOSITED DOLOMITES IN GRAVITY FLOW DEPOSITS

If the Sandholdt dolomite concretions were not fully formed until they reached burial depths of several tens to hundreds of meters, as argued earlier, it seems at first glance difficult to reconcile this with the common occurrence of dolomite as clasts in gravity flow deposits and with the field evidence that concretions were uncovered by erosion on the sea floor (Fig. 7). Individual gravity flow events would not be expected to erode the sea floor to depths of more than a few meters or, at a maximum, of a few tens of meters (personal communications from M. Fields and T. Nilsen, 1984), values less than the indicated depths of formation for the concretions.

Several alternative sources for the dolomite clasts thus deserve consideration:

(1) These clasts could represent dolomites which were parts of older Monterey sections which were uplifted, subaerially eroded and reworked into a deep marine environment. Graham (1976), for example, has shown that Saucesian elements of the Monterey were eroded from the Lockwood High south of the Arroyo Seco in pre-middle Miocene time; similar stripping of older Miocene sediments occurred on the Gabilan Mesa shelf to the north and northeast. It this had occurred, however, a mixture of redeposited dolomite clasts with a variety of other clast types would be expected. With the exception of unit 14 (Figs. 6 and 7), the reworked dolomite clasts occur only with reworked clasts of laminated mudrocks and sandstones, in other words with elements indigenous to the

Sandholdt.

At several localities in southern California, the Monterey or the overlying Sisquoc Formation contains breccias with dolomite chert, and porcelanite clasts which have borings made by pholad bivalves (e.g., the Monterey Formation at Gaviota State Park Beach, Bramlette, 1946; Isaacs, 1981; the Sisquoc Formation north of Mussel Rock, Santa Maria Basin). Clasts of these kinds thus were reworked in shallow marine zones where they became colonized by boring bivalves, a relationship compatible with subaerial exposure and erosion of older parts of the Monterey. We noted no such organic borings in redeposited dolomite clasts in the Arroyo Seco. We are led to conclude that, with the possible exception of unit 14 (Figs. 6 and 7) where dolomite clasts occur together with more allochthonous components, all of the redeposited dolomites were derived from submarine erosion of Sandholdt sediments.

(2) If individual gravity flows were incapable of exhuming concretions buried a hundred meters or more beneath the sea floor, it is still possible that repeated gravity flow events could progressively strip away their overburden. Many of the gravity flow beds in the Arroyo Seco section do not contain reworked dolomite clasts (Fig. 4), though they commonly contain ripup clasts of the hemipelagic mudrocks. These beds may thus be records of incremental erosion of slope sediments originally deposited farther upslope, incremental erosion which would eventually expose dolomite concretions to erosion on the sea floor. This kind of progression might explain instances like that portrayed in Figure 7 where a debris flow deposit (Bed 12C) apparently eroded down to the level of a large concretion (C). Estimated maximum temperature of formation of this concretion is 26°C (Sample AS-5 in Table 1). Using the methods applied earlier (Table 2), we may estimate the maximum depth of formation as between about 180 and 300 meters. Though these depths are rough approximations, a substantial burial depth is nonetheless indicated, one that would seem to preclude uncovering by only one cycle of erosional scour.

(3) Erosion to depths of several hundred meters is common in submarine canyons and the upper, highly channelized portion of submarine fans (Normark, 1978). One possibility, therefore, is that many of the dolomite concretions originally formed in upper slope sediments which were subsequently deeply incised by erosion, exposing the concretions in the walls of canyons and channels where they could be undercut and transported downslope in gravity flows.

(4) Another possibility is that large scale slumping or sliding of slope sediments may have brought previously more deeply buried concretions to or near the sea floor. As noted previously, slump folds are present in the Arroyo Seco section and slump breccias containing reworked dolomite concretions have been observed by the authors in the Monterey Formation at Naples Beach west of Santa Barbara (Isaacs, 1981). Slumping might have resulted from the same stresses which initiated gravity flows, or it might also be due to the impact of gravity flows on hemipelagic slope sediments (cf. Moore et al., 1982). Even if it did not completely exhume the concretion, sea floor slumping locally could have removed substantial sediment overburden; this would leave concretions which formed at sub-bottom depths of 100 meters or more within a few meters of the sea floor and thus susceptible to erosion and transport by gravity flows. Moore et al. (1982) have documented an example of these processes in Pleistocene fan deposits southeast of Baja California. There, the impact of a debris flow, which was derived from an adjacent shelf, caused massive sliding of mid-fan hemipelagic sediments to a depth of at least 75 meters below the sea floor.

In summary, subaerial erosion is not a likely source for the majority of dolomite clasts in the Arroyo Seco, nor does simple submarine scour at the base of gravity flows seem adequate by itself to account for their origin. We believe their explanation lies in a combination of multiple cycles of gravity flow scour with other slope processes, such as channelized erosion of upper slope sediments and large scale slumping or sliding.

CONCLUSIONS

Our field observations in the Arroyo Seco section demonstrate that dolomite concretions of substantial size formed in hemipelagic slope sediments relatively early in diagenesis. Limited isotopic data support this conclusion and further suggest that concretion development occurred at moderate temperatures (22 to 35°C) in the upper few hundred meters of sediment within a few hundred thousand years after their host sediment was deposited. Because erosion accompanying single gravity flows was unlikely to uncover the concretions, our data indicate more complex combinations of slope processes which may have included multiple cycles of erosional scour, channeling, and large scale slumping.

The early genesis of brittle dolomites made them susceptible to fracture-inducing stresses over much longer time spans than, for example, porcelanites or cherts which typically became embrittled through silica phase changes only after several millions years of burial diagenesis. Monterey dolomites therefore commonly exhibit intense, fracturing, and the earliest formed fractures may serve as conduits for vertical movement of fluids. In the Arroyo Seco area, they served also as planes of weakness along which sediment was injected during gravity flow transport, in some instances leading to fragmentation of the concretions.

ACKNOWLEDGEMENTS

Robert Compton and William Dickinson first introduced us to the geology of the Arroyo Seco area and willingly shared information with us during the subsequent years. For discussions, sharing of data and comments on the manuscript we express gratitude to Warren Addicott, Howard Ellis, Michael Field, Miriam Kastner, Karl Mertz, Ken Pisciotto, Tor Nilsen, and Don Zenger. Field and laboratory expenses were supported by grants from the Petroleum Research Fund (PRF #13953-AC2), the Chevron Oil Field Research Co., and the Faculty Research Committee of the University of California, Santa Cruz, and we wish gratefully to acknowledge the donors to these funds.

REFERENCES CITED

Baker, P. A. and Kastner, M., 1981, Constraints on the formation of sedimentary dolomite: Science, v. 213, p. 214-216.

Bramlette, M. N., 1946, The Monterey Formation of California and the origin of its siliceous rocks: U.S. Geol. Survey Prof. Paper 212, 57 p.

Bukry, D., 1981, Pacific Cost coccolith stratigraphy between Point Conception and Cabo Corrientes, D.S.D.P. Leg 63, in Orlofsky, S., ed., Initial Reports DSDP, v. 63: Washington, U.S. Govt Printing Office, p. 445-472.

Burnett, W. C., 1977, Geochemistry and origin of phosphorite deposits from off Peru and Chile: Geol. Soc. Am. Bull, v. 88, p. 813-823.

Burnett, W. C., 1980, Apatite-glauconite associations off Peru and Chile: paleooceanographic implications: J. Geol. Soc. London, v. 137, p. 757-764.

Compton, R. R., 1966, Analysis of Pliocene-Pleistocene deformation and stresses in northern Santa Lucia Range, California: Geol. Soc. America Bull., v. 77, p. 1361-1379.

Curtis, C. D., Petrowski, C., and Oertel, G., 1972, Stable carbon isotope ratios within carbonate concretions: a clue to place and time of formation: Nature, v. 235, p. 98-100.

Durham, D. L., 1974, Geology of the southern Salinas Valley area, California: U.S. Geol. Survey Prof. Paper 819, 111 p.

Friedman, I. and Murata, K. J., 1979, Origin of dolomite in Miocene Monterey Shale and related formations in the Temblor Range, California: Geochim. Cosmochim. Acta, v. 43, p. 1357-1365.

Friedman, I. and O'Neil, J. R., 1977, Compilation of stable isotope fractionation factors of geochemical interest: U.S. Geol. Survey Prof. Paper 440-K, 12 p.

Fritz, P. and Smith, D. C. W., 1970, The isotope composition of secondary dolomites: Geochim. Cosmochim. Acta, v. 34, p. 1161-1173.

Goldsmith, J. R. and Graf, D. L., 195, Structural and compositional variations in some natural dolomites: Jour. Geology, v. 66, p. 678-693.

Graham, S. A., 1976, Tertiary sedimentary tectonics of the central Salinian block of California: Ph.D. Thesis, Stanford University, Stanford California, 510 p.

Griggs, D. and Handin, J., 1960, Observations on fracture and a hypothesis of earthquakes, in Griggs, D. and Handin, J., eds., Rock Deformation: Geol. Soc. America, Memoir 79, p. 347-364.

Hein,, J. R., O'Neil, J. R. and Jones, M. G., 1979, Origin of authigenic carbonates in sediments from the deep Bering Sea: Sedimentology, v. 26, p. 681-705.

Hudson, J. D., 1978, Concretions, isotopes, and the diagenetic history of the Oxford clay (Jurassic) of central England: Sedimentology, v. 25, p. 339-370.

Irwin, H., Curtis, C. D. and Coleman, M., 1977, Isotopic evidence for source of diagenetic carbonates formed during burial of organic-rich sediments: Nature, v. 269, p. 209-213.

Isaacs, C. M., 1981, Field trip guide for the Monterey Formation, Santa Barbara Coast, California, in Isaacs, C. M., ed., Guide to the Monterey Formation in the California Coastal Area, Ventura to San Luis Obispo: Pacific Section AAPG Public., v. 52, p. 55-71.

Isaacs, C. M., 1984, The Monterey--key to the offshore California boom: Oil and Gas Journal, Jan. 9, 1984, p. 75-81.

Kastner, M. and Baker, P. A., 1982, A new insight into the origin of sedimentary dolomite: 1982/83 Yearbook of Science and Technology, McGraw Hill, p. 406-408.

Kelts, K. and McKenzie, J. A., 1982, Diagenetic dolomite formation in Quaternary anoxic diatomaceous muds of Deep Sea Drilling Project Leg 64, Gulf of California, in Curray, J. R., Moore, D. G., et al., Initial Reports DSDP, v. 64: Washington, U.S. Govt. Printing Office, p. 553-569.

Kulm, L. D., Schrader, H., Resig, J. M., Thornburg, T. M., Masias, A., and Johnson, L., 1981, Late Cenozoic carbonates on the Peru continental margin: lithostratigraphy, biostratigraphy, and tectonic history, in Kulm, L. D. et al., eds., Nazca Plate: Crustal Formation and Andean convergence: Geol. Soc. America, Memoir 145, p. 469-507.

Mertz, K. A., Jr. and Garrison, R. E., 193, Notes on the Sandholdt Member of the Monterey Formation in the Arroyo Seco area, Monterey County, California, in Ingle, J. C. and Garrison, R. E., eds., Field guide to the Monterey Formation, Northern Santa Lucia Mountains and Monterey Area, Monterey County, California: Pacific Section SEPM, p. 89-96.

Moore, D. G., Curray, J. R. and Einsele, G., 1982, Salado-Vinorama submarine slide and turbidity current off the southest tip of Baja California, in Curray, J. R., Moore, D. G., et al., Initial Reports DSDP, v. 64: Washington, U.S. Govt. Printing Office, p. 1071-1082.

Murata, K. J., Friedman, I. I., and Madsen, B. M., 1967, Carbon-13-rich diagenetic carbonates in Miocene formations of California and Oregon: Science, v. 156, p. 1484-1485.

Murata, K. J., Friedman, I. and Madsen, B. M., 1969, Isotopic composition of diagenetic carbonates in marine Miocene formations of California and Oregon: U.S. Geol. Survey Prof. Paper 614-B, 24 p.

Murata, K. J., Friedman, I. and Cremer, M., 1972, Geochemistry of diagenetic dolomites in Miocene marine formations of California and Oregon: U.S. Geol. Survey Prof. Paper 724-C, 12 p.

Normark, W. R., 1978, Fan valleys, channels, and depositional lobes on modern submarine fans: characters for recognition of sandy turbidite environments: Amer. Assoc. Petroleum Geologists Bull., v. 62, p. 912-931.

Pisciotto, K. A., 1978, Basinal sedimentary facies and diagenetic aspects of the Monterey Shale, California. Ph.D. thesis, Univ. of Calif., Santa Cruz, California, 450 p.

Pisciotto, K. A., 1981a, Review of secondary carbonates in the Monterey Formation, California, in Garrison, R. E., et al., eds., The Monterey Formation and related siliceous rocks of California. Pacific Sec., Soc. Econ. Paleont. Mineral., 273-283.

Pisciotto, K. A., 1981b, Distribution, thermal histories, isotopic compositions and reflection characteristics of siliceous rocks recovered by the Deep Sea Drilling Project, in J. D. Warme et al., eds., The Deep Sea Drilling Project: A Decade of Progress: SEPM Spec. Publ. No. 32, p. 129-147.

Pisciotto, K. A. and Mahoney, J. J., 1981, Isotopic survey of diagenetic carbonates, DSDP Leg 63, in Yeats, R. S., Haq, B. U., et al., Initial Reports of the Deep Sea Drilling Project, v. 63: Washington, U.S. Govt. Printing Office, p. 595-609.

Raiswell, R., 1971, The growth of Cambrian and Liassic concretions: Sedimentology, v. 17, p. 147-171.

Redwine, L., 1981, Hypothesis combining dilation, natural hydraulic fracturing, and dolomitization to explain petroleum reservoirs in Monterey Shale, Santa Maria area, California, in Garrison, R. E., et al., eds., The Monterey Formation and related siliceous rocks of California. Pacific Sec., SEPM Public., p. 221-248.

Roehl, P. O., 1981, Dilation brecciation--a proposed mechanism of fracturing, petroleum expulsion and dolomitization in the Monterey Formation, California, in Garrison, R. E., et al., eds., The Monterey Formation and related siliceous rocks of California. Pacific Section SEPM Public., p. 285-315.

Seilacher, A., 1969, Fault-graded beds interpreted as seismites: Sedimentology, v. 13, p. 155-159.

Spotts, J. H. and Silverman, S. R., 1966, Organic dolomite from Point Fermin, California: Am. Mineralogist, v. 51, p. 1144-1155.

Thorup, R. R., 1943, The type locality of the Vaqueros Formation: Calif. Div. Min. Geol. Bull. 118, p. 463-466.

Turner, F. J. and Weiss, L. E., 1963, Structural analysis of metamorphic tectonites: New York, McGraw Hill Book Co., 545 p.

Veeh, H. H., Burnett, W. C. and Soutar, A., 1973, Contemporary phosphorite on the continental margin off Peru: Science, v. 181, p. 845-847.

Walker, R. G., 1978, Deep-water sandstone facies and ancient submarine fans--models for exploration for stratigraphic traps: Amer. Assoc. Petroleum Geologists Bull., v. 62, p. 932-966.

Williams, L. A., 1982, Lithology of the Monterey Formation (Miocene) in the San Joaquin Valley of California, in Williams, L. A. and Graham, S. A. (eds.), Monterey Formation and Associated Coarse Clastic Rocks, Central San Joaquin Basin, California: Pacific Section SEPM Public., p. 17-35.

Woodruff, F., Savin, S. and Douglas, R. G., 1981, Miocene stable isotope record: a detailed Pacific Ocean study and its paleoclimatic implications: Science, v. 212, p. 665-668.

ORIGIN OF DOLOMITES IN THE MONTEREY FORMATION: PISMO AND HUASNA BASINS, CALIFORNIA

R.I. Kablanow
University of Wyoming
Dept. of Geology/Geophysics
P.O. Box 3006
Laramie, WY 82071

R.C. Surdam
University of Wyoming
Dept. of Geology/Geophysics
P.O. Box 3006
Laramie, WY 82071

D. Prezbindowski
Amoco Production Co.
Research Center
P.O. Box 591
Tulsa, OK 74102

ABSTRACT

The Monterey Formation in the Pismo and Huasna Basins is characterized by four lithofacies (siliceous, calcareous, phosphatic, and siliciclastic). Each lithofacies contains dolomite as matrix, cements, concretions, replacement features, and as fracture fillings. Some dolomites apparently represent a single diagenetic event, whereas others clearly result from multiple diagenetic events.

Due to its complex diagenetic history the dolomite exhibits a wide range of isotopic compositions ($\delta^{13}C$ = -25.5 to +16.8 ‰ and $\delta^{18}O$ = -7.2 to +1.2 ‰, relative to PDB). In the various facies of the Pismo Basin the dolomite replacing microfossils typically exhibits $\delta^{13}C$ values of about -15.0 ‰ and an estimated isotopic temperature of formation (T.F.) of 50°C ($\delta^{18}O$ values from -3 to -4; using a water $\delta^{18}O$ composition of 0.0). In the Huasna Basin the dolomite in the calcareous facies exhibits a range of compositions and T.F. ($\delta^{13}C$ = -7.3 to +1.1‰ and $\delta^{18}O$ = -2.0 to -4.0, T.F. of 40 to 50°C). The siliceous facies from both basins contains dolomite, as matrix and framework grain replacements, with the following compositions: $\delta^{13}C$ = +6.5 to +12.0 and T.F. of 35 to 50°C ($\delta^{18}O$ = -7.0 to +0.8). The above isotopic compositional differences are interpreted as resulting primarily from variations in burial rates for different lithofacies.

Other conclusions drawn from the isotopic data pertain to concretion growth and the spatial distribution of high temperature dolomite. Concretions and replacement structures from both basins and from various lithofacies show very large variations in isotopic composition ($\delta^{13}C$ = -7.6 to +16.8 ‰) and T.F. (from 26 to 52°C, $\delta^{18}O$ = +1.2 to -4). Some concretionary growth occurred at or near the sediment-water interface, whereas other concretionary structures grew at greater depth and higher temperatures. All dolomite samples characterized by high temperatures of formation (>70°C) are spatially associated with regionally significant fluid conduits, either faults or large sand bodies.

INTRODUCTION

The widely distributed marine Monterey Formation (Middle to Late Miocene) typically contains significant amounts of dolomite. The dolomite populations contained in the Monterey Formation underwent complex diageneic histories, as reflected in the petrography and geochemistry of the units. The purpose of this study is to document the textural and geochemical nature of the diagenetic dolomite in the Monterey Formation from the Huasna and Pismo Basins and evaluate their origin. Methods used in conducting this study were x-ray diffraction, cathode luminescence, petrographic analysis, and stable isotopic geochemistry.

The Pismo and Huasna Basins are located in the central California Coast Range, San Luis Obispo County, north of the western Transverse Range (Fig. 1). These basins are contained in fault bounded, but reasonably coherent tectonic blocks that lie between the San Andreas and Hosgri fault systems (Fig. 2). Hall (1981) describes these basins as pull-apart structures that formed along the San Luis Obispo Transform. In brief, this transform was near the mid-Miocene continental margin and joined the early mid-Miocene triple junction with the crest of the East Pacific Rise (Fig. 2; for more detail see Hall, 1981). These pull-apart structures may have been initiated when the Mendocino triple junction migrated past the San Luis Obispo area about 20-22 m.y.b.p. (Dickinson and Snyder, 1979). Hall (1981) estimates this migration at 20-28 m.y.b.p. Zandt and Furlong (1981) proposed a model to characterize tectonic activity associated with the triple junction. According to their model, the transition from subduction of the Gorda Plate beneath the North American Plate to transform motion between the Pacific and North American Plates resulted in thinning of the lithosphere and a subsequent increase in heat flow and volcanic activity.

The Pismo and Huasna Basins are relatively simple folds that served as depositional basins during the Miocene and Pliocene. The Monterey Formation in the Pismo basin ranges from 300 to 600 meters thick along the limbs of the fold and ranges up to 1500 meters in thickness at the center of the fold (Fig. 3). The Monterey Formation in the Huasna Basin averages about 1,500 meters in thickness (Fig. 4).

A volcanic high forms the western margin of the Pismo Basin (Surdam and Stanley, 1981a). The sediments deposited on the western side of the volcanic high are actually form the eastern margin of the next basin, offshore Santa Maria. The spatial distribution of lithofacies among these three basins varies greatly, from shallow water siliciclastic deposition along the eastern margin of the Huasna to starved-basin marine (calcareous/phosphatic) deposition along the eastern margin of the basin which lies to the west of the Pismo Basin.

The two significant lithofacies in the Monterey Formation in the Pismo Basin are the siliceous and carbonate-phosphatic facies, with the former being by far the most voluminous lithofacies. In fact, exclusive of the lowermost portion of the formation, most of the Monterey in the Pismo Basin east of the volcanic high can be placed in the siliceous lithofacies.

The carbonate-phosphatic lithofacies is developed on the west side of the volcanic high and consists of a mixture of intercalated carbonate, phosphate and siliceous rocks. The carbonate rocks consist of laminated dolomicrite, dolomitized foraminiferal packed biomicrite and fine-grained dolomitic turbidite sandstone. The phosphatic rocks consist of

in Garrison, Robert E.; Kastner, Miriam and Zenger, Donald H., eds., 1984, **Dolomites of The Monterey Formation and Other Organic-Rich Units:** Pacific Section S.E.P.M., Vol. 41, p. 103-114.

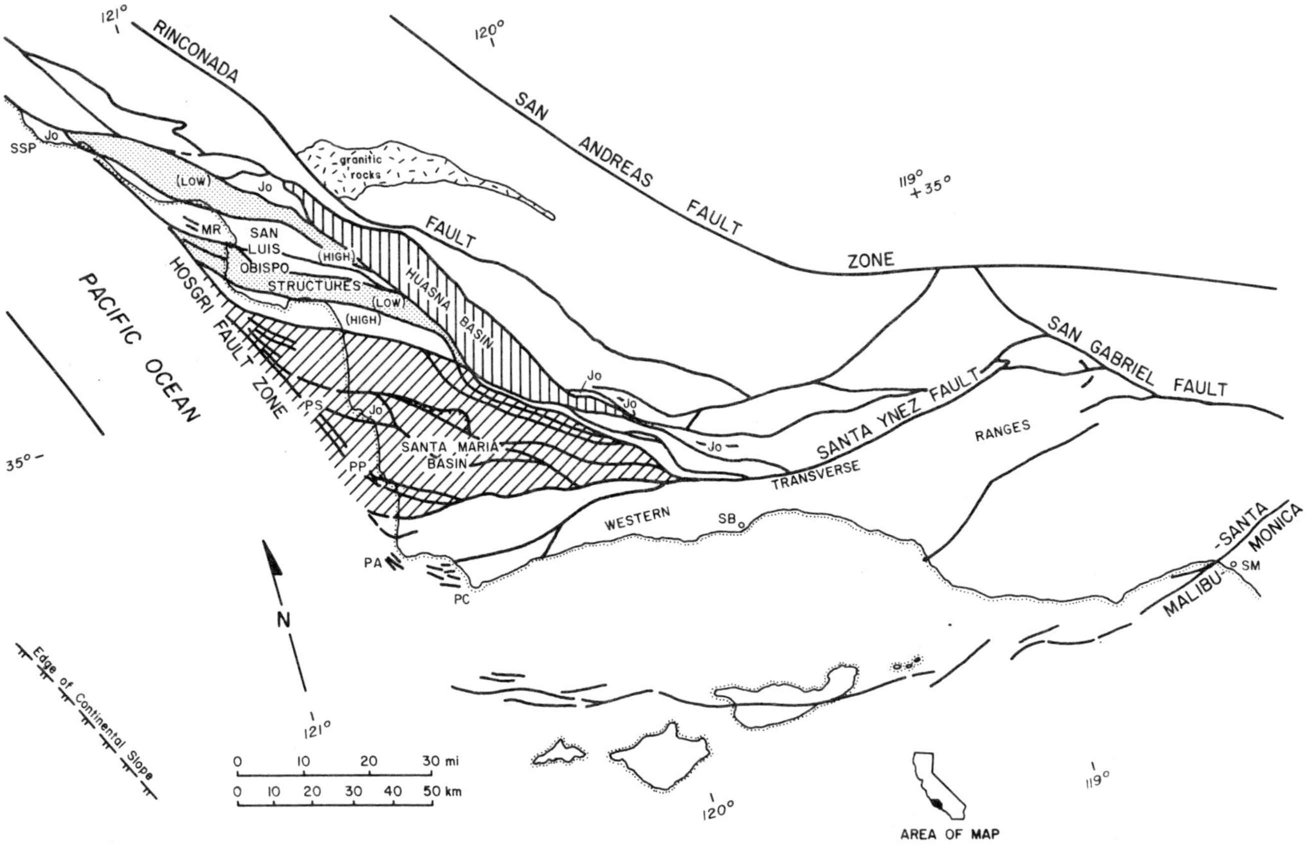

Figure 1. Generalized geologic map of tne southwestern Coast Ranges and the western Transverse Ranges, California, depicting three structural regions: (a) San Luis Obispo structures, Pismo Basin (stippled pattern and intervening blank areas), consisting of horst-graben-like pull-apart structures between faults (b) Huasna basin (vertically ruled pattern), a pull-apart structure. (c) Santa Maria basin (diagonally ruled pattern). The distance between the midpoints of stratigraphic sequences near San Simeon Point (SSP) and Point Sal (PS) is 95 km. Abbreviations are Morro Rock (MR), Purisima Point (PP), Point Arguello (PA), Point Conception (PC), Santa Barbara (SB), and Santa Monica (SM). Jurassic ophiolite is Jo. Simplified from Hall (1981).

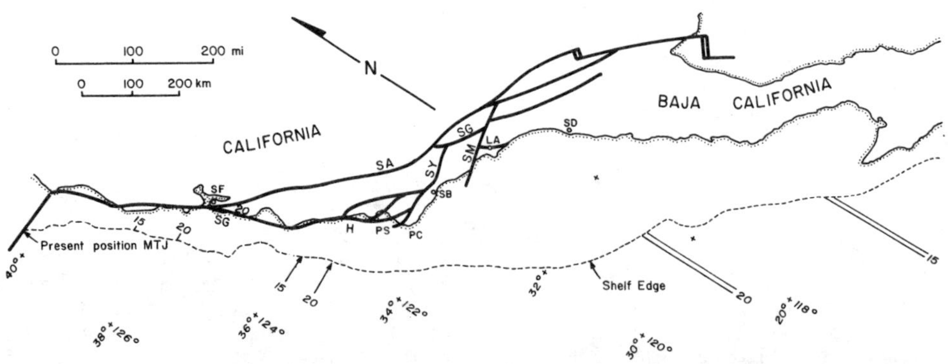

Figure 2. Sketch map of coastal California and Baja California, showing approximate prior positions of the Mendocino fracture zone with respect to interior North America (Atwater and Molnar, 1973), depicted as single-line arrows along the continental slope at 15 and 20 m.y.b.p. The 15-and 20-numbered single lines represent the general locations of the Mendicino triple junction at 15 and 20 m.y.b.p., with the slivers between the San Andreas fault (SA), San Gregorio-Hosgri Fault (SG-H), and the continental slope (Dickinson and Snyder, 1979). The double solid lines are prior positions of the East Pacific Rise crest (Atwater and Molnar, 1973; Dickinson and Snyder, 1979). Abbreviations are San Francisco (SF). Point Sal (PS), Point COnception (PC), Santa Barbara (SB), Los Angeles (LA), and San Diego (SD). Faults are San Andreas (SA), Santa Ynez (SY), San Gabriel (SG), and Santa Monica-Malibu Coast (SM) faults. After Hall, (1981).

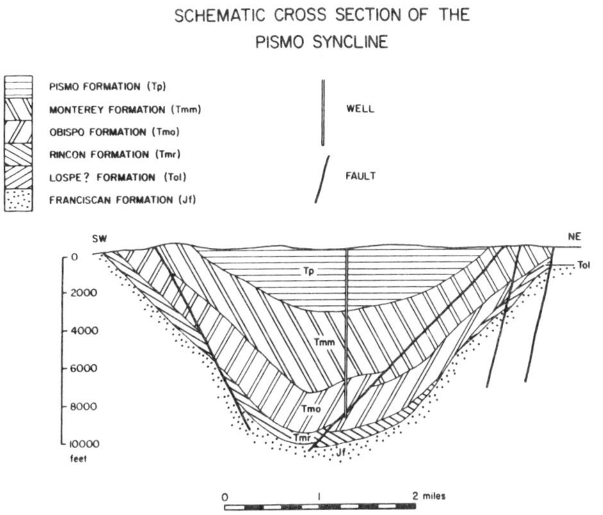

Figure 3. Schematic structure section of the Pismo Basin.

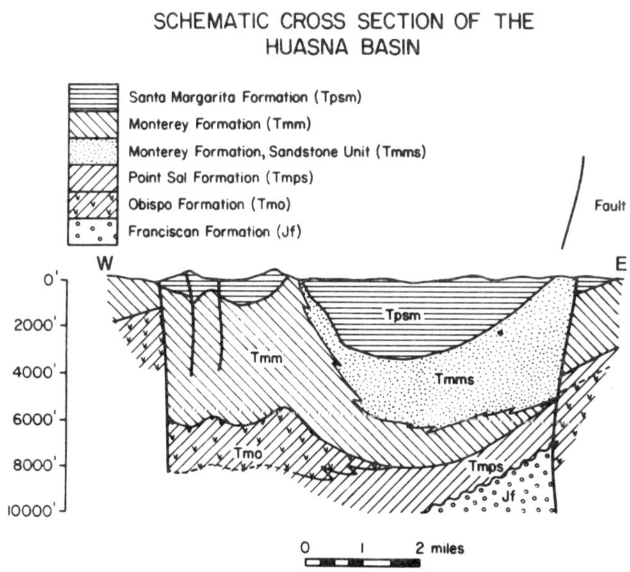

Figure 4. Schematic structure section for the Huasna Basin. Note the facies change form siliciclastic sediments on the east interfingering into siliceous sediments on the west.

laminated to pebble phosphates, whereas the siliceous rocks consist of laminated porcelanites. For a more detailed description of this lithofacies see Surdam and Stanley (1981a).

Kablanow and Surdam (1983) describe the lithofacies in the Huasna Basin in some detail. The Monterey Formation in the north and east side of Huasna Basin is a siliciclastic unit nearly 1,500 meters thick. This unit is mainly fine to medium grain sandstone, locally course grained, with usually a high matrix content. As in the Pismo Basin, the siliceous lithofacies also dominates in the Huasna Basin, and most is commonly present as porcelanite. The silica matrix in the porcelanite consists of both opal-CT and quartz. The siliceous facies has varying amounts of detrital quartz or feldspar. The calcareous facies is exposed below the procelanite along Lopez Lake and is cut off by the West Huasna Fault. It also occurs at the bottom of the Porter D-2 well in the south end of the basin, indicating that it may be a continuous unit below the porcelanite, at least on the western and southern portions of the basin (see Figure 3A, Kablanow and Surdam, 1983).

DOLOMITES IN THE MONTEREY FORMATION

All of the lithofacies in the Monterey Formation contain dolomites as concretions, layers, pervasive cement, matrix or framework grain replacement features, and fracture fillings. Dolomitization occurred predominately as multiple diagenetic events as Figures 5 and 6 illustrate. Therefore, due to their complex diagenetic history, dolomites from the Monterey Formation commonly exhibit extreme isotopic variability (Murata, Friedman and Madison, 1969). Table 1 demonstrates the scatter in isotopic data from the Pismo and Huasna Basins. Reasonable interpretations of isotopic data from dolomites in these two basins can be made only if the dolomite is studied within each lithofacies and correlated to thin section observations.

Dolomite In The Calcareous-Phosphatic Lithofacies

Dolomite in the calcareous-phosphatic facies occurs predominately as matrix and framework grain replacement, and less commonly concretions. The extent of dolomitized matrix is highly variable, ranging from scattered to pervasive micritic dolomite. However, most of the original calcite has been dolomitized, 75-95% diagenetic dolomite (analyzed by point counting). Microfossils and calcite framework grains have been extensively replaced by dolomite. Typically, where microfossil replacment was observed, the matrix is nearly pure dolomite (Fig. 7) and dolomite crystals are larger in the microfossil replacement than in the matrix. The original calcareous framework grains have been extensively replaced by dolomite ($\delta^{13}C$ = -4.3 to -17.2; $\delta^{18}O$ = +1.2 to -4.2; Fig. 8), and as can be seen most of the megacrystalline dolomite is euhedral. Some dolomite in the calcareous-phosphatic facies occurs as fracture and void fillings. Void filling is probably an extension of the matrix dolomitization.

Surdam and Stanley (1981b) describes thin fine-grained detrital, dolomitic turbiditic sandstones which are scattered throughout the carbonate facies. These beds are typically cross-stratified at the base and horizontally laminated at the top. The dolomite grains are rounded to euhedral in shape. Examination of the dolomite rhombs by cathode luminescence reveals that they are rounded grains with syntaxial overgrowths. The rounding probably occurred during transport.

Table 2 shows the isotopic compositions of the dolomite from the calcareous-phosphatic lithofacies in the Huasna and Pismo Basins. The Huasna Basin is not a starved basin facies. Carbon isotope composition range from -1.1 to -17.2 and oxygen isotope composition range from +1.2 to -4.2.

Sample	Lithofacies	Dolomite Type	$\delta^{13}C$ (o/oo PDB)	$\delta^{18}O$ (o/oo PDB)	$\delta^{18}O$ (o/oo SMOW)	Estimated Temperature of Formation (°C) 1	2	3	Estimated Zone of Formation
PISMO									
NSB-2	calcareous	concretion	- 4.3	+1.2	31.6	26	38	39	II
NSB-6	calcareous	replacement	-17.2	-3.1	27.1	48	59	60	IV
NSB-7	calcareous	replacement	-13.7	-4.1	26.1	55	63	64	IV
N-2530	siliceous	cement	+ 5.1	-2.1	28.2	43	54	55	III
E-4510	siliceous	cement	+10.6	-2.0	28.3	42	53	54	III
E-4985	siliceous	cement	+12.0	-1.2	29.1	37	49	50	III
E-5652	siliceous	cement	+ 4.8	-3.0	27.3	49	58	60	III
PO-23	calcareous	replacement	-14.1	-4.0	26.2	54	64	65	IV
PO-31	phosphatic	replacement	-13.1	-0.1	30.2	32	44	44	II
P-8	siliceous	cement	+ 1.7	-7.0	23.2	75	81	83	IV
P-22A	siliceous	concretion	+ 4.4	-2.8	27.5	46	57	59	III
P-22B	siliceous	concretion	+ 4.3	-3.0	27.3	47	57	60	III
P-22C	siliceous	concretion	+ 5.2	-2.3	28.0	44	54	58	III
HUASNA									
LL-3	calcareous	replacement	+ 1.1	-4.0	26.3	53	63	63	III
LL-14	calcareous	replacement	- 7.3	-2.0	28.3	42	53	54	II
LL-23	siliceous	concretion	- 4.2	-2.2	28.1	41	54	57	II
LL-63	siliceous	concretion	- 7.6	-1.1	29.2	37	48	50	II
LL-94A	siliceous	concretion	- 4.3	+0.8	31.2	27	39	40	II
ASE-5	siliceous	replacement	- 0.4	-7.2	23.0	76	82	85	IV
ASE-5C	siliceous	calcite	- 3.0	-5.2	25.1	43	43	40	?
ASE-14	siliceous	concretion	- 3.3	-5.4	24.8	63	72	73	IV
LRI-10	siliclastic	cement	+16.2	-0.9	29.4	36	47	49	III
LRII-1	siliclastic	cement	+ 1.8	-4.6	25.6	57	67	69	III
LRII-26	siliclastic	cement	+10.9	-3.8	26.5	53	63	64	III
LRII-36	siliceous	replacement	+ 2.3	-7.0	23.2	75	82	83	IV
LRII-40	siliclastic	cement	- 1.7	-5.5	24.8	64	73	75	IV
LC-1	siliclastic	cement	+14.9	-3.1	27.2	48	58	60	III
R-20	siliclastic	concretion	+13.4	-0.1	30.3	32	44	45	III
HR-1	siliclastic	cement	+16.8	-0.9	29.4	36	47	49	III
ST-3270	siliclastic	replacement	+11.2	-3.6	26.7	51	62	62	III
HC-2760	siliceous	replacement	+ 2.9	-6.5	23.7	71	79	81	IV

Table 1. Lithofacies, type of dolomite, estimated temperature of fortmation, and estimated zone of formation. Under estimated T.F.: column 1 is calculated from curve C and column 2 calculated from curve D of Pisciotto and Mohoney (Figure 7; 1981). Column 3 was calcuated from Northrup and Clayton (1966). Zones of microbial decay of organic matter as described by Pisciotto and Mahoney (1981). The one calcite sample (ASE-5c) analyzed is fracture filling material. The conversion of $\delta^{18}O$ $^{o}/oo$ (PDB) $\delta^{18}O$ $^{o}/oo$ SMOW was calcuated according to Faure (1977).

Dolomite In The Siliceous Lithofacies

Dolomite in the siliceous facies occurs predominately as cement and concretions, and as framework grain replacement. The cement occurs as a patchy (Fig. 9) to ubiquitous fine grained micritic dolomite (20-40% dolomite analyzed by joint; Fig. 10). In outcrop, concretions usually occur as nodules (Fig. 11) of nearly pure dolomite, and also as discontinuous layers or lenses. Baker and Kastner (1981) postulate that in organic rich sediments, dolomitization can occur without a solid carbonate precursor present and suggest that the most important bicar-

Table 2. Isotopic composition of dolomites in the calcareous-phosphatic lithofacies.

Sample	$\delta^{13}C$ (o/oo PDB)	$\delta^{18}O$ (o/oo PDB)
PISMO		
NSB-2	- 4.3	+1.2
NSB-6	-17.2	-3.1
NSB-7	-13.7	-4.1
PO-23	-14.1	-4.0
PO-31	-13.0	-0.1
HUASNA		
LL-3	+1.1	-4.0
LL-14	-7.3	-2.0

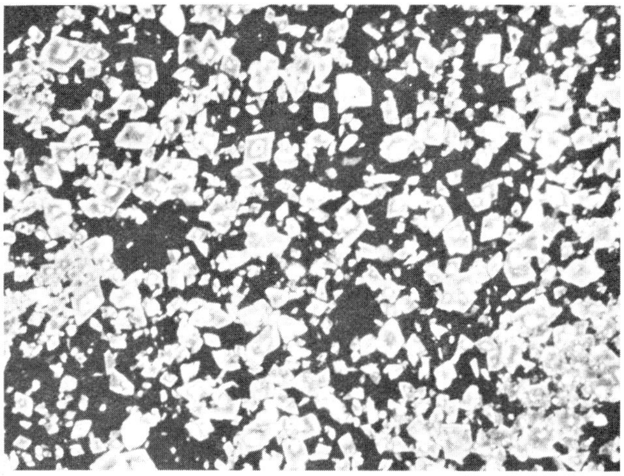

Figure 5. Cathode luminescence photomicrograph of authigenic dolomite showing evidence of multiple diagenetic events, i.e. zoned euhedral crystals. 10x magnification.

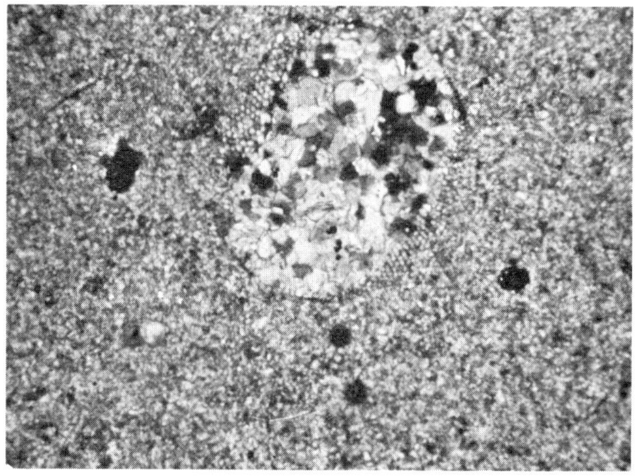

Figure 7. Photomicrograph, crossed nicols, showing pervasive dolomitization in calcareous-phosphatic facies. Note the crystallite size difference between the microfossil and the matrix. 20x magnification.

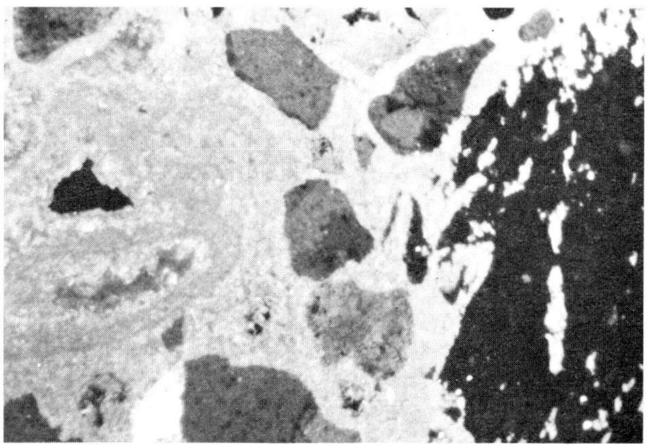

Figure 6. Cathode luminescence photomicrograph of authigenic dolomite, i.e. the light colored matrix material, showing zoning and crystal growth into voids. Note some framework grain replacement with authigenic dolomite to the right of photograph. 10x magnification.

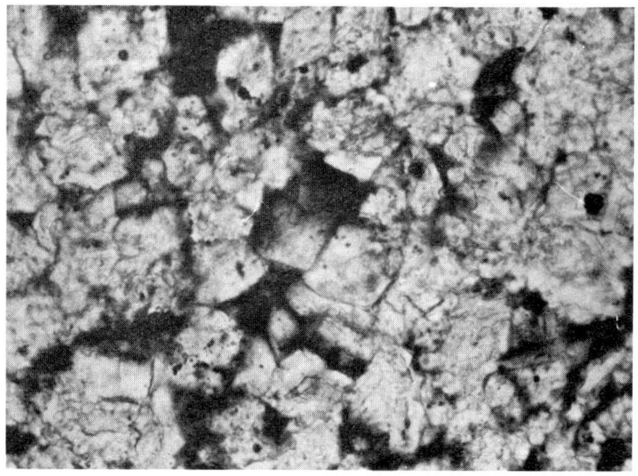

Figure 8. Photomicrograph plain light, of pervasive dolomitized calcite framework grains. Matrix, i.e. the dark material, is organic matter. From the phosphatic facies. 20x magnification.

bonate sources for this reaction are either sulfate reduction or methanogenesis ($\delta^{13}C$ = -25°/oo and +15°/oo relative to PDB, respectively; Irwin and Curtis, 1977) and that calcium supply is the limiting factor for dolomite formation during progressive diagenesis. Organic matter oxidation by Fe III reduction was found to be an additional source of bicarbonate in the shallow suboxic diagenetic systems (Froelich et al., 1979). The processes of sulfate reduction or methanogenesis can lead to the precipitation of carbonates with distinct stable isotopic compositions as indicated above (Irwin and Curtis, 1977).

Table 3 shows the isotopic compositions of the dolomite types from the siliceous facies in both basins. Carbon isotope values range from -7.6 to +12.0 and oxygen isotope values range from +0.8 to -7.2.

Figure 9. Photomicrograph, crossed nicols, showing the patchy or scattered nature of the diagenetic dolomite in the siliceous facies. 40x magnification.

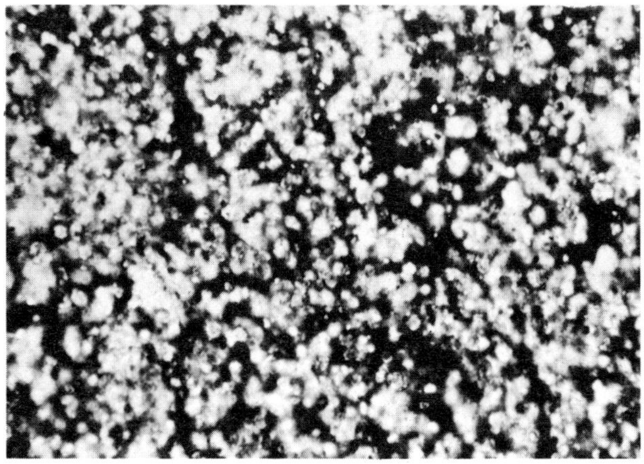

Figure 10. Photomicrograph, crossed nicols, showing ubiquitous nature of some diagenetic dolomite in the siliceous facies. 20x magnification.

Table 3. Isotopic composition of dolomite in the siliceous facies.

Sample	$\delta^{13}C$ (o/oo PDB)	$\delta^{18}O$ (o/oo PDB)
PISMO		
N-2530	+ 5.1	-2.1
E-4510	+10.6	-2.0
E-4985	+12.0	-1.2
E-5652	+ 4.8	-3.0
P-8	+ 1.7	-7.0
P-22A	+ 4.4	-2.8
P-22B	+ 4.3	-3.0
P-226	+ 5.2	-2.3
HUASNA		
LL-23	- 4.2	-2.2
LL-63	- 7.6	-1.1
LL-94A	- 4.3	+0.8
ASE-5	- 0.4	-7.2
ASE-14	- 3.3	-5.4
LRII-36	+ 2.3	-7.0
HC-2760	+ 2.9	-6.5

Dolomite In The Siliciclastic Lithofacies

Dolomite in the siliciclastic facies occurs predominately as cement, but it can also occur as concretions and framework grain replacements, and ranges from 10 to 20% of the rocks analyzed. Some of the dolomite cement can be con- sidered a replacement feature, because it typically replaces primary calcite cement. In some rocks, the primary calcite cement may have formed immediately after deposition, thus preventing extensive compaction (Fig. 12). Diagenetic processes have subsequently removed this early carbonate cement, resulting in very high porosity and permeability (Fig. 13). Consistent with the distribution of concretions in other facies in the Monterey Formation, the concretions in the siliciclastic facies occur mainly as large dolomitic nodules.

Replacement of feldspar framework grains is not common, but does occur locally within these units (Fig. 6), typically along the margins of feldspar grains. This is not surprising, because the margins of grains are the location of the highest surface free energy and the place where compaction stress is concentrated.

Table 4 illustrates the isotopic composition of the dolomite types from the siliciclastic facies. Carbon isotope composition ranges from -1.7 to +16.8 and oxygen isotope composition ranges from -0.9 to -5.5.

Figure 11. Photograph of large dolomite concretion in the siliceous facies. Note rock hammer for scale.

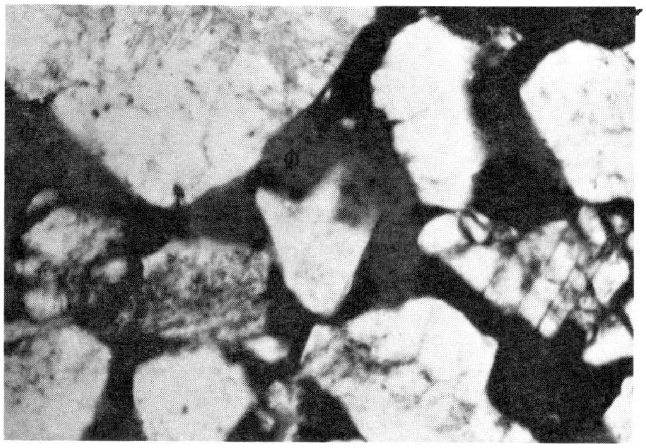

Figure 13. Photomicrograph, crossed nicols, showing rock with high porosity and permiability resulting from diagenetic processes that has removed the early calcareous cement in the siliciclastic facies. 10x magnification.

Table 4. Isotopic composition of dolomite in the siliciclastic facies.

Sample	$\delta^{13}C$ (o/oo PDB)	$\delta^{18}O$ (o/oo PDB)
HUASNA		
LRI-10	+16.2	-0.9
LRII-1	+ 1.8	-4.6
LRII-26	+10.9	-3.8
LRII-40	- 1.7	-5.5
LC-1	+14.9	-3.1
R-20	+13.4	-0.1
HR-1	+16.8	-0.9
ST-3270	+11.2	-3.6

STABLE ISOTOPIC GEOCHEMISTRY

The rocks analyzed in this study were chosen as representative samples for each lithofacies. The samples were examined by means of X-ray diffraction, cathode luminescence, conventional petrographic microscopy, and light stable isotopic analysis (oxygen and carbon). The samples are ordered dolomite with some rhombs exhibiting zoning and/or syntaxial overgrowths. $\delta^{13}C$ and $\delta^{18}O$ analyses were carried out using the phosphoric acid technique first documented by McCrea (1950). Before reaction, the samples were vacuum roasted at 360°C for a period of 1 hour to remove volatile components. All stable

Figure 12. Photomicrograph, crossed nicols, of dolomite cement in the siliciclastic facies. 10x magnification.

isotopic data unless otherwise indicated, are referenced to the PDB standard. Analitical error is no greater than 0.1°/oo. No attempt was made to separate various diagenetic generations of dolomite, therefore isotopic values for a particular sample are average values for the carbonate fraction. However, based on the results of X-ray diffraction, only those samples that contained one carbonate phase were chosen. The lithologic, mineralogic and textural settings for the dolomites have been discussed previously. It should be noted, however, that with the exception of the siliciclastic lithofacies, all other samples are from rocks which are relatively organic-rich. For example, rocks of the siliceous facies typically contains 1-5 wt % organic carbon (average TOC 3 wt %), whereas those of the calcareous-phosphatic lithofacies typically contains 5-20 wt % organic carbon (average TOC 8 to 10 wt %).

Stable isotopic data has been an important source of information used in delineating the diagenetic history of authigenic carbonates. This is particularly true of recent investigations of dolomite occurrances in organic rich sediments (Supko et al., 1974 and Pisciotto and Mahoney, 1981). Even though frequently used as a investigative tool, we still have an incomplete understanding of the relationships of the stable isotopic composition of an authigenic carbonate to the diagenetic process responsible for its formation.

Since many factors can influence the stable isotopic composition of a diagenetic carbonate e.g., temperature, volumes and compositions of different carbon and oxygen reservoirs, equilibrium and nonequilibrium processes, reaction rates etc., precise interpretations of the data may not be feasible. However, general insights into the diagenetic history of a carbonate can be obtained by careful comparisons to experimental and controlled field studies.

Carbon and oxygen stable isotopic data of dolomites from the Pismo and Huasna Basins have a wide range of compositions (see Table 1 and Figure 14). Similar stable isotopic compositional ranges have been reported from other studies of authigenic dolomites recovered from DSDP coring operations (see Pisciotto and Mahoney, 1981). Combined with information obtained from interstial water and gas studies

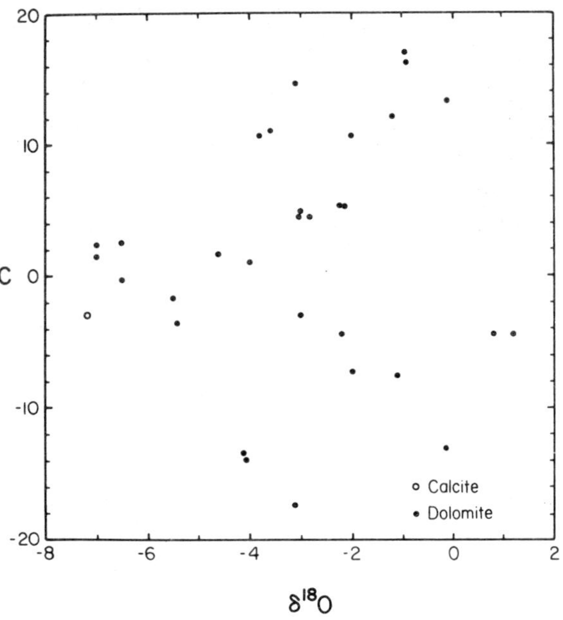

Figure 14. Carbon and oxygen stable isotopic cross plot: Monterey Formation, Pismo and Huasna Basins, California.

of shallow, organic rich sediment systems e.g., Froelich et al., (1979) and Galimov and Simoneit, (1981), a plausible diagenetic history of the dolomites studied can be advanced.

Oxygen Stable Isotopes

Estimates of the average temperature of formation of dolomites in the Monterey Formation can be determined from the oxygen isotopic composition of the dolomite. However, there are a number of factors which contribute to the uncertainty of these calculated temperatures. First, the oxygen isotopic fractionation between dolomite and water at sedimentary temperatures is not well established (Land, 1980). Second, the oxygen stable isotopic composition of the interstial water will change during burial diagenesis. These changes have been documented in several studies (Pisciotto and Mahoney, 1981; Gieskes et al., 1982). The typical $\delta^{18}O$ shift observed for interstial waters recovered from depths up to 900 meters sub bottom is approximately -0.5°/oo per 100 meters of burial. This shift can be attributed to preferential removal of ^{18}O from the interstial water by carbonate precipitation. However, ^{18}O enrichment of interstial waters with shallow burial has also been documented (Gieskes et al., 1982). Although this enrichment trend was attributed to hydrothermal alteration of igneous sills and/or sediments associated with the sills, an assumption that ^{18}O depletion in interstial waters will always occur is not valid. ^{18}O enrichment of interstial waters with depth in sedimentary basins has been commonly observed (Clayton et al., 1966). Since we have no reliable information regarding the ^{18}O evolution of the interstial water in these sediments during burial, we have chosen to calculate all temperatures of dolomite formation and based on a interstial water $\delta^{18}O$ composition of 0°/oo (SMOW). Third, during formation dolomite records temperature and $\delta^{18}O$ variations in the water responsible for precipitation. This variable $\delta^{18}O$ signal is homogenized to an average value during whole rock analysis. The experimental dolomite-water and the protodolomite-water fractionation curves presented by Pisciotto and Mahoney (Fig. 7, 1981) were used to determine temperatures of formation for the dolomite. The protodolomite-water fractionation expression (curve C of Fig. 7, Piciotto and Mahoney, 1981) was used for a comparison of the temperatures of formation for dolomite from the Pismo and Huasna Basins with those determined for dolomites from organic-rich sediments (from DSDP Leg 63, Pisciotto and Mahoney, 1981). Estimated temperatures of formation are presented in column 1, for T.F., Table 1. Also shown in Table 1 are the temperatures of formation derived from the dolomite-water fractionation expression (column 2 T.F.), and a third temperature of formation estimated from a different set of fractionation curves for dolomite-water (Northrup and Clayton, 1966) are shown in column 3 for T.F. Pisciotto and Mahoney (1981) discuss the limitations and constraints of temperatures derived from dolomite fractionation expressions. Nonetheless the relative temperature of formation of samples from this study compared to dolomites from DSDP Leg 63 are intriguing. The range of temperatures from Leg 63 was 10°C to 50°C using curve C (Fig. 7, Pisciotto and Mahoney, 1981); the range of temperatures for samples from the Monterey Formation are 26°C to 76°C using the same fractionation expression. Thus, the range in temperature of formation determined for the dolomites in the Monterey Formation are 15 to 25°C higher than the range determined for Neogene dolomites in organic-rich siliceous sediments from offshore California and Baja California (Pisciotto and Mahoney, 1981).

This shift to slightly higher temperatures of formation for dolomite from the Monterey Formation may be explained by three factors. First recall that the stable isotopic temperatures calculated for dolomite from the Monterey are <u>average</u> temperatures for all stages of dolomite formation. Secondly, Kelts and McKenzie (1982) have shown that dolomites from a single dolomitic layer forming in Quaternary anoxic diatomaceous muds from the Gulf of California can form over a wide range of depth and temperature which will also yield an <u>average</u> T.F. For one layer they were able to document dolomite formation over a depth range of 80 to 440 meters probably representing a 1 to 2 million year burial history (Kelts and McKenzie, 1982). Using curve C (from Fig. 7, Piciotto and Mahoney, 1981) the temperatures of formation for this layer range from 18°C to 75°C. The <u>average</u> temperature of formation of this layer would be between 18° and 75°C depending on where the sample was taken. Third, the evolution of the $\delta^{18}O$ composition of the interstitial water by interaction with the sediment may have been different in the two systems. The dolomites in the Monterey started forming at very shallow depths, probably comparable to the Quaternary dolomites described by Kelts and McKenzie (1982). Because they have had a much longer and deeper burial history during which recrystallization and solution reprecipitation occurred at higher temperatures (Land, 1980), their temperatures of formation are higher than those exhibited by the more recent dolomites. In summary, the oxygen isotopic data for dolomites from the Monterey are compatible with early diagenetic dolomite formation close to the sediment-water interface and continued dolomite formation and/or recrystallization during progressive burial, up to temperature in excess of 70° to 80°C.

Carbon Stable Isotopes

Previous investigations have established that carbon isotopes of authigenic carbonates precipitating within the various zones of microbial metabolic processes can assist in documenting the types and importance of organic-inorganic diagenetic reactions (Irwin et al., 1977; Pisciotto and Mahoney, 1981).

Pore water studies in pelagic sediments of the eastern equatorial Atlantic have documented the importance of microbial processes in early diagenesis (Froelich, et al., 1979). The simplest model of organic diagenesis during early burial consist of organic matter being oxidized by the oxidant yielding the greatest free energy change per mol of organic carbon oxidized (Froelich et al., 1979). When the initial oxidant is depleted, oxidation will proceed with the next most efficient oxidant. The oxidants most commonly documented in order of decreasing energy production are: O_2 > manganese oxides nitrate > iron oxides > sulfate (Froelich et al., 1979). The net effect of organic matter oxidation by O_2 and nitrate reduction is to increase total alkalinity promoting carbonate dissolution. While organic oxidation by manganese, iron oxide, and sulfate reduction will decrease total alkalinity promoting carbonate mineral precipitation (Froelich et al., 1979). Organic matter oxidation by sulfate reduction appears to be the most influential processes affecting the pH of the diagenetic system during shallow burial (Nissenbaum et al., 1972). Irwin et al. (1977) and others including Pisciotto and Mahoney (1981: see Figure 9) have simplified the microbial metabolic zones and divided them into four major zones in order of increasing depth: the uppermost and thinnest zone of oxidation (I); the zone of sulfate reduction (II); zone of methanogenesis (III); the zone of thermocatalytic decarboxylation (IV).

The carbon contained in authigenic dolomite can originate from one or more of the following sources:

1) sea water bicarbonate ($\delta^{13}C \approx 0$)
2) volcanogenic CO_2 ($\delta^{13}C \approx -7.0$ ‰) producing carbonates with a $\delta^{13}C$ range of +1 to +5
3) dissolution of biogeneic calcite and aragonite ($\delta^{13}C \approx 0$)
4) dissolution of detrital carbonate material ($\delta^{13}C$ variable)
5) CO_2 derived from metablolic oxidation of organic material ($\delta^{13}C \approx -18$ to -23 ‰). However, nonequilibrium oxidation of organic material has been shown to produce CO_2 with a +20.5 ‰ δC^{13} composition (Murata et al., 1969)
6) CO_2 derived from oxidation of CH_4 ($\delta^{13}C < -25.$ ‰)
7) carbon exchange between CH_4 and CO_2 leading to an ^{18}O enrichment in the CO_2 ($\delta^{13}C$ of the CO_2 tending toward enriched values)
8) fermentation of organic acids CH_3COOH $CH_4 + CO_2$ ($\delta^{13}C$ of CO_2 tending toward enriched values)
9) preferential reduction of ^{12}C rich CO_2 by methane forming bacteria using available hydrogen (Nissenbaum, 1972) $CO_2 + 4H_2$ $CH_4 + 2H_2O$ ($\delta^{13}C$ of CO_2 tending toward enriched values)
10) Thermocatalytic decarboxylation ($\delta^{13}C$ of CO_2 ≈ -10 to -25) (Irwin et al., 1977).

Because the ^{13}C composition of a dolomite reflects the composition of the bicarbonate at the time of precipitation it yields some insight into the dominate diagenetic process supplying carbon. Note however that various diagenetic reactions can produce HCO_3 with similiar $\delta^{13}C$ compositions. Table 1 lists the average $\delta^{13}C$ values for dolomite from the Huasna and Pismo Basins. Also in Table 1 (last column) the dolomites have been assigned to one of the four microbial metabolic zones. These are general assignments based on both $\delta^{13}C$ values as well as the average temperatures of formation based on $\delta^{18}O$ values. Consistent with the $\delta^{18}O$ values, the $\delta^{13}C$ values are compatible with initiation of diagenesis at or near the sediment water interface with continued diagenesis down to and including the zone of thermocatalytic decarboxylation. At elevated temperatures the $\delta^{13}C$ values tend to converge near zero (Fig. 15), probably reflecting the averaging effect of these processes.

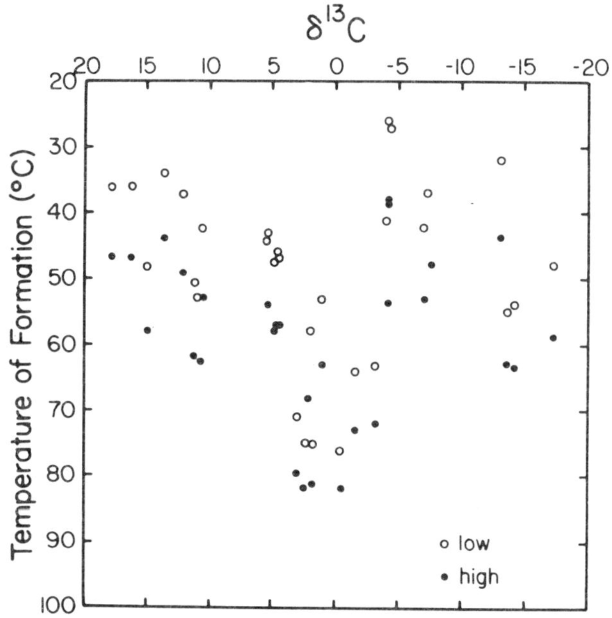

Figure 15. Plot of temperature of formation versus $\delta^{13}C$. Open circles (low) are temperatures calculated from curve C and closed circles (high) are temperatures calculated from curve D, in Figure 7 of Pisciotto and Mahoney, 1981.

Discussion

As suggested by Pisciotto and Mahoney (1981) sedimentation rate may be the most important factor influencing the character of authigenic carbonates in organic-rich seidments. It was noted that sedimentation rate controls preservation of organic matter and extent of oxygen and sulfate diffusion from overlying sea water. Just as important is the control sedimentation rates (burial) have on vertical fluid movement (compaction expulsion) and reaction times in specific diagenetic environments.

We believe that the major isotopic variation of dolomite in the different lithofacies can be interpreted as a result of varying rates of burial that in turn control the time-space relationships of the various diagenetic processes. For example, the calcareous-phosphatic facies represents starved-basin deposition characterized by very low rates of burial; whereas the siliceous lithofacies represents higher but still moderate rates of burial. The siliciclastic facies represents high rates of burial, with the highest burial rate represented by the coarse-grained siliciclastic rocks. Table 5 suggests possible sedimentation rates for the various lithofacies in the Pismo and Huasna Basin, taken from modern and ancient environments. Figure 16 shows the $\delta^{13}C$ values of the dolomite plotted against interpreted sedimentation rates. Consistent with the dolomites from more recent organic-rich diatomaceous sediments (Pisciotto and Mahoney, 1982) there is a distinct correlation between the $\delta^{13}C$ values of dolomite in the Monterey and the rates of sedimentation. This correlation explains much of the isotopic variation observed in the different lithofacies. As Pisciotti and Mahoney (1981) suggest; slow rates of deposition allow more complete oxidation and anaerobic sulfate reduction, producing ^{13}C depleted CO_2 that could be incorporated into precipitating carbonates; more rapid rates of sedimentation allow the organic rich sediments to pass through the oxidizing and sulfate reduction zones rapidly. Thus promoting methanogenic processes and the production of ^{13}C enriched carbonates. This is a plausible explanation for the isotopic variations between the calcareous-phosphatic rocks (i.e. slow rates of sedimentation) and the siliceous siliciclastic rocks (i.e. moderate to high rates of sedimentation). At still higher rates of burial (coarse grain sandstone; i.e. clastic sediments in Figure 16) much of the dolomitization probably took place in the initial part of the zone of thermocatalytic decarboxylation, resulting in a ^{13}C depletion of the bicarbonate, shifting it away from enriched values toward the normal marine range.

Table 5. Possible sedimentation rates for the different lithofacies in the Pismo and Huasna Basin.

Calcareous-phosphatic Facies: 0.5 to 2.5 m/my: for starved basin deposition (Gorsline and others, 1968)

5 to 10 m/my: for Mid to Late Miocene in the Pacific (Whitman and Davis, 1979)

Siliceous Facies: 80 m/my: from varve estimates, average maximum compacted thickness (Pisciotto and Garrison, 1981)

20-50 m/my: for equitorial Pacific Ocean (Reading, 1979)

25 to 200 m/my: for California continental borderland (Pisciotto and Mahoney, 1981)

Siliciclastic Facies: 100-300 m/my: for Mid Miocene sandstone from the Salinas Valley Basin (Graham, 1976)

500-1,200 m/my: for California offshore fans and basins (Blatt, Middleton and Murray, 1980)

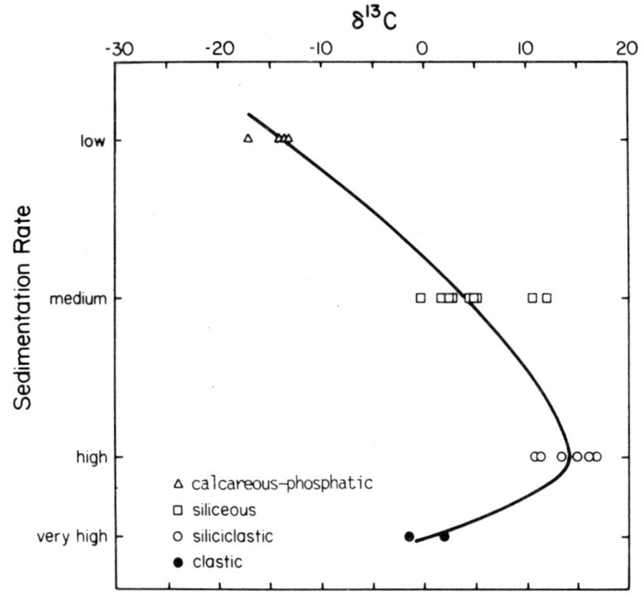

Figure 16. Interpreted sedimentation rates versus $\delta^{13}C$. Refer to Table 5 for estimated sedimentation rates for each lithofacies. Clastic refers to course-grained sandstone units.

COMPARISON BETWEEN DOLOMITE IN THE MONTEREY FORMATION AND MODERN ORGANIC RICH DEPOSITONAL ENVIRONMENTS

Numerous studies have been conducted in recent years on organic rich depositional environments with particular interest in interstitial water chemistry and dolomitization (e.g., Presley and Kaplan, 1968; Nissenbaum, Presley and Kaplan, 1972; Sayles, Manheim and Waterman, 1973; Presley, Culp, Petrowski and Kaplan, 1973; Kelts and McKenzie, 1979; Pisciotto and Mahoney, 1981; Kulm, Suess and Thornburg, 1982; Suess, Kulm and Killingley, 1982, Kastner and Baker, 1983). These papers focus on the interaction of framework grains, organic matter, interstitial fluid, and authigenic minerals (mainly dolomites) that form during shallow diagenetic processes. In addition to these modern depositional environment studies, several experimental studies have been conducted to evaluate the process of dolomitization in marine environments (e.g., Baker, Kastner, Byerlee and Lockner, 1980; Baker and Kastner, 1981; Kastner, Keene and Gieskes, 1977; Kastner and Baker, 1983). In most of the modern, organic-rich environments studied, the carbon isotopic composition of the dolomites have variability similar to the dolomite from the Monterey Formation.

The experimental work of Baker and Kastner (1981) on dolomitization suggested that dissolved sulfate is a very effective inhibitor of calcite dolomitization, even when concentrations are less than 5 percent of sea water value (28m μ; Kastner and Baker, 1983). They suggested that sulfate inhibition is a possible explanation as to why dolomitization is so rare in open marine environments. Where anoxic environments are established due to high organic content in the sediments, sulfate reduction takes place, and the sulfate inhibitor is removed, allowing dolomitization. Baker and Kastner (1981) suggest that this explains why early diagenetic dolomite is so common in organic-rich sediments, but rare in sediments with low organic content.

Other investigators have suggested that organic acids derived from the decomposition of organic matter help promote dolomitization of pre-existing calcite by facilitating the entry of Mg ions into the calcite structure (Kitano and Kanamori, 1966). Numerous other studies have proposed different dolomitization processes to explain dolomite occurrances in DSDP material (see Supko et al., 1974). Dolomite most certainly can form in a variety of diagenetic environments by different processes. However, the strong link between organic diagenesis and dolomite formation documented in these samples favors Baker and Kastner's (1981) model of dolomitization. Later modification (recrystallization and dolomite cementation) of this dolomite most certainly occurred during burial, but did not greatly modify the established whole rock stable isotopic character.

Our investigation of authigenic dolomites from the Monterey Formation from the Pismo and Huasna Basins indicates that it formed at shallow burial depths and at low temperature, and that in some cases it continued to form with progressive burial. Exceptions exist with samples inferred to be associated with conduits (i.e. located adjacent to faults or in sandstone units adjacent to faults), or for late hydrothermal fluid movement (temperatures greater than 70°C). As Table 1 (column 1 for T.F.) illustrates, these samples have the highest calculated temperatures of formation. Different rates of sedimentation of the various lithofacies can explain the wide variance in carbon isotope values and temperatures of formation derived from oxygen isotope values in the Monterey dolomites.

A major problem which remains to be solved is the identification of the source of Mg^{++} for dolomitization. Is the diagenetic alteration of volcanic material capable of supplying the Mg^{++} as suggested by Hein et al. (1979), for dolomites found in sediments from the Bering Sea? Or is the Mg^{++} supplied from interstial and/or recirculated sea water? At present the source of the Mg^{++} incorporated into the volumetrically important dolomites ($\approx$ 75-85 vol. % in the calcareous-phosphatic facies, $\approx$ 30% in the siliceous facies and $\approx$ 15% in the siliciclastic facies) remains uncertain. We hope that further work will provide an answer.

In conclusion, the work on dolomite from more recent anoxic organic-rich diatomaceous sediments from offshore California and Baja California forms an excellent model for the origin and distribution of dolomites in the Miocene Monterey Formation. The rate of sedimentation (burial) is one of the most significant parameter when trying to explain the origin and distribution of dolomite in the Monterey Formation.

ACKNOWLEDGEMENTS

We are greatful to Marathon Oil Company Research Laboratory, Littleton, Colorado and Amoco Production Research Laboratory, Tulsa, Oklahoma for conducting the isotopic determinations for us. We are greatful to Dr. Robert Reitsema, Marathon Research Laboratory for his assistance. Some of the samples used in this study were provided by Phillips Petroleum Company, Denver, Colorado, we appreciate their cooperation. Kablanow thanks Phillips Petroleum Company for support and Surdam thanks the NSF (Grant EAR-8207570) for support.

REFERENCES

Baker, P.A. and Kastner, M., 1981, Constraints on the formation of sedimentary dolomite: Science, v. 213, p. 214-216.

Baker, P.A., Kastner, M., Byerlee, J.D., and Lockner, D.A., 1980, Pressure solution and hydrothermal recrystallization of carbonate sediments - an experimental study: Marine Geology, v. 38, p. 185-203.

Blatt, H., Middleton, G., and Murray, R., 1980, Origin of sedimentary rocks: Prince-Hall, Inc., New Jersey, 782 p.

Claypool, G.E. and Kaplan, I.R., 1974, The origin and distribution of methane in marine sediments. in Kaplan, I.R. (ed.), Natural Gases in Marine Sediments: Mar. Sci., v. 3, p. 99-139.

Clayton, R.N., Friedman, I., Graf, D.L., Mayeda, T.K., Ments, W.F. and Shimp, N.R., 1969. The origin of saline formation waters, I. Isotopic composition, J. Geophys. Res., v. 71, p. 3869-3882.

Dickinson, W.R. and Snyder, W.S., 1979, Geometry of triple junctions related to San Andreas Transform: Jour. Geophysical Research, v. 84, no. B2, p. 561-572.

Faure, G., 1977, Principles of isotope geology: John Wiley and Sons, Inc., 464 p.

Froelich, P.N., Klinkhammer, G.P., Bender, M.L., Luedtke, N.A., Heath, G.R., Cullen, D., Dauphin, P., Hammond, D., Hartman, B. and Maynard, V., 1979, Early oxidation of organic matter in pelogic sediments of the eastern equatiorial Atlantic: suboxic diagenesis, Geochimica et Cosmochin. Acta, v. 43, p. 1075-1090.

Galimov, E.M., and Simoneit, B.R.T., 1982, Geochemistry of interstitial gases in sedimentary deposits of the Gulf of California, Deep Sea Drilling Project Leg 64: Initial Reports of the Deep Sea Drilling Project, v. 64, Wash. (U.S. Govt. Printing Office) p. 781-787.

Gieskes, J.M., Elderfield, H., Lawrence, J.R., Johnson, J., Meyers, B. and Campbell, A., 1981, Geochemistry of interstial waters and sediments, Leg 64, Gulf of California: Initial Reports of the Deep Sea Drilling Project, v. 64, Wash. (U.S. Govt. Printing Office) p. 675-694).

Gorsline, D.S., Drake, D.E., and Barnes, P.W., 1968, Holozene sedimentation in Tanner Basin, California continental borderland: Geological Society Am. Bull., v. 79, p. 659-674.

Graham, S.A., 1976, Tertiary sedimentary tectonics of the central Salinian block of California: Ph.D. Thesis, Stanford University, 510 p.

Hall, C.A., 1981, San Luis Obispo Transform fault and Miocene rotation of the Western Transverse Ranges, California: Journ. Geophysical Research, v. 86, no. B2, p. 1015-1031.

Hein, J.R., O'Neil, J.R. and Jones, M.G., 1979, Origin of authigenic carbonates in sediments from the deep Bering Sea: Sedimentology, v. 26, p. 681-705.

Ingle, J.C., 1980, Cenozoic paleobathymetry and depositional history of the selected sequence within the southern California continental borderland. Cushman Foundation spec. pub. 19, p. 163-195.

Irwin, H., Curtis, C. and Coleman, M., 1977, Isotopic evidence of source of diagenetic carbonaes formed during burial of organic-rich sediments: Nature, v. 269, p. 209-213.

Kablanow, R.I., and Surdam, R.C., 1983, Diagenesis and Hydrocarbon generation in the Monterey Formation, Huasna Basin, California, in Isaacs, C.M. and others, eds., Petroleum Generation and Occurrence in the Miocene Monterey Formation, California: Society Economic Paleontologists and Mineralogists, Pacific Section, Spec. Publ., p. 53-68.

Kastner, M. and Baker, P.A., 1983, A new insight into the origin of sedimentary dolomite: in Yearbook of Science and Technology, McGrew Hill, p. 406-408.

Kastner, M., Keene, J.R., and Gieskes, J.M., 1977, Diagenesis of siliceous oozes-I. Chemical controls on the rate of opal-A to opal-CT transformation-an experimental study: Geochim. et. Cosmochem. Acta, v. 41, p. 1041-1059.

Kelts, K., and McKenzie, J.A., 1982, Diagenetic dolomite formation in quaternary anoxic diatomaceous muds of Deep Sea Drilling Project Leg 64, Gulf of California: Initial Reports of the Deep Sea Driling Project, v. 64, Wash. (U.S. Govt. Pringing Office) p. 553-569.

Kitano, Y. and Kanamori, N., 1966, Sythesis of magnesian calcite at low temperatures and pressures: Geochim. J., v. 1, p. 1-10.

Kulm, L.D., Suess, E., and Thornburg, T.M., 1982, Mechanism of dolomitization of Peru convergent margin sediments: Depositional and tectonic history: EOS v. 63, no. 45, p. 1000.

Land, L.S., 1980, The isotopic and trace element geochemistry of dolomite: The state of the Art: in Concepts and Models of Dolomitization, eds. Zenger, D.H., Dunham, J.B. and Ethington, R.L., SEPM Special Publication No. 28, p. 87-110.

Marata, K.J., Friedman, I., and Madsen, B.M., 1969, Isotopic composition of diagenetic carbonates in marine Miocene formations of California and Oregon. U.S. Geol. Sur. Prof. Paper, 614-B.

McCrea, J.M., 1950, On the isotopic chemistry of carbonates and a paleotemperature scale: Jour. Chem. Physics, v. 18, p. 849-857.

Northrup, D.A., and Clayton, R.M., 1966, Oxygen-isotope fractionations in systems containing dolomite: Jour. Geology, v. 74, p. 174-196.

Nissenbaum, A., Presley, B.J., and Kaplan, I.R., 1972, Early diagenesis in a reducing fjord, Saanich Inlet, British Columbia-I. Chemical and isotopic changes in major components of interstitial water: Geochim. et Cosmochem. Acta, v. 36, pp. 1007-1027.

Pisciotto, K.A., 1978, Basinal sedimentary facies and diagenetic aspects of the Monetery Shale, California: Santa Cruz, California, University of California, Ph.D. Thesis, 450 p.

Pisciotto, K.A., and Mahoney, J.J., 1981, Isotopic survey of diagenetic carbonates, Deep Sea Driling Project Leg 63: Initial Reports of the Deep Sea Drilling Project, v. 63, Wash. (U.S. Govt. Printing Office) p. 595-609.

Pisciotto, K.A., and Garrison, R.E., 1981, Lithofacies and depositional environments of the Monterey Formation, California, in the Monterey and Related Siliceous Rocks of California: Society of Economic Paleontologists and Mineralogists, p. 97-122.

Presley, B.J., Culp, J., Petrowski, C. and Kaplan, I.R., 1973, Interstitial water studies. Leg 15 - Major ions Br, Mn, NH_3, Li, B, Sr, and C^{13}: Initial Reports of the Deep Sea Drilling Project, v. 20, Wash. (U.S. Govt. Printing Office) p. 805-810.

Presley, B.J., and Kaplan, I.R., 1968, Changes in dissolved sulfate, calcium and carbonate interstitial water of near-shore sediments: Geochim. et. Cosmochem. Acta, v. 32, pp. 1037-1048.

Reading, H.G., Sedimentary environments and facies: Elsevier, New York, 557 p.

Sayles, F.L., Manheim, F.T., and Waterman, L.S., 1973, Interstitial water studies on small core samples. Leg 15: Initial Reports of the Deep Sea Drilling Project, v. 20, Wash. (U.S. Govt.Printing Office) p. 783-804.

Suess, E., Kulm, L.D., and Killingly, J.S., 1982, Mechanism of dolomitization of Peru convergent margin sediments: Isotope and mineral record: EOS, v. 63, no. 45, p. 1000.

Supko, P.R., Stoffers, P. and Coplen, T.B., 1974, Petrography and geochemistry of Red Sea dolomite: Initial Reports of the Deep Sea Drilling Project, v. 23, Wash. (U.S. Govt. Printing Office) p. 871-878.

Surdam, R.C., and Stanley, K.O., 1981a, Diagenesis and migration of hydrocarbons in the Monterey Formation, Pismo Basin, California, in Garrison, R.E. and others, eds., the Monterey Formation and Related Siliceous Rocks of California: Society of Economic Paleontologists and Mineralogists, Pacific Section, Spec. Publ., p. 317-327.

Surdam, R.C., and Stanley, K.O., 1981b, Stratigraphic and sedimentologic framework of the Monterey Formation, Pismo Syncline, California, in Isaacs, C.M., ed., Guide to the Monterey Formation in the California Coastal Area, Ventura to San Luis Obispo: American Association Petroleum Geologist, Pacific Section, v. 52, p. 83-91.

Whitman, J.M. and Davis, T.A., 1979, Cenozoic ocean sedimentation rates: how good are the data?: Marine Geology, v. 30, p. 269-284.

Zandt, G., and Furlong, K.P., 1982, Evolution and thickness of the lithosphere beneath coastal California: Geology, v. 10, no. 7, p. 376-381.

SOLUTION POROSITY OF DOLOMITES IN THE CALCAREOUS - PHOSPHATIC FACIES OF THE MONTEREY FORMATION, PISMO BASIN, CALIFORNIA

R.C. Surdam
University of Wyoming
Dept. of Geology/Geophysics
P.O. Box 3006
Laramie, WY 82071

D. Prezbindowski
AMOCO Production Co.
Research Center
P.O. Box 591
Tulsa, OK 74102

R.I. Kablanow
University of Wyoming
Dept. of Geology/Geophysics
P.O. Box 3006
Laramie, WY 82071

INTRODUCTION

In the calcareous-phosphatic facies of the Miocene Monterey Formation in the Pismo Basin, there is significant development of enhanced porosity due to solution of dolomite (see Fig. 1 and 2). The calcareous-phosphatic facies has been described in detail by Surdam and Stanley (1981). This facies consists of interbedded foraminiferal and coccolith oozes that have been extensively dolomitized and phosphatic dolostone (see Fig. 3). The dissolution of dolomite is particularly prominent in the phosphatic dolostones. These rocks average 10 wt. % organic carbon, but can have up to 20 wt. % TOC and 15% P_2O_5.

SEDIMENTOLOGY

During deposition of the Monterey Formation in the Pismo Basin large amounts of siliciclastic material derived from the continent to the east were trapped in the Huasna Basin (Fig. 4; see also, Kablanow and Surdam, 1983). The facies relationships in the Monterey of the Huasna Basin described by Hall and Corbato (1967) are compatible with the nearshore basin model of Pisciotto (1978, Fig. 1-12). However, some of the siliciclastic material was transported to the west and deposited in the eastern and central portions of the subsiding Pismo Basin (Fig. 4). As a consequence a thick sequence of silty diatomaceous sediments accumulated in the central portion of the Pismo Basin. These lithofacies suggest distal basin deposition.

In contrast, the Monterey sequence west of the topographic high (volcanic ridge) is much thinner and is characterized by more diverse lithologies. The topographic high served as a barrier to continental debris, so that west of the high the basin was starved of siliciclastic material. In addition, judging from the extremely high organic content of the sediments west of the high (up to 20 wt. %) this portion of the basin was effectively silled. Almost all the lithologies west of the high exhibit abundant evidence that they were the products of resedimentation. Current directional features suggest that much of the calcareous, phosphatic, and diatomaceous material was originally deposited or formed in situ on the slopes of the topographic high and was later swept downslope into the lower portions of the basin. Thus, the western portion of the Pismo Basin fits the offshore basin facies model of Pisciotto (1978, Fig. 1-12). Following the

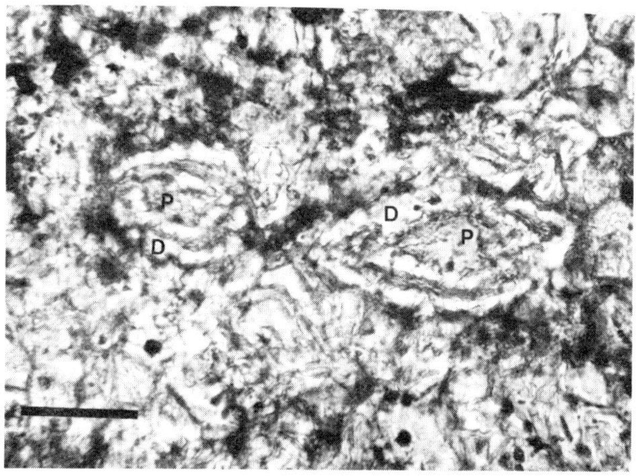

Figure 1 Photomicrograph showing dissolution textures in the dolomite rhombs from the calcareous - phosphatic facies. D = dolomite, P = intragranular solution porosity. 2.0 cm scale bar = 40 microns.

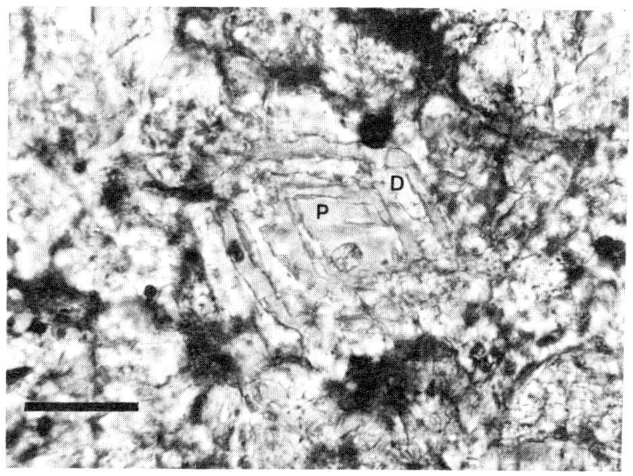

Figure 2 Photomicrograph of intragranular solution porosity (P) in authogenic dolomite (D). Note (in both figure 1 and 2) that solution occurs mainly along zones reflecting compositional differences. In some cases, dissolution cross-cuts these zones. 2.0 cm scale bar = 40 microns.

in Garrison, Robert E.; Kastner, Miriam and Zenger, Donald H., eds., 1984, **Dolomites of The Monterey Formation and Other Organic-Rich Units:** Pacific Section S.E.P.M., Vol. 41, p. 115-117.

Figure 3 Interbedded foraminiferal and coccolith dolostone and phosphatic dolostone.

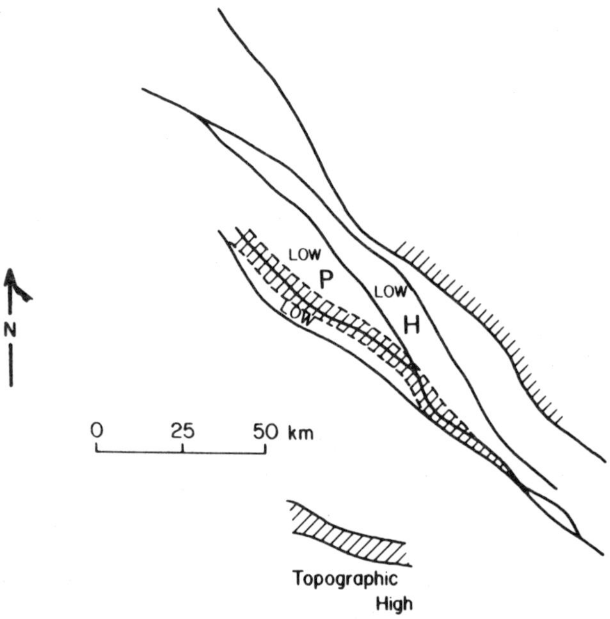

Figure 4 Schematic palinspastic map of the Pismo area during mid-Miocene, showing the Huasna Basin to the northeast (low, H) and the two parts of the Pismo Basin (low, P) separated by the volcanic topographic high. Modified from Hall, 1981.

suggestions of Pisciotto (1978) and Garrison, Stanley, and Horan (1979), it is speculated that the phosphatic and dolomitic material originally formed at the intersection of the upper boundary of the oxygen minimum with the slope of the topographic high, whereas the foraminiferal and diatomaceous material probably represents hemipelagic sedimentation. For more details see Surdam and Stanley (1981).

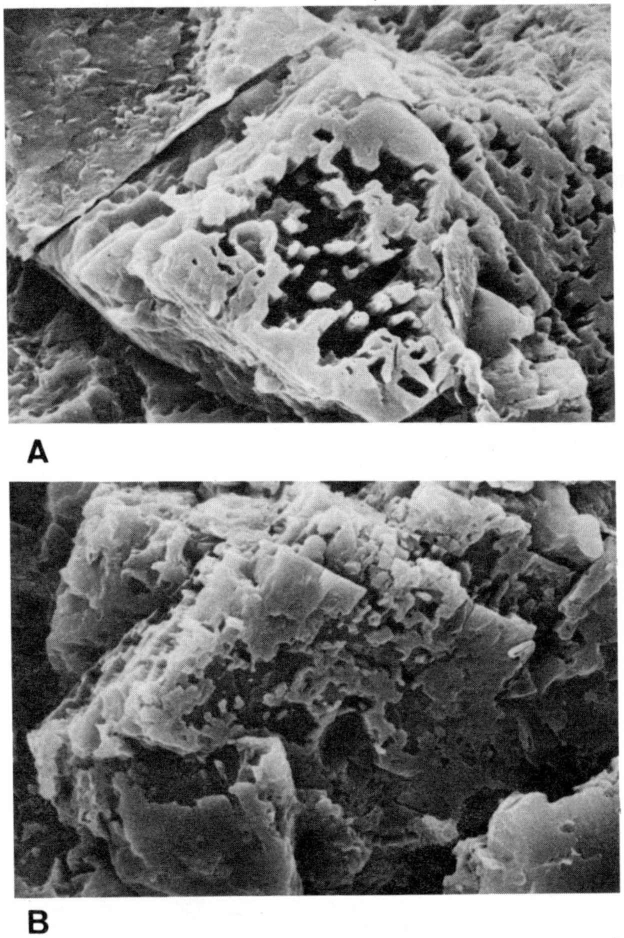

Figure 5 SEM photomicrographs showing solution textures in dolomite rhombs. Note in both A and B that dissolution has selectively etched the central portions of the grains. A is 3000x, B is 2600x.

DIAGENESIS

Kablanow, Surdam and Prezbindowski (this volume) have shown that isotopic evidence strongly supports the hypothesis that dolomitization of the calcareous-phosphatic rocks started near the sediment-water interface and continued during early burial. They suggest that much of the dolomitization in this facies formed over an average temperature range of 26 to 55°C (see Kablanow, Surdam and Prezbindowski, this volume, Table 1). During the generation of carboxylic acids (temperature >80°C) these early formed dolomite grains partially dissolved (see Figures 5, and 6). The generation of carboxylic acids is a natural consequence of kerogen maturation and takes place over the temperature range of 80-120°C (Surdam, Boese, and Crossey, 1984). Carbonates are highly soluble in these short chained organic acids, providing the P_{CO_2} is not elevated (Surdam, Boese, and Crossey, 1984). Thus, the dissolution of the dolomites probably took place over a relatively small temperature range between approximately 80°C and the beginning of thermal decarboxylation (elevated P_{CO_2}) and/or hydrocarbon generation.

Figure 6 SEM photomicrograph, closeup of Fig. 5A. Note the selective dissolution. 8600x.

CONCLUSIONS

Dissolution of dolomite in the Monterey Formation has at least two very significant consequences. First, as can be seen in the figures, dissolution is compositionally selective. Thus, in dolomites that have undergone some dissolution there will be a compositional bias introduced into any compositional determination, such as isotopic geochemistry. More importantly, the calcareous rocks of the Monterey Formation may contain significant amounts of enhanced intergranular and intragranular porosity. This so-called secondary porosity coupled with the fracture porosity commonly present in the calcareous rocks may be responsible for the outstanding reservoir characteristics of some calcareous rocks in the Monterey Formation.

ACKNOWLEDGEMENTS

We are grateful to Phillips Petroleum Company for support and the NSF (Grant EAR-8207570) for support.

REFERENCES

Garrison, R.E., Stanley, R.G., and Horan, L.J., 1979, Middle Miocene sedimentation on the southwestern edge of the Lockwood High, Monterey County, California: in Graham, S.A. (ed.), Tertiary and Quaternary geology of the Salinas Valley and Santa Lucia Range, Monterey County, California, Pacific Coast Paleogeography Field Guide 4, Pacific Section, Society Economic Paleontologists and Mineralogists, p. 51-65.

Hall, C.A., Jr., 1981, San Luis Obisbo transform fault and Miocene rotation of the Western Transverse Ranges, California: Journ. Geophysical Research, v. 86, No. B2, p. 1015-1031.

Hall, C.A. and Corbato, C.E., 1967, Stratigraphy and structure of Mesozoic and Cenozoic rocks, Nipomo quadrangle, Southern Coast Ranges, California: Geological Society of Am. Bull., v. 7, p. 559-582.

Kablanow, R.I. and Surdam, R.C., 1983, Diagenesis and hydrocarbon generation in the Monterey Formation, Huasna Basin, California, <u>in</u> Isaacs, C.M. and others, eds., Petroleum Generation and Occurrence in the Monterey Formation, California: Society Economic Paleontologists and Mineralogist, Pacific Section, Special Publ., p. 53-68.

Pisciotto, K.A., 1978, Basinal sedimentary facies and diagenetic aspects of the Monterey shale, California: Santa Cruz, California, University of California, Ph.D. thesis, 450 pp.

Surdam, R.C., and Stanley, K.O., 1981, Stratigraphic and sedimentologic frame work of the Monterey Formation, Pismo Syncline, California, <u>in</u> Isaacs, C.M., ed., Guide to the Monterey Formation in the California Coastal Area, Ventura to San Luis Obispo: American Association Petroleum Geologist, Pacific Section, v. 52, p. 83-91.

Surdam, R.C., Boese, S., and Crossey, L.J., 1984, The chemistry of secondary porosity: <u>in</u> McDonald, D. and Surdam, R.C., eds., Clastic Diagenesis: the Key to Porosity and Permiability Development: in press.

RELATIONSHIP OF DIAGENETIC DOLOMITE TO DEPOSITIONAL FACIES IN DIATOMACEOUS STRATA, UPPER SISQUOC FM., CASMALIA HILLS, SANTA BARBARA CO., CALIFORNIA.

Nuel C. Henderson, Jr., G. Allan Crawford, John B. Dunham
Union Oil Company of California
Brea, California 92621

Robert H. Osborne
Dept. Geol. Sciences
Univ. Southern California
Los Angeles, California 90089

ABSTRACT

Study of nine cores from the Casmalia Hills, Santa Barbara County, CA, indicates that dolomite occurrence is related to diagenesis of fine-grained diatomaceous sediment and is, in part, facies controlled. Strata observed in the cores represent borderland basin and slope sedimentation.

The following upward succession of lithostratigraphic units occurs: (A) alternating laminated and nonlaminated opal-CT bearing claystone; (B) diatomaceous and opal-CT bearing mudstone (indistinctly laminated to thin bedded) with minor thin (<15 cm thick) dolomite nodules; (C) thick pebbly mudstone or conglomerate with minor diatomaceous mudstone; (D) diatomaceous mudstone with thin pebbly mudstone or conglomerate; (E) diatomaceous mudstone, mudstone with resedimented chert clasts, and chert-rich sandstone; (F) diatomaceous mudstone, chert-rich sandstone, and mudstone with a molluscan fauna. Thick pebbly mudstone or conglomerate, which occurs in two of the cores, appears to be, in part, laterally equivalent to thin pebbly mudstone (unit #D) in other cores and may represent channelized debris flow deposits. This sequence of rock types is interpreted to represent an upward succession from restricted basin or slope (anoxic to dysaerobic) to unrestricted basin or slope (dysaerobic), and finally to an aerobic slope environment. Units A through D were apparently deposited within an oxygen minimum zone; however, the rocks show evidence of a gradual increase in oxygen within the environment through time as indicated by increased bioturbation.

Dolomite has three types of occurrences in the cores from the Sisquoc Formation:

1. Nodular ferroan dolomite with relatively heavy oxygen-isotopic values (-1.7 to +5.1‰, PDB) that are heaviest in the center of a nodule and become progressively lighter toward the margins. This relationship suggests an accretionary growth of the dolomite at burial depths ranging from 0 to less than 1 km. Relatively heavy carbon isotopic values (+15 to +32.5‰, PDB), suggest that the carbon in the dolomite formed as a result of bacterial fractionation of organic carbon during methanogenesis. Variation in $\delta^{13}C$ is minor within a single dolomite nodule (<3‰, PDB) but differs significantly among nodules (range +15 to +32‰, PDB). This variation among nodules suggests isolation of carbon-isotope reservoirs.

2. Clasts of ferroan dolomite that closely resemble the ferroan dolomite nodules in both color and texture, and have a similiar isotopic composition. These clasts were probably derived from the ferroan dolomite nodules.

3. Disseminate, nonferroan dolomite is characterized by oxygen-isotopic values ranging from -1.2 to -8.4‰, PDB, which suggest that this dolomite formed at burial depths greater than the nodular ferroan dolomite. Carbon-isotopic values ranging from -4.5 to +12.5‰, PDB, may indicate CO_2 formation from degradation of organic matter by both abiotic and bacterial processes, including bacterial methanogenesis, or may be due to the effects of higher temperature of dolomite formation at greater burial depths.

The nodular ferroan dolomites are restricted to the opal-CT bearing and the diatomaceous mudstones (unrestricted basin and lower slope facies). The dolomite clasts occur in pebbly mudstone or conglomerate or in the associated disturbed mudstone (lower slope). The nonferroan disseminated dolomite occurs in all lithologies (except sandstone) but is most abundant in the opal-CT bearing claystone (restricted basin facies).

INTRODUCTION

The discovery of dolomite within deep-water organic rich sediment (Kelts and McKenzie, 1982; Pisciotto and Mahoney, 1981) has focused attention on deep-water environments as sites of early-diagenetic dolomite formation. This paper describes the occurrence of dolomite in several distinctive diatomaceous lithofacies present in nine cores from the late Miocene to middle Pliocene (late Mohnian to early Delmontian, G. Blake, 1983, written com.) Sisquoc Formation in the Casmalia Hills, Santa Barbara County, California (Figs. 1 & 2). The cores, taken by Union Oil Co. of Calif. from the upper Sisquoc Fm, are on the average 150 m (500 ft) long and are spaced on a 150 m (500 ft) grid, allow a detailed lithofacies analysis that is generally not possible from outcrop studies due to the weathering characteristics of diatomaceous rocks, and provide close lateral control of the lithofacies defined.

Following a brief review of the geologic and tectonic history of the Casmalia Hills area, this paper first discusses rock types and lithofacies noted in the study cores and their environmental interpretations. Secondly, the types of dolomite encountered and their inferred origins is reported. Finally the association between dolomite types and lithofacies is discussed.

in Garrison, Robert E.; Kastner, Miriam and Zenger, Donald H., eds., 1984, **Dolomites of The Monterey Formation and Other Organic-Rich Units:** Pacific Section S.E.P.M., Vol. 41, p. 119-140.

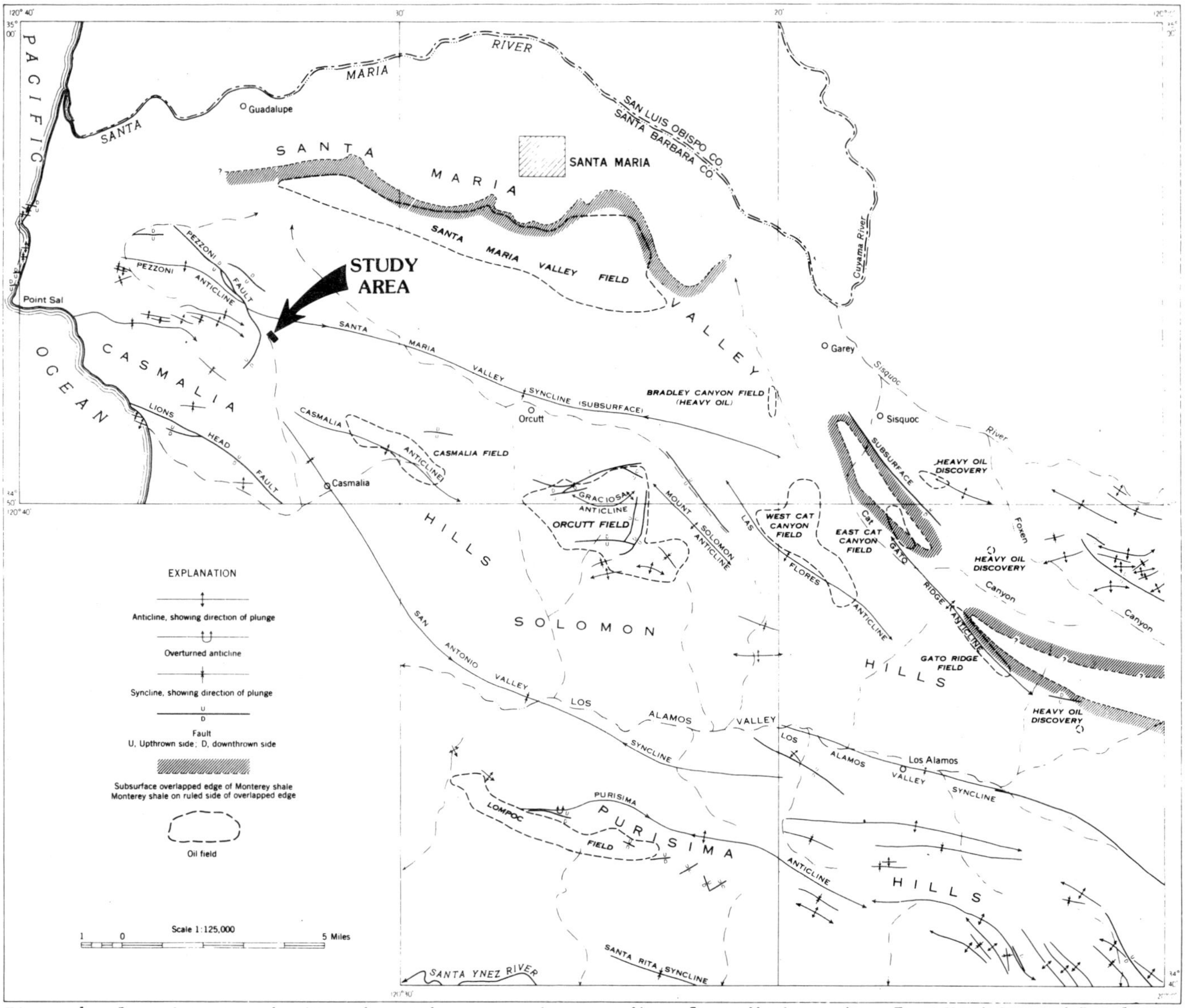

Figure 1. Location map showing the study area and surrounding Santa Maria region. From Woodring and Bramlette (1950).

GEOLOGIC AND TECTONIC HISTORY

During the late Miocene, anticlinal ridges began to form in the basin of the Santa Barbara Embayment, and continued to develop through the Pliocene resulting in the formation of smaller basins with intervening highs (Woodring and Bramlette, 1950; Porter, 1932). The Sisquoc Formation of the Casmalia Hills represents an accumulation of fine-grained diatomaceous sediment within one of these smaller basins (Woodring and Bramlette, 1950).

The Sisquoc Fm conformably overlies the Monterey Fm (Woodring and Bramlette, 1950) in the Casmalia Hills, and is similiar to that unit as it represents a continuation of diatom-rich hemipelagic basinal sedimentation. The Sisquoc Fm differs from the Monterey Fm, in that it contains more terrigenous silt and clay, and is not as well laminated or stratified as the Monterey (Isaacs, 1981b).

By mid-Pleistocene, the Sisquoc Fm in the Casmalia Hills area had become oil-saturated due to leakage from the underlying Monterey Fm, and had been uplifted to the surface (Farley and Wilson, 1983). Middle-to-late Pleistocene marine terrace deposits, devoid of oil, truncate the oil-saturated Sisquoc rocks (Farley and Wilson, 1983; Woodring and Bramlette, 1950). A further discussion of the geologic setting and stratigraphy of the Sisquoc Fm in the Casmalia Hills is given in Henderson (1983).

METHODS

Data on lithofacies and dolomite occurrences were obtained from detailed core descriptions and thin-section petrography. The cored intervals were correlated along lines of cross-section (Fig. 3, 4) drawn from a geologic map of the area prepared by D.A. Oriola and B.R. Baca (Union Oil of Calif, unpublished report). A down-plunge profile section of the formation (Fig. 5) was used for lateral and vertical comparison of the lithostratigraphic units encountered in the cored intervals.

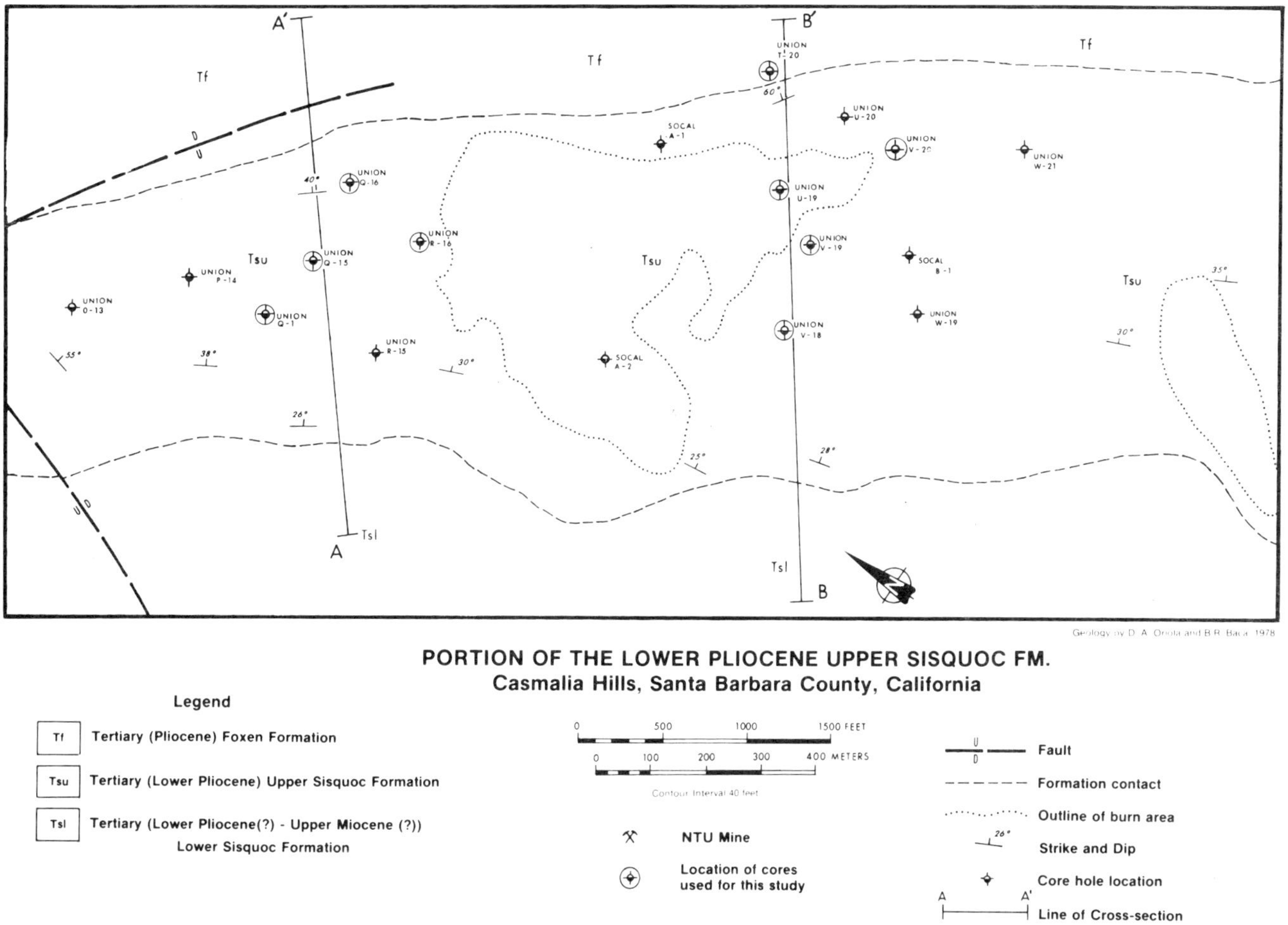

Figure 2. Simplified geologic map of the study area showing corehole localities and lines of cross-section.

Core descriptions were compiled with the aid of a binocular microscope. The fine-grained nature of the rocks presented some difficulty in estimating percentages of constituents. Petrographic data are biased toward dolomite nodules, in part because of interest in the dolomite, but also because of the difficulty in preparing thin-sections from soft, fine-grained rocks.

ROCK TYPES

Rocks of the upper Sisquoc Fm have four main components: siliciclastic silt and sand, clay, diatoms, and dispersed dolomite. Significant amounts of organic matter, pyrite, and locally, phosphate, glauconite, and opal-CT also are present. This mix of material, most finer than 0.062 mm and isotropic or weakly anisotropic, makes classification difficult. The following lithologic classification, modified from Potter and others (1980), is used: Claystone - clay-rich rock containing less than 25% siliciclastic sand and silt; Mudstone - clay-rich rock containing more than 25% siliciclastic silt and sand; Sandstone - rock containing more than 50% siliciclastic silt and sand. Dolomite - is the term used for rocks containing in excess of 50% the mineral dolomite. The modifier 'diatomaceous' is used for rocks containing from 10 to 50% diatoms. Table 1 summarizes the classification scheme used in this paper.

TABLE 1.- Claystone and mudstone classification used for this study.

Percent sand and silt size constituents	0 - 25%	25 - 50%	>50%
Clay-rich fine-grained rock	CLAYSTONE	MUDSTONE	CLAYEY SANDSTONE OR SILTSTONE
Diatom content between 10 and 50%	DIATOMACEOUS CLAYSTONE	DIATOMACEOUS MUDSTONE	DIATOMACEOUS SANDSTONE OR SILTSTONE
Containing opal-CT, but no free silica	OPAL-CT BEARING CLAYSTONE	OPAL-CT BEARING MUDSTONE	OPAL-CT BEARING SANDSTONE OR SILTSTONE

Nine rock types are distinguished :

1) Claystone and diatomaceous claystone. Pyrite is a common constituent. These rocks are locally dolomitic, containing as much as 40% disseminated dolomite.

2) Mudstone and diatomaceous mudstone. Locally, these rocks are dolomitic, containing as much as 15% disseminated dolomite. Pyrite is a common component. Some mudstones contain randomly dispersed, medium to very coarse sand-size angular chert grains.

3) Opal-CT bearing claystone. Diatoms are absent, or are present only as partially dissolved robust fragments. Their former presence is indicated

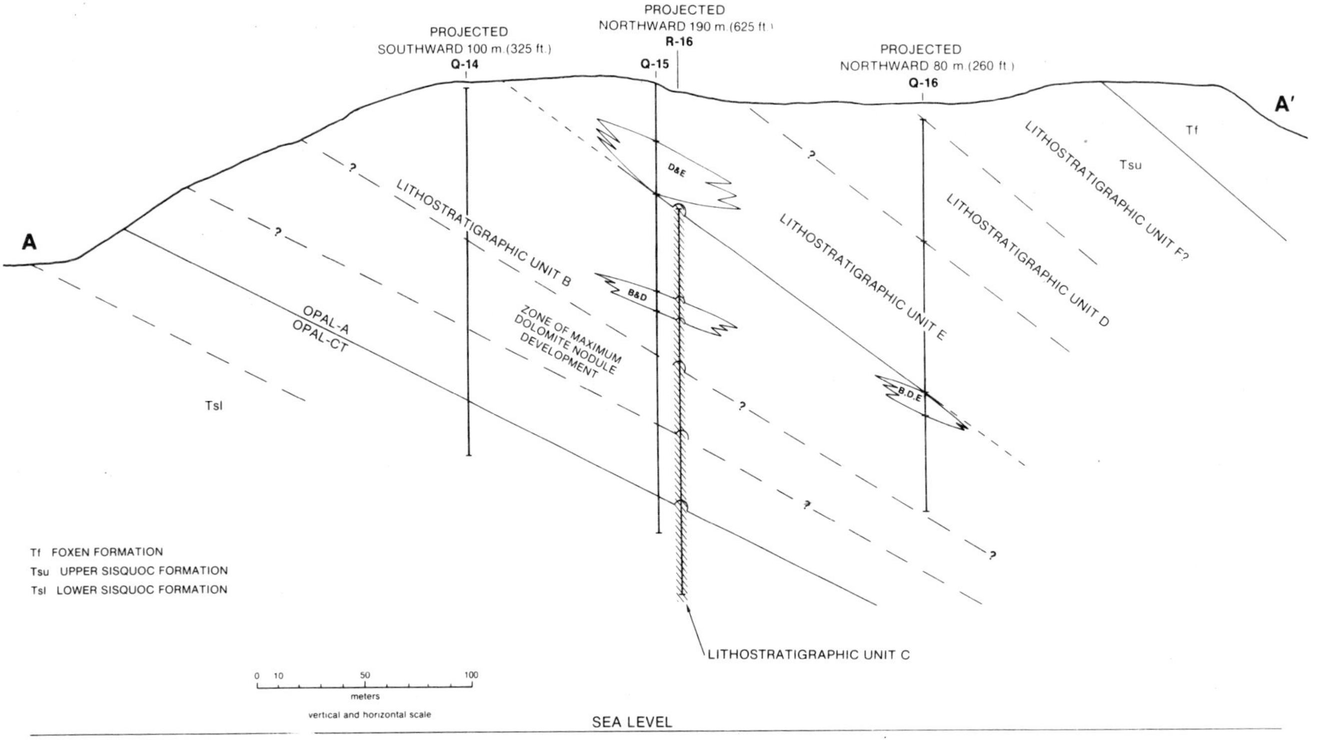

Figure 3. Cross-section A-A′. Position of cross-section shown in Figure 2. Letter combinations, such as 'D and E' indicate a zone containing rock characteristic of the lithostratigraphic units D and E, etc.

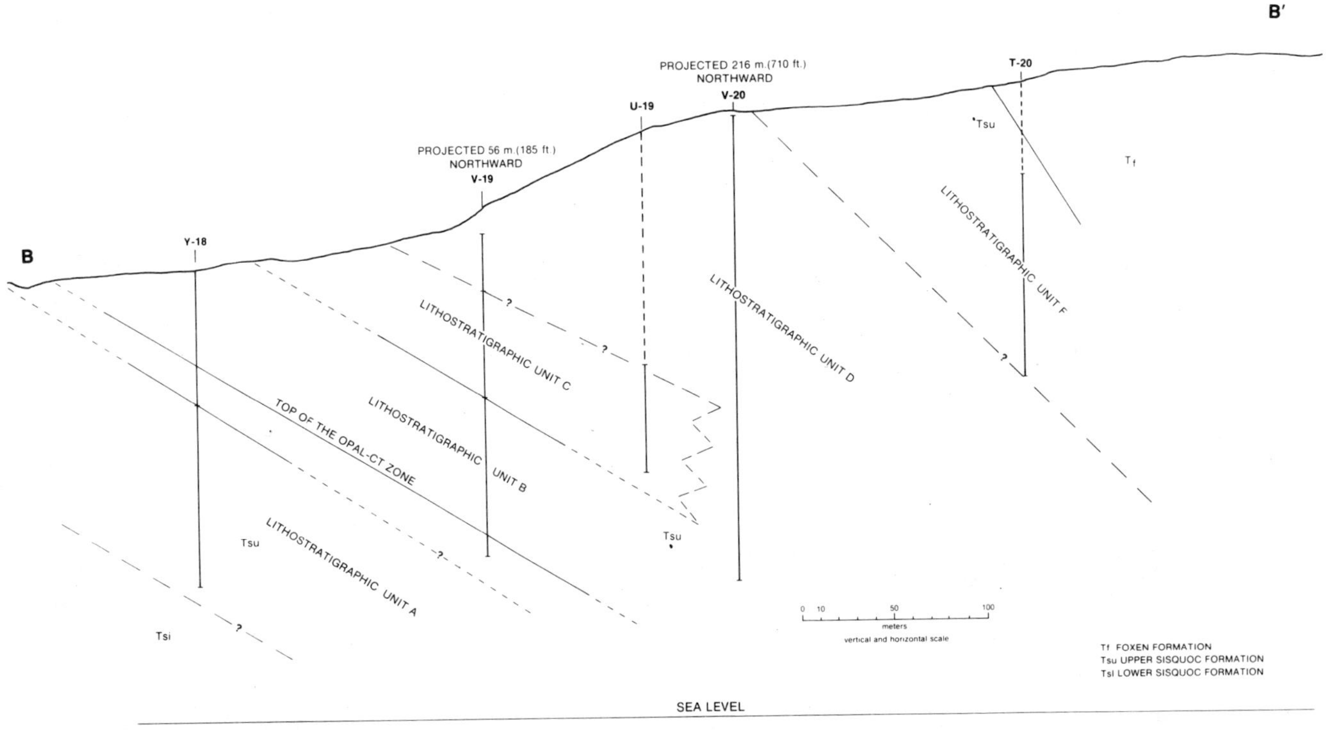

Figure 4. Cross-section B-B′. Position of cross-section shown in Figure 2.

by distinct molds, pyrite fillings or replacements of frustules, and the opal-CT is detected by X-ray diffraction (Isaacs (1980) notes that silica does not exhibit significant mobility during transformation of phases). These claystones commonly contain as much as 30% disseminated dolomite. Pyrite, and locally glauconite, are commonly present.

4) Opal-CT bearing mudstone. Diatoms are absent, or present as poorly preserved robust fragments. The former presence of diatoms is indicated by distinct molds, pyrite filling or replacement of frustules, and opal-CT detected by X-ray diffraction. Pyrite is a common constituent and the rocks are

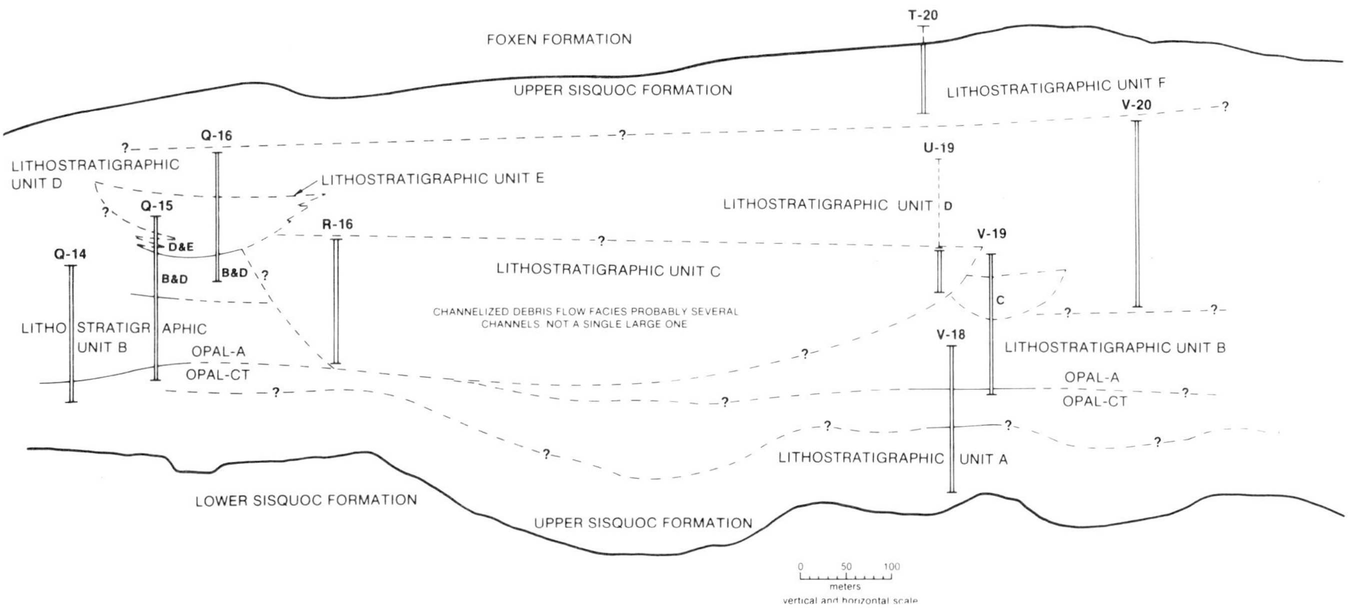

Figure 5. Computer-generated composite cross-section (down-plunge section) through the upper Sisquoc Fm, Casmalia Hills. The cross-section was constructed perpendicular to dip and takes into account topographic relief.

locally dolomitic. Dolomitic varieties commonly contain less than 5% disseminated dolomite.

5) Pebbly mudstone. Matrix-supported pebbly rocks with most clasts composed of mudstone or diatomaceous mudstone. A minor portion, commonly less than 5%, of the clasts are composed of dolomite. Clasts are usually rounded to well rounded, and range in size from 0.3 to 2.0 cm. Clasts up to 70 cm in size are less common. Matrix locally contains as much as 25% sand.

6) Pebbly conglomerate. Clast-supported pebbly rocks which are similiar in all other respects to pebbly mudstone. Locally, rare thin beds of conglomerate occur that are composed of dolomite and/or chert clasts.

7) Sandstone. Sandstone commonly occurs as thin beds of poorly-sorted, medium sand composed predominantly of quartz and feldspar with pyrite as a minor component and dolomite in trace quantities. Less commonly the sand is composed of chert grains. Interstices contain silt and clay.

8) Dolomite. Dolomite as a rock type occurs as nodules from centimeters to tens of centimeters in thickness. In core samples, the dolomite commonly appears to be bedded. However, the dolomite 'beds' cannot be correlated between even closely spaced coreholes. Variable thicknesses of dolomite in core, difficultly of correlation of individual 'beds', and the nature of Sisquoc dolomite occurences in coastal outcrops (Fig. 12e), suggests that the dolomite occurs as nodules. Most are tightly cemented, light gray to white, and contain abundant diatoms, common pyrite and fecal(?) pellets, rare siliciclastic silt and sand grains, and locally foraminifera and mollusc shell fragments. Rarely, volcanic glass shards are present.

9) Vitric tuff. Vitric tuff is a rare rock type composed predominantly of well-preserved volcanic glass shards with minor mudstone clasts and diatomite. Pyrite is a common constituent.

Diatomaceous rocks that have undergone the opal-A to opal-CT transition are referred to here as 'opal-CT bearing' rather than by the field term porcelanite. Although these rocks are more brittle and less friable than the rocks still in the opal-A state, they lack the matte luster associated with porcelanite.

LITHOSTRATIGRAPHY OF THE UPPER SISQUOC FM IN THE CASMALIA HILLS

Within the study area, the nine basic rock types defined above combine to form six distinctive lithostratigraphic units in the Sisquoc Fm. Identification of these lithostratigraphic units is based on unique assemblages of rock type, sedimentary structures and textures, and fossil content.

Roughly in upward succession, the six lithostratigraphic units are:

A) Opal-CT bearing claystone,
B) Opal-CT bearing or diatomaceous mudstone and minor dolomite,
C) Thick pebbly mudstone and minor diatomaceous mudstone,
D) Diatomaceous mudstone and mudstone-clast pebbly mudstone or conglomerate,
E) Diatomaceous mudstone, chert-clast mudstone and chert sandstone,
F) Diatomaceous mudstone, chert-rich sandstone and chert-sand bearing mudstone.

Lithostratigraphic Unit A

Unit A consists predominantly of opal-CT bearing claystone with minor interbedded opal-CT bearing mudstone and sandstone. The claystone varies from irregular medium laminated and very-thin bedded, to mottled or massive. Laminated to very-thin bedded intervals are approximately 20 m thick; rocks become less-well laminated upward with each succeeding laminated zone. Burrows are abundant in the nonlaminated

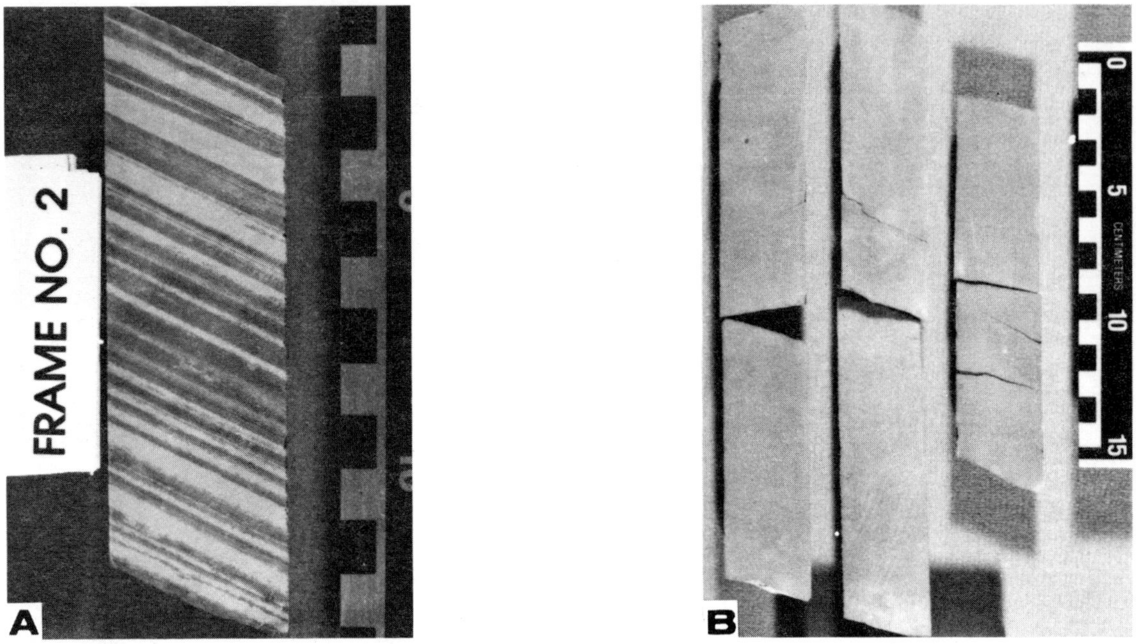

Figure 6. Prominant rock types characteristic of lithostratigraphic unit A. A. Core photo shows thin to thick laminated diatomaceous mudstone. This rock type comprises greater than 50% of unit A. The distinct laminae represent deposition under the most intensely dysaerobic conditions during deposition of the upper Sisquoc Fm. Scale in centimeters. Core V-18, 141.02m. B. Massive, bioturbated claystone, core V-18, 110.5 to 110.8m. This rock type makes up approximately 40% of unit A. Top of interval is in the upper left corner. Scale in centimeters.

zones. Silt- and sand-size quartz and feldspar increase from an estimated 5% at the base to 20% near the top of the unit. Muscovite, glauconite, and volcanic glass shards are locally present. Phosphatized shell fragments are common in parts of the facies. Dolomite most commonly occurs in unit A as disseminated, anhedral to euhedral nonferroan crystals. Figures 6A and B show the prominant rock types in unit A.

Lithostratigraphic Unit B

Unit B

Unit B consists predominantly of diatomaceous mudstone (opal-CT bearing mudstone below the opal-A to opal-CT boundary), with minor, but prominent dolomite nodules, and diatomaceous, silty-sandy claystone. The rocks of this facies are indistinctly layered, and highly burrowed. Locally, bioturbation is so intense that the rock has a massive appearance (Fig. 7).

Silt and sand size quartz and feldspar with minor (<1%) glauconite constitute 3 to 30% of the rock. The more clayey rock also contains muscovite(?) flakes. Volcanic glass shards are locally abundant.

Dolomite occurs in unit B most prominently as discrete, well-cemented nodules of ferroan dolomite, and less commonly as disseminated anhedral to euhedral nonferroan crystals, and rarely as disseminated ferroan rhombs.

Lithostratigraphic Unit C

Unit C consists of very thick, mudstone-clast pebbly mudstone and conglomerate beds (with thicknesses of well over 10 m) with minor interbedded diatomaceous mudstone. The thick mudstone-clast pebbly units are composed largely of diatomaceous mudstone clasts and may have concentrations of dolomite or mud clasts with similiar color and texture, in beds ranging from tenths of a meter to approximately a meter thick. Large mudstone clasts (over 30 cm in diameter), commonly contain whole articulated pelecypods. Figure 8 shows examples of rocks and bedding typical of unit C.

The matrix of the pebbly units consists largely of mud. Sand-size quartz and feldspar occur in variable quantities in the matrix ranging from near 0 to 25%. Foraminifera are locally abundant.

Dolomite occurs in unit C as dolomite clasts, and as disseminated crystals in partially dolomitized mudstone.

Lithostratigraphic Unit D

Unit D is characterized by the occurrence of thin mudstone-clast pebbly mudstone or conglomerate units and thicker diatomaceous mudstone beds, often locally disturbed (Fig. 9). Clasts in unit D are diatomaceous mudstone and dolomite, the former being the dominant clast type. Mudstone clasts are generally subrounded to rounded and commonly have millimeter thick clay rims suggestive of accretionary overgrowth acquired by movement at the sea floor. The mudstone clasts range in size from approximately 0.2 to 30.0 cm in diameter; approximately 95% are are finer than 2 to 3 cm. The larger clast sizes occur in the thicker pebbly units.

The matrix of the pebbly units is predominantly mud. Sand content, which ranges from a trace to approximately 20%, consists of very fine to medium-grained quartz and feldspar. Foraminifera tests are locally present in the matrix of the pebbly units.

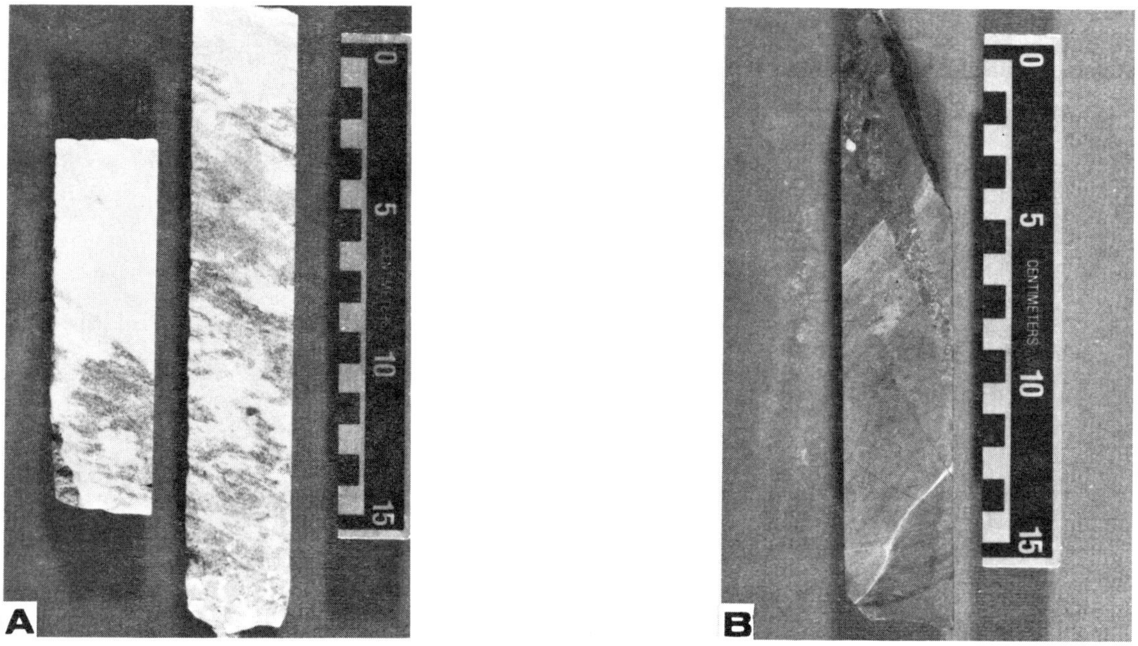

Figure 7. Rock types characteristic of lithostratigraphic unit B. A. Core photo shows indistinctly layered to massive, bioturbated opal-CT bearing mudstone. Core V-18, 76.24 to 76.56m. B. Sedimentary dike composed of mudstone-clast pebbly conglomerate, which cross-cuts indistinctly bedded diatomaceous mudstone. Indistinctly bedded diatomaceous mudstone like that shown in this photo is the dominant lithology in unit B. Core Q-14, 20.56m.

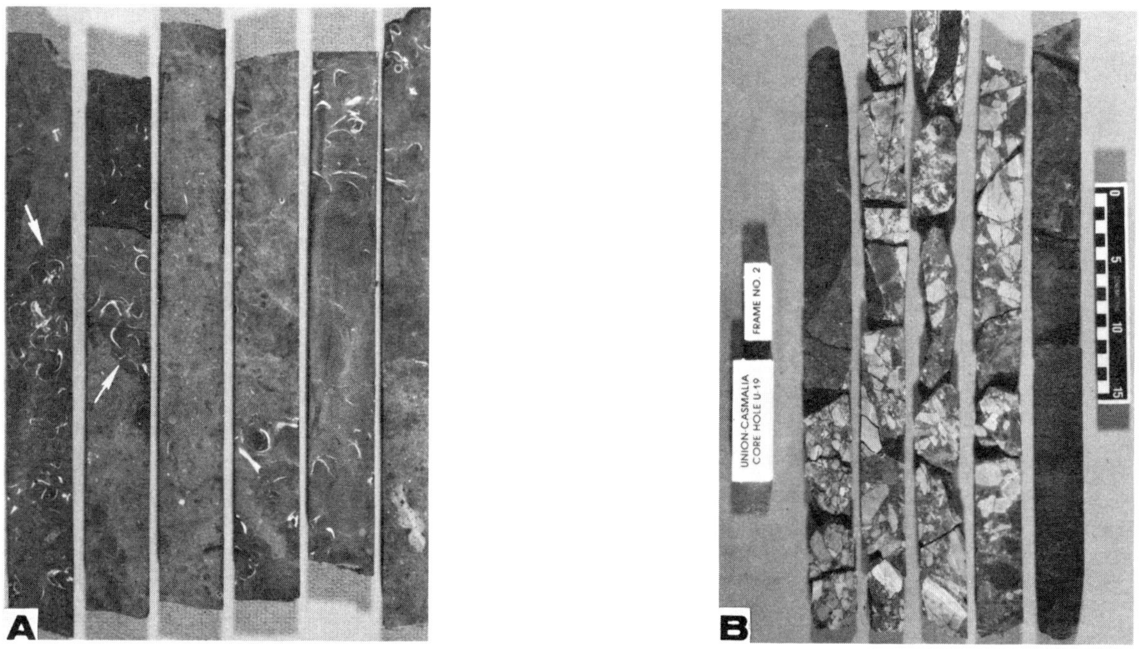

Figure 8. Rock types of lithostratigraphic unit C. A. Pebbly mudstone; the mudstone clasts contain articulated and disarticulated pelecypods. Finer mudstone clasts are subrounded to rounded. Core U-19, 107.72 to 109.12m. Top of interval is in upper left corner. Arrows indicate top and bottom of one large mudstone clast that bears pelecypod shells. Core is approximately 3.5 cm wide. B. Coarse breccia; angular to subangular clasts consist largely of diatomaceous mudstone. Core U-19, 127.86 to 129.73m. Top of interval is in upper left corner.

The pebbly units rarely display strong internal organization such as clast alignment or imbrication. The largest mudstone clasts commonly occur near the middle of the pebbly unit. Some thin pebbly units cut across bedding in the mudstone and may represent sedimentary dikes.

Ribbon-like distributions of phosphatic grains occur in the pebbly units and associated diatomaceous mudstone interbeds. The phosphate appears to have replaced or cemented mudstone clasts. Some clasts have siliciclastic grains as a nucleus and concentric-banded phosphate at the margin. In diatom-

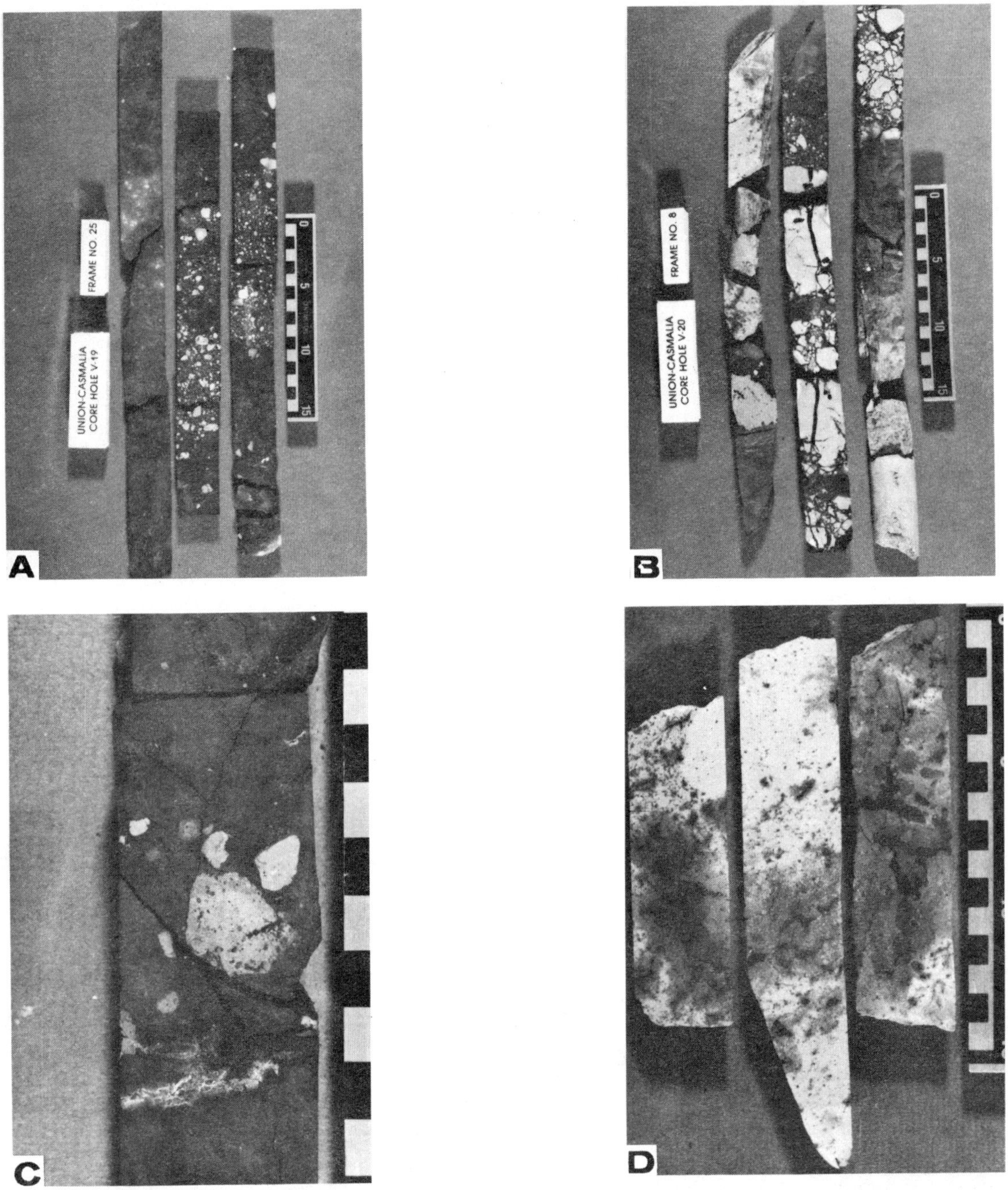

Figure 9. Rock types of lithostratigraphic unit D. A. Pebbly mudstone, containing a high proportion of dolomite clasts, overlain and underlain by disturbed diatomaceous mudstone. Core V-19, 5.7 to 6.4m. Top of interval at upper left. Scale in centimeters. B. Dolomite breccia; similarity of color and texture of the angular to subrounded dolomite clasts indicate brecciation of a single dolomite bed with only minor movement of the clasts. Core V-20, 146.04 to 147.15m. Top of interval at upper left corner. Scale in centimeters. C. Dolomite clasts at base of pebbly mudstone. Clasts have varied color and texture, suggesting an origin from separate dolomite nodules. Core V-20, 119.65m. Scale at right in centimeters. D. Disturbed diatomaceous mudstone with mudstone filling disruptions in the rock. The diatomaceous mudstone is overlain by mudstone-clast pebbly conglomerate (not shown). Commonly, the disrupted areas in such disturbed mudstone is filled with fine pebbly mudstone. Core V-20, 180.20m, 179.21m, and 176.73m core depths from left to right.

aceous mudstone the phosphatic clasts most commonly occur in small burrow-like, semi-vertical concentrations, close to the contact with adjacent pebbly units.

The diatomaceous mudstone interbedded with the pebbly units is usually mottled to indistinctly stratified and typically has disturbed texture. The disturbed textures are the result of soft sediment deformation, particularly injection.

Figure 10. Chert-clast-rich rocks of lithostratigraphic unit E. A. Core photo showing distribution of chert grains in chert-bearing mudstone. The grains are dispersed in diatomaceous mudstone and concentrated in burrows. Core Q-16, 75.56 to 75.81m. Top of interval in upper left. Scale in centimeters. B. Sandstone composed of poorly sorted, angular chert fragments. The sandstone is massive, matrix poor, tar impregnated, and is in sharp contact (not shown) with adjacent lithologies. Core Q-16., 68.77m.

Pyrite is widely distributed through the rocks of unit D. Dolomite occurs as detrital clasts in thin concentrations in pebbly units (Fig. 9). Disseminated dolomite crystals occur in mudstone clasts and matrix.

Lithostratigraphic Unit E

Unit E is characterized by abundant diatomaceous mudstone, with subordinate chert-sand bearing mudstone, chert-pebble mudstone, chert-rich sandstone, and rare chert-clast conglomerate. Diatomaceous mudstone in the unit commonly contains thin zones of Chondrites. Locally the diatomaceous mudstone has a disrupted texture, due to injection of mud and/or fine-pebbly mudstone. The disrupted zones commonly occur near or adjacent to conglomerate or pebbly mudstone. Chert clasts in the diatomaceous mudstone typically display an irregular distribution, appearing to preferentially fill large burrow traces (Fig. 10a).

Chert-clast pebble mudstone units contain angular to subrounded chert clasts less than 1 cm in size supported in a mud matrix. These rocks contain large pelecypods and gastropods that appear to be indigeneous.

Chert-rich sandstone is associated with chert-clast pebble mudstone units occurring within the pebbly mudstone. The chert fragments are angular to subrounded. Bedding is massive, showing no prominent sedimentary structures (Fig. 10b).

Dolomite is present in the unit as rounded detrital ferroan dolomite clasts and as disseminated nonferroan crystals in the matrix. The crystals occur throughout unit E, but are most common in the diatomaceous, chert-sandy mudstone, where they locally comprises nearly 50% of the rock.

Lithostratigraphic Unit F

Unit F consists of diatomaceous, chert-bearing sandy mudstone, chert-rich sandstone and minorchert-sandy mudstone (Fig 11a). Diatomaceous mudstone and claystone also occur (Fig. 11b). The strata are bioturbated and large, thin-walled pelecypods (Macoma aff M. brota?) appear to be indigenous. Foraminifera are common.

Bioturbation is prevelant in unit F, but some thin beds with clay-rich laminae do occur. Sandstone beds typically have an erosional contact with underlying mudstone units, and frequently contain mudstone rip-up clasts near the base of the bed. The sandstone units grade upward to diatomaceous chert-sandy mudstone.

Dolomite is locally present in the disseminated, nonferroan form. Rarely, large monocrystalline dolomite grains with abraided margins occur, which may be a detrital form that was not derived from the Sisquoc Fm.

INTERPRETATIONS OF ENVIRONMENTS AND SEDIMENTARY PROCESSES

The sedimentary sequence in the Casmalia Hills cores shows a vertical change from laminated claystone, to bioturbated, indistinctly bedded mudstone, chaotic pebbly mudstone and conglomerate with diatomaceous mudstone intervals, and chert-sand-bearing mudstones with a molluscan fauna. This overall sequence suggests an upward change from a nearly anoxic basin or outer slope environment to an aerobic upper slope or outer shelf environment.

Lithostratigraphic unit A is interpreted to have been deposited slowly in a deep, isolated basin, receiving little siliciclastic sand and silt; hemi-

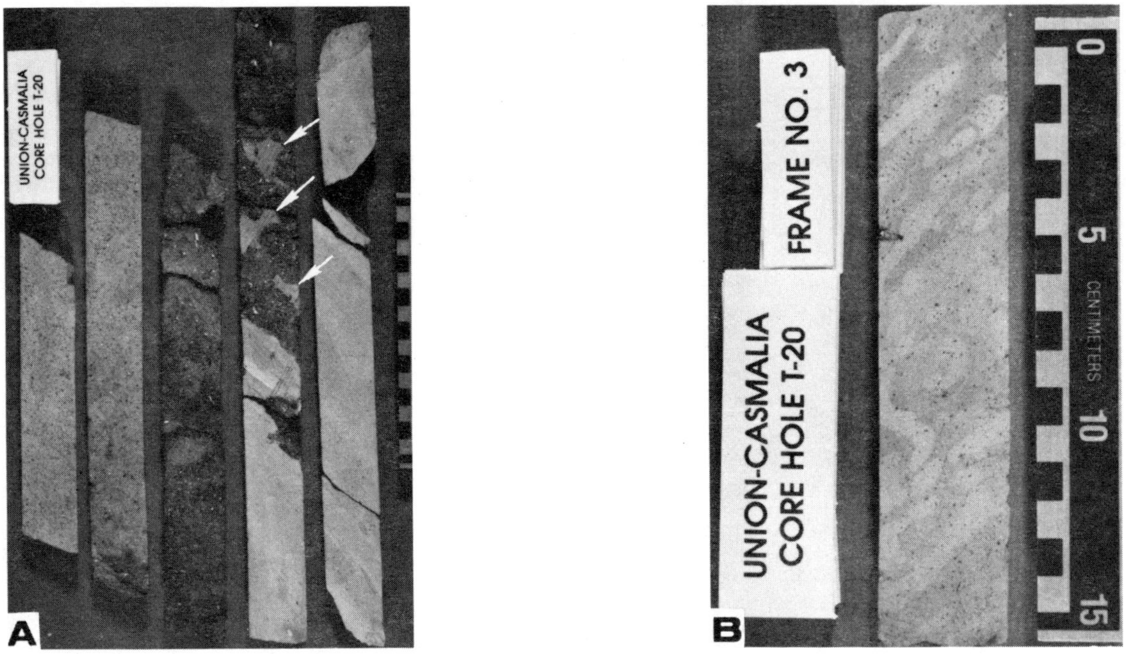

Figure 11. Characteristics of rocks in lithostratigraphic unit F. A. Chert-rich sandstone overlying diatomaceous mudstone (at right) with mudstone rip-up (?) clasts (arrow) near the base of the sandstone. The sandstone fines upward and grades abruptly to chert-bearing mudstone. Core T-20, 86.50 to 87.47m. Top of interval in upper left. Scale in centimeters. B. Bioturbated, chert-sand bearing diatomaceous mudstone. Core T-20, 90.37m. This is the dominant rock type in unit F.

pelagic sedimentation was the dominant process. Deposition was under low oxygen conditions, the waters varied from nearly anaerobic (intensely dysaerobic) to dysaerobic. A deep-water origin is inferred for unit A based on: 1) benthic foraminifera indicative of an upper bathyl, marine environment with water depths of 150 to 500 m (G. Blake, written com. 1983); 2) the presence of a thick section of fine-grained strata, locally laminated, which implies accumulation under low energy conditions that persisted for an extended period of time; and 3) the stratigraphic position between the Monterey Fm and Sisquoc lithostratigraphic units that both contain deep-water molluscs and benthic foraminiferia.

The presence of an oxygen minimum zone may be inferred from the occurrence of preserved laminae and the foraminiferal assemblages (G. Blake, written com, 1983). Anaerobic conditions in the water column were episodic as indicated by interbedded laminated and bioturated strata. Biógenic structures are small (<1 mm) and probably related to Chondrites. The lack of mega-invertebrates in highly-bioturbated strata and the interbedded laminated strata suggest dysaerobic conditions within the depositional basin. The more heavily bioturbated, peloidal intervals represent dysaerobic conditions (dissolved oxygen content of less than 1.0 ml/1, but more than 0.1 ml/1) under which a burrowing infauna developed, but a calcified epifauna was restricted (Byers, 1977).

Lithostratigraphic unit B consists of indistinctly layered, bioturbated, diatomaceous mudstone with rare pebbly mudstone. The depositional setting is interpreted to be similar to that of unit A (basinal to lower slope setting) but in a slightly more oxygenated environment, allowing for continuous development of a burrowing infauna. Widely-spaced, pebbly-mudstone interbeds suggest periodic deposition from gravity-driven mass flows.

The pebbly mudstone and conglomerate of lithostratigraphic units C and D mark the onset of major gravity-driven mass-transport and deposition on the lower slope. Many of the thicker pebbly units of lithostratigraphic unit C probably represent debris-flow deposits as indicated by their poor sorting and grading, abundant mud matrix, and lack of sedimentary structures indicative of other sedimentary processes (Field, 1981).

The debris flow deposits in unit C, are interpreted to be channelized based on: 1) the limited lateral extent of unit C, 2) the thick, amalgamated pebbly units with minor interbeds of hemipelagic diatomaceous mudstone, and 3) the occurrence of mudstone clasts containing thick-walled pelecypods (Fig. 8A) at stratigraphic levels equivalent to dysaerobic strata of unit B (Figs. 3 and 5).

The limited lateral extent of unit C is graphically represented in Figure 5. Unit C comprises the whole 125 m of core R-16, but despite the proximity of cores Q-14 and Q-15 (Fig. 2) unit C is not encountered in these cores. Farther to the southeast, unit C occurs in cores V-19 and U-19, but not in the nearly stratigraphically equivalent core V-20.

Many of the thick pebbly sections of unit C in cores R-16 and U-19 appear to be amalgamations of several flow deposits or surges. The repetitious deposition of pebbly mud with little intervening hemipelagic mud, and the limited lateral extent of such deposition suggests that the debris flows followed a preferred pathway down the slope.

Lithostratigraphic unit D, is laterally equivalent to unit C, but extends higher in the stratigraphic section than the known occurrences of unit C (Fig 5). Unit D, like unit C, is composed of pebbly mudstone and conglomerate interbedded with diatomaceous mudstone and claystone. In contrast to unit C, however, it contains a much higher proportion of mudstone and claystone and fewer and thinner pebbly mudstone and conglomerate units.

Some of the pebbly units of unit D may have been deposited by thin, low-viscosity debris flows as suggested by concentration of dolomite clasts at the base of some pebbly units. The lower abundance of pebbly and conglomerate units argues for depositon away from the channel setting of unit C. An open slope or overbank origin is suggested for unit D. Other pebbly units may have been emplaced as sedimentary dikes. This is suggested by: 1) the occurrence of pebbly units that cross-cut diatomaceous mudstone strata (Fig. 7b), 2) the occurrence of pebbly units that are only slightly thicker than the average clast size within the unit, and 3) the disrupted texture of interbedded mudstones (Fig. 9d).

Lithostratigraphic unit E marks the initial influx of chert debris onto the slope. The chert-clasts probably were derived from the Monterey Formation. Woodring and others, (1943) and Woodring and Bramlette (1950) describe a clastic wedge of chert-rich sediment in the Upper Sisquoc Fm, derived from the Monterey Fm that extends from Corralillos Canyon, which is 4.8 km to the northwest of the study area, southeastward to within 3.2 km of the study area. The coarse, angular nature of the chert clasts suggests minimal transportation of the clasts to the depositional site.

Unit E, which was only encountered in two cores, is overlain and underlain by mudstone and pebbly mudstone of unit D (Fig 5). The limited lateral extent of the chert sand suggests that the lithostratigraphic unit represents a channel.

The depositional origin of the chert-rich sandstones in unit E is unclear. The sandstones are massive, poorly sorted and consist of angular to subrounded, fine-sand to granule size chert. The sandstone beds are up to 1.5 m thick (too thick to have been deposited by grain flows (Lowe, 1976)), lack the muddy matrix to sustain a debris flow, display no cross-bedding or sorting to indicate deposition by traction currents, and have no grading suggestive of turbidite deposition. The sandstone beds are associated with chert-clast pebbly mudstone, and may represent clastic dikes.

The chert-sand bearing diatomaceous mudstone units in unit E represents continued hemipelagic sedimentation of silt, clay, and diatom debris. The coarser chert sand may represent biogenic redistribution of thin sandy turbidites(?) or thin traction current deposits. Chert grains are locally concentrated in burrows or dewatering structures.

Lithostratigraphic unit F, which is characterized by alternating mudstone, claystone, and chert-sand bearing mudstone with interbedded thin, normally-graded, chert-rich sandstone, represents the establishment of aerobic conditions in the depositional environment as indicated by the development of an indigenous shelled molluscan fauna. Diatoms become less abundant upward and eventually disappear toward the top of the unit.

Large-scale mass transport of sediment was not an important process involved in the deposition of unit F as indicated by the lack of pebbly mudstone or conglomerate units associated with the debris flows in units C and D. Thin-bedded, graded sandstone, which may be of turbidite origin, is the only mass flow deposits.

The present data suggests an upper slope setting for unit F. The paleodepths derived from the analysis of the benthic foraminifera (G. Blake, written com, 1983) indicate depths from 100 to 500 m (outer shelf to upper bathyal), which are not greatly different from the upper bathyal (150-500 m) paleodepth determination from lithostratigraphic unit A. Unit F, overlies units C and D which contain pebbly mudstone and conglomerate units interpreted to represent mass flow deposits, characteristic of a slope environment.

Summary of Lithostratigraphic Description and Interpretation

The alternating laminated and burrowed claystone of unit A was deposited slowly on a stable slope by hemipelagic settling of silt, clay and planktonic organisms, predominantly diatoms. The sediments accumulated under low oxygen conditions that varied from dysaerobic to nearly anaerobic (intensely dysaerobic).

The distinctly thin-bedded to massive mudstone, locally diatomaceous or opal-CT bearing, of unit B was deposited on the continental slope under dysaerobic water conditions. The sedimentation rate was increased by additional silt and very fine sand entering the environment, and possibly by an increase in biologic productivity in the surface waters.

Lithostratigraphic units C and D resulted from combined sediment mass flow and hemipelagic processes. The high sedimentation rate and slope angle initiated submarine slumping and mass flow of the sediment resulting with mudstone-clast pebbly mudstone or conglomerate deposited by debris flow. Hemipelagic diatomaceous mudstone accumulated between debris flows. Unit C represents channelized debris flows marked primarily by limited lateral extent of numerous, thick pebbly mudstone and conglomerate units and volumetrically less important interbedded mudstone. The thinner pebbly mudstone and conglomerate of unit D may represent overbank deposits related to unit C, thin open slope debris flows, or injections of pebbly conglomerate generated by rapid dewatering of debris flow sediments.

Lithostratigraphic unit E, laterally equivalent to unit D may represent an adjacent channelized depositional environment that received sand and pebble size chert fragments derived from subaerial(?) erosion of the Monterey Fm. The chert-clast-rich sediment may have been transported by debris flow. The channel(?) of unit E may have been the site of enough current action to oxygenate the bottom water sufficiently to allow the development of pelecypods and gastropods.

The upper Sisquoc Fm is capped by lithostratigraphic unit F, which is characterized by alternating sections of mudstone, locally claystone, and chert-sandy mudstone with interbedded thin, normally graded, chert-rich sandstone. The chert-rich strata include indigenous pelecypods. Diatoms decrease in abundance and disappear up-section. The slope sediment became more stable, possibly as a result of decreased sedimentation rate, decreased slope angle and a change in sediment characteristics. Increased stability of the slope sediment is indicated by the absence of thick mass-flow deposits. The thin, poorly sorted, graded sandstones may represent turbidite deposition. Unit F is thought to be deposited on an open, unchanneled upper slope or outer shelf.

DOLOMITE

The core material from the upper Sisquoc Fm of the Casmalia Hills provides an opportunity to study the origin of dolomite in fine-grained diatomaceous sediment, and to compare dolomite genesis in several lithostratigraphic units.

Dolomite Occurrence

Ferroan and nonferroan dolomite in a variety of habits (nodules, clasts, and disseminated crystals) occurs in the observed cores.

Nodules

Ferroan dolomite nodules, which range in thickness from 1 to 150 cm, consist of an interlocking network of subhedral to anhedral crystals, ranging in size from 3 to 50 μm, averaging 10-20 μm. The dolomite contains unaltered diatoms, pyrite and rare siliciclastic silt and sand. Commonly the siliciclastic grains are partially replaced by dolomite. Nodular dolomite grades abruptly into the surrounding diatomaceous mudstone (Fig. 12a). The margins of the nodules consist of isolated euhedral ferroan dolomite rhombs 10 to 40 μm in size, averaging 30 μm. The nodules most commonly have a bioturbated to massive appearance and only rarely will be distinctly layered (Fig. 12b). Based on examination of similar dolomite in outcrops of the Monterey Fm, the nodular dolomites probably have a lateral continuity of a few meters (Fig. 12E).

Clasts

Dolomite clasts appear to have been derived from the breakup of nodular dolomite. Subangular to subrounded dolomite clasts with uniform texture and color, which are not mixed with clasts of different lithologies, represent brecciation of individual dolomite beds (Fig. 9b). Subrounded to rounded dolomite clasts with varying colors and textures occur as localized inhomogeneous concentrations in a mudstone matrix (Fig. 9c). These units appear to represent debris flow deposits. Less common, but not rare, are dike-like features containing dolomite clasts that occur in lithostratigraphic unit C. This type of dolomite concentration consists of poorly-sorted, rounded, inhomogeneous clasts with little mud in the matrix. The mudstone clasts and matrix that do occur are commonly partially dolomitized. The matrix of these dolomite concentrations is tar impregnated.

Disseminated Crystals

Disseminated dolomite occurs in claystone and mudstone clasts and matrix. Most disseminated dolomite is nonferroan, occurring as discrete anhedral to euhedral crystals ranging from 2 to 40 μm, and averaging 6 to 10 μm in size.

Formation of Dolomite

The process of ferroan dolomite cementation of diatomaceous strata is inferred from the study of ferroan dolomite nodules grading into the underlying and overlying diatomaceous mudstone. Figure 14 illustrates the transition from diatomaceous mudstone to the well-indurated dolomite. Petrographic observations suggest the following sequence:

1) Carbonate begins to nucleate in diatom-rich strata. The rhombs are euhedral, about 30 μm in size, which is somewhat larger than in the well-cemented beds. These crystals probably represent growth into open pores; no carbonate precursor is evident.

2) As cementation continues, the crystals impinge on one another, leading to an anhedral fabric. Although diatomaceous strata with a low silt/sand content is the preferred host for dolomite cementation, it appears that the diatom frustules are not directly involved in the nucleation of the dolomite as illustrated in Figure 14b and c.

3) The final stage of cementation engulfs the diatom frustules and produces an anhedral fabric (Fig. 14c).

Stable Isotope Analysis

Carbon and oxygen stable isotopic analyses were performed to evaluate probable sources of carbon in the dolomite and the timing of dolomite formation. Both isotopes yield important insights concerning the diagenetic environment in which the dolomite formed (Pisciotto, 1981b; Murata and others, 1969; and Kelts and McKenzie, 1982). Twenty-one dolomite or dolomitic samples were analyzed for stable carbon and oxygen-isotopic ratios.

The results of the carbon and oxygen stable isotopic analysis are presented in Table 2. Stable isotope analysis was done by Global Geochemistry following the method of McCrea (1950).

Carbon-Isotopic Ratios

The $^{13}C/^{12}C$ ratio in dolomite reflects the bicarbonate pool from which the dolomite formed (Claypool and Kaplan 1974). By determining the $\delta^{13}C$ value in dolomite the source of carbon can be estimated. In most platform dolomites the $\delta^{13}C$ values fall in a range from -4 to +4‰, PDB, indicative of a normal marine source for the carbon (Land, 1980). However, in organic-rich dolomites, such as those from the Monterey Fm, carbon isotopic values can range from very negative to very positive (typically -20 to +20‰, PDB; Murata and others, 1969; Murata and Larson, 1975; Friedman and Murata, 1979; Pisciotto, 1981b). The very positive values are indicative of carbon sources from the anaerobic bacterial methanogenesis zone.

The heavy carbon isotopic values in the nodular dolomites from the Sisquoc Fm reflect formation of carbon in the zone of bacterial methanogenesis (Irwin and others, 1977). In this zone bacteria selectively concentrate ^{13}C in the carbon dioxide and ^{12}C in methane during breakdown of organic matter.

Dolomites in the Sisquoc Fm fall into two distinct groups based on carbon-isotopic ratios (Table 2). Disseminated dolomite has $\delta^{13}C$ values ranging from -4.5 to +5.8‰, PDB (one value of +12.58) and dolomite nodules have $\delta^{13}C$ values ranging from +15 to +32.5‰, PDB (one value of +4.25‰).

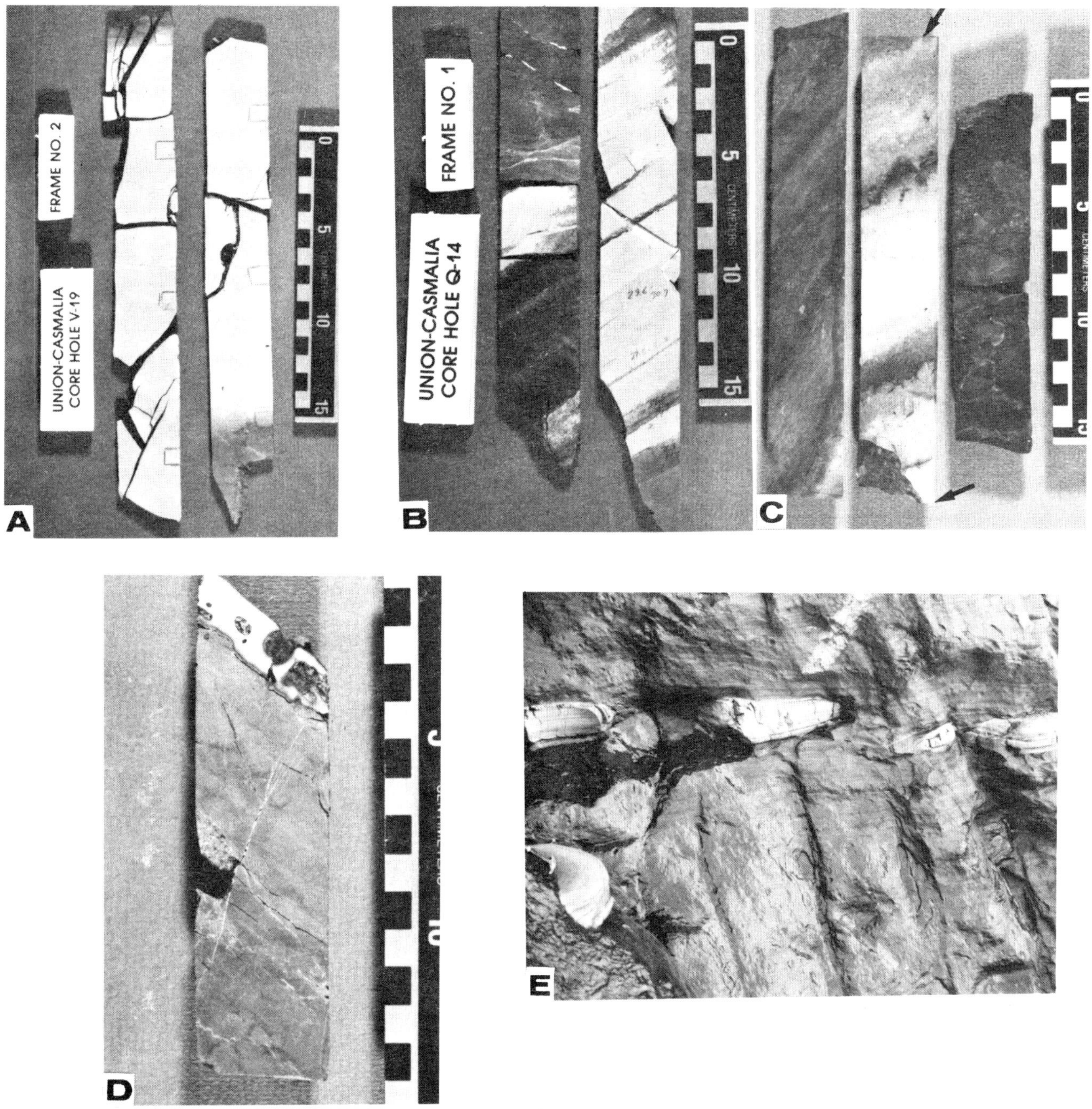

Figure 12. Nodular dolomites. A. Massive, bioturbated dolomite nodule common to lithostratigraphic unit B. The nodule is thoroughly dolomitized in the light gray portion, and grades abruptly to diatomaceous mudstone at the margins. The nodule was sampled at 7 cm intervals for stable isotopic analysis. Sample sites are marked. Lithostratigraphic unit B, core V-19, 108.26 to 108.83m. Top of interval in upper left. Scale in centimeters. B. Distinctly bedded dolomite nodule. This is the only distinctly bedded dolomite noted. Five samples taken from the nodule were analyzed for stable carbon and oxygen-isotopic ratios. The mean $\delta^{13}C$ value was +32.3‰, PDB. Lithostratigraphic unit B, core Q-14, 131.62 to 132.16m. Top of interval in upper left. Scale in centimeters. C. The vertical symmetry of the dolomite nodule, marked with arrows, is suggestive of accretionary growth. Lithostratigraphic unit C, core V-19, 27.12 to 27.48m. Top of interval in upper left. Scale in centimeters. D. Indistinctly layered, bioturbated diatomaceous mudstone with a thin dolomite nodule that contains undolomitized burrows (?). Burrow (?) traces are parallel to bedding. Lithostratigraphic unit B, core V-19, 90.34m. E. Outcrop photograph showing the discontinuous, nodular nature of dolomite in Sisquoc Fm exposures along the coast, north of Point Conception. Photo taken by John Dunham.

Figure 13. Disseminated dolomite. A. Anhedral to euhedral disseminated nonferroan dolomite (arrow). This type of dolomite is most abundant in lithostratigraphic unit A, but is also found in all other lithostratigraphic units. The white grains (S) are siliciclastic silt; some dolomite crystals are indicated by a (D). Dark gray material is the clayey matrix. Lithostratigraphic unit A, core V-18, 141.9m; thin section photomicrograph, crossed nichols. Field of view is approximately 0.3 mm in width. B. SEM photograph of euhedral, ferroan dolomite rhomb. Lithostratigraphic unit B, core V-18, 37.9m.

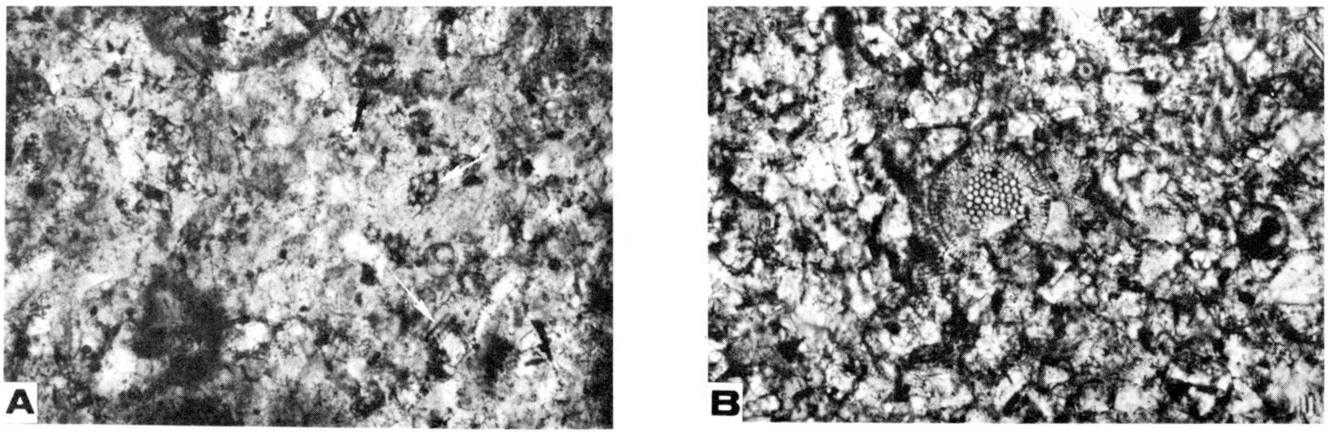

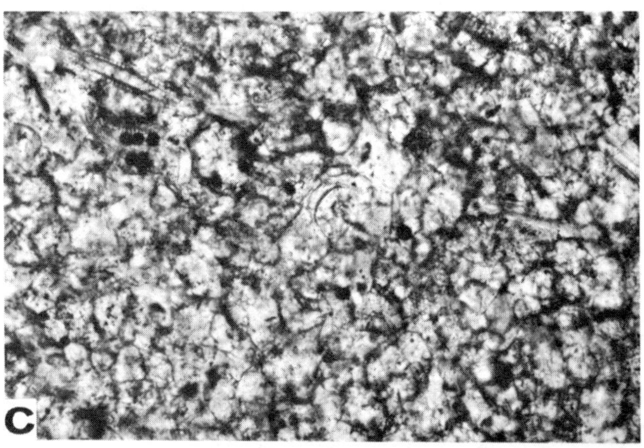

Figure 14. Photomicrographs showing characteristics of nodular dolomites in diatomaceous mudstone. All photos were taken with plane light. Field of view in all photos is approximately 0.3 mm. A. Scattered euhedral ferroan rhombs in mudstone matrix. This is characteristic of the outer edges of the nodular dolomite. Arrows indicate dolomite rhombs. Matrix is clay-rich and contains diatom fragments. Lithostratigraphic unit B, core V-19, 108.7m. B. Incompletely cemented (dolomite) zone in nodular dolomite with with preserved diatoms. The dolomite appears to fill pore spaces between diatoms. Some diatom frustules contain pyrite. Core V-19, 108.7m. C. Completely cemented dolomite nodule with scattered diatom frustules, some of which contain pyrite. Dolomite has filled all pore space. Core V-19, 108.7m.

TABLE 2.- Results of carbon and oxygen stable isotope analyses. Temperature estimates for dolomite formation based on equation from Shepard and Schwarcz (1970), assuming an initial oxygen-isotopic composition of seawater at 0.0‰, SMOW. All isotopic values reported are in per mil relative the PDB scale. The margin of error for the $\delta^{13}C$ values is ± 0.10‰ and for the $\delta^{13}O$ values ± 0.20‰ (P. Mankiewicz, 1984, oral com.). Disseminated dolomite is abbreviated as 'dissem'.

Nonferroan Dolomite						
Spl#	Core#	Depth/ meters	Occur	$\delta^{13}C$	$\delta^{18}O$	Est. Temp (°C)
1.	V-18	141.9	dissem	4.53	-1.25	44
2.	V-18	47.3	dissem	- 0.06	-8.34	88
3.	V-20	53.0	dissem	12.58	-3.03	54
4.	T-20	87.2	dissem	- 4.48	-6.14	73

Ferroan Dolomite						
Spl#	Core#	Depth/ meters	Occur	$\delta^{13}C$	$\delta^{18}O$	Temp (°C)
5.	V-18	37.9	dissem	5.81	-0.12	39
6.	V-18	75.4	nodule	4.25	1.13	33
7.	V-19	27.4	nodule	23.35	5.12	16
8.		27.3	nodule	23.02	2.95	25
9.		20.5	nodule	23.68	4.78	18
10.	V-19	108.32	nodule	15.55	-1.48	46
11.		108.37	nodule	17.83	2.23	28
12.		108.45	nodule	15.92	4.50	19
13.		108.55	nodule	16.23	4.30	19
14.		108.70	nodule	18.06	1.74	30
15.		108.82	nodule	14.98	-1.72	47
16.	Q-14	131.71	nodule	31.93	-0.64	41
17.		131.80	nodule	32.36	0.48	36
18.		131.82	nodule	32.22	0.51	36
19.		131.93	nodule	32.50	0.80	34
20.		131.97	nodule	32.00	-1.35	45
21.	V-20	119.70	clast	23.38	4.19	20

Oxygen-Isotopic Ratios

In this study oxygen isotopes have been analyzed to determine relative temperatures and depth of burial of dolomite formation. Temperatures were determined by assuming that the interstitial fluids in the sediment at the time of deposition of the Sisquoc Fm resembled the composition of normal sea water, and calculating the temperature from the measured $\delta^{18}O$ value of the dolomite using the equation given by Shepard and Schwarcz (1970) for oxygen isotopic fractionation between dolomite and water. By assuming a geothermal gradient, the depth of burial during formation of various types of dolomite in the Sisqouc Fm was calculated.

The use of oxygen isotopes to determine paleotemperatures presents several difficulties:

1) The oxygen-isotopic composition of a carbonate is a function of both temperature and the oxygen-isotopic composition of the solution from which the carbonate formed. If neither factor is known, one must be assumed to calculate the other.

2) The oxygen isotope fractionation factor between dolomite and water has been experimentally determined at high temperature, and that fractionation factor must be extrapolated from high temperature experimental results to surface temperature conditions. It has also been determined indirectly by analyzing naturally coexisting, supposedly cogenetic dolomite and calcite to determine the difference in the oxygen-isotopic composition (Land, 1980), and then comparing it with the known calcite-water oxygen isotope fractionation factor. The fractionation factor has not been experimentally determined at earth surface temperatures.

3) Dolomite may re-equilibrate with fluids through partial dissolution and precipitation several times after formation. During dissolution and precipitation of protodolomite and dolomite, the oxygen isotopes may interact with the surrounding interstitial water, mutually modifying the carbonate and water oxygen-isotopic compositions (Land, 1980).

The composition of interstitial fluids from which the dolomite in the Sisquoc Fm formed is assumed to have been close to that of normal sea-water. This assumption is based on the following characteristics of the rocks:

1) The sediment was deposited in a marine environment with seawater incorporated into the pore spaces.

2) Sedimentation occurred in a deep-marine environment where the possibility of infiltrating meteoric waters of contrasting oxygen-isotopic composition that could modify the oxygen-isotopic composition of the pore fluids was minimal. An influx of fresh meteoric water cannot be entirely ruled out, however. Uplift of adjacent areas, as indicated by chert clasts derived from the Monterey Fm that are found in the Sisquoc sediments, may have provided a catchment area for freshwater recharge into the area. However, based on isotopic trends consistent with no input of fresh water and the low permeability of diatomaceous and clay-rich sediments, this possibility is considered unlikely.

3) Diagenetic reactions involving clays that could alter the oxygen-isotopic composition of pore waters (smectite-illite transformation) have not occurred based on XRD analyses of the clay minerals.

4) Silica diagenesis had not begun during dolomite formation as indicated by the majority of the dolomite occurring in unaltered diatomaceous (opal-A) mudstone, and therefore did not affect the oxygen-isotopic composition of the pore waters. Some of the latest dolomites may have formed during or after the transformation of opal-A to opal-CT. The waters released by this transformation would be isotopically heavier than sea water (Pisciotto, 1981a).

5) There is no evidence that magmatic water, with contrasting oxygen-isotopic composition, has influenced the sediment. Igneous rocks are virtually absent except for occasional volcanic glass shards and a single 3 m thick vitric tuff.

Results

Stable carbon and oxygen isotope analysis, and selective staining for iron-bearing carbonates show two distinct populations of dolomite (Table 2): 1) ferroan dolomite with $\delta^{13}C$ values of +4.25 to +32.50‰, and $\delta^{18}O$ values of -1.72 to +5.12‰, PDB, and 2) nonferroan dolomite with $\delta^{13}C$ values of -4.48 to +12.58‰, PDB, and $\delta^{18}O$ values of -8.34 to -1.25‰, PDB. The ferroan dolomites are nodular or occur as clasts but rarely as isolated rhombs. The nonferroan dolomite occurs as disseminated crystals.

The $\delta^{18}O$ values and estimated temperatures are listed in Table 2. These results also are shown with $\delta^{13}C$ values plotted against $\delta^{18}O$ values in Figure 15.

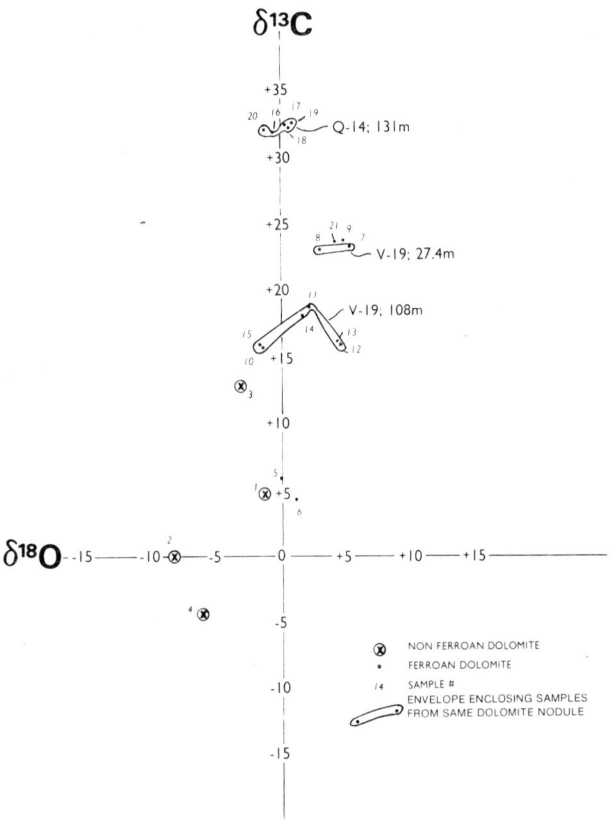

Figure 15. Plot of carbon vs oxygen isotopic values (PDB) for all dolomites analyzed in this study. Note the distinct differences in isotopic populations between the nonferroan disseminated dolomites and the bedded ferroan dolomites.

Discussion of Results

The results of the stable carbon and oxygen isotopic analysis demonstrate that there are significant differences between nodular ferroan dolomite and disseminated, nonferroan dolomite in the upper Sisquoc Fm. Although there is some overlap of $\delta^{13}C$ values of both dolomite types, ferroan dolomite tends to have values greater than +15‰, PDB, whereas nonferroan dolomite tends to have lower values, averaging +3.2‰, PDB. The ferroan and nonferroan dolomite also are differentiated by the $\delta^{18}O$ values, with ferroan dolomite $\delta^{18}O$ values generally being heavier than those of the nonferroan dolomite (Table 2).

Oxygen Isotopes

The more positive $\delta^{18}O$ values of the nodular ferroan dolomite suggests that this dolomite type formed at shallower burial depths than the disseminated nonferroan dolomite, which have more negative $\delta^{18}O$ values. This is supported by petrographic data that indicates the disseminated dolomite post-dates the nodular dolomite. Based on the oxygen-isotopic data and the assumption that the dolomite formed from sea water, temperatures for the nodular ferroan dolomite range from 16 to 47°C and for the disseminated dolomite from 39 to 88°C.

Depth of Burial, Temperature, and Pore Water Composition: Pisciotto (1981a) estimated present geothermal gradients in the Casmalia Oil Field, to the southwest of the study area, to range from 21 to 63°C/km. If this range of geothermal gradients is similar to that during the Pliocene, it would suggest nodular ferroan dolomite, which oxygen isotope analyses indicate formed at a maximum of about 47°C, developed at maximum burial depths of between 750 m and 2200 m. Maximum burial depths of 1380 m to 4100 m would be required to achieve maximum temperatures of approx. 88°C indicated by the $\delta^{18}O$ values of the disseminated dolomite.

The upper Sisquoc Fm probably was not buried to these depths, as indicated by reconstruction of burial depths based on stratigraphic and structural data and by the probable depth of burial and temperature of formation of the upper boundary of the opal-A:opal-CT transition zone. By using oxygen-isotopes, Pisciotto (1981a) calculated the temperature range at which diatomaceous sediment of the Monterey Fm transformed from opal-A to opal-CT. This data puts constraints on the depth of burial of the upper Sisquoc strata in the Casmalia Hills, which contain the Opal-A:Opal-CT boundary. Pisciotto's (1981a) isotopic data indicated that the transformation took place over a temperature range of 18 to 56°C, a range that encompasses the 38 to 54°C Pisciotto calculated from present geothermal gradients, and reconstructed maximum burial depth in the Santa Maria region.

The temperature (or depth of burial) at which the opal-A to opal-CT transformation occurs, in part, depends on the composition of the host sediment containing the opal-A (Murata and Nakata, 1974; Kastner, et al., 1977; and Isaacs, 1981). Kastner, et al., (1977), experimentally demonstrated that the presence of reactive clays retards the transformation rate. Isaacs, (1981), showed the effect of sediment composition on the relative temperature requirment for the silica transformations in the rocks of the Monterey Fm along the Santa Barbara coast. Nearly pure diatomites transform to opal-CT at relatively low temperature, whereas very impure diatomaceous sediment requires a higher temperature (or greater burial depths) to initiate the silica transformation. As the upper Sisquoc Fm consists of clay-rich diatomaceous rocks, a temperature of approximately 56°C, the highest temperature determined by Pisciotto (1981a) will be assumed to have been required to effect the opal-A to opal-CT transition in the upper Sisquoc Fm. This assumption requires that the top of the opal-CT zone reached the temperature of 56°C at the point of maximum burial, and that overlying sediment was at a lower temperature. This conflicts with the calculated temperatures of Sisquoc dolomite formation. The oxygen-isotopic ratio for a disseminated, nonferroan dolomite sample taken from just below the top of the opal-CT zone (Core V-18, 47.3 m), indicates an approximate temperature of formation of 88°C. A second disseminated dolomite sample from approximately 330 m above the opal-CT zone (core T-20, 87.2 m), indicates an approximate temperature of formation of 73°C. Both of these calculated, approximate temperatures of dolomite formation are well above the maximum temperature

(56°C) calculated by Pisciotto (1981a) to be required for the opal-A to opal-CT transition.

For purposes of comparison, an independant estimate of the temperature of dolomite formation was made using geothermal gradients from nearby oil fields and reconstructing the burial depths from stratigraphic and structural data. An estimate of the stratigraphic thickness of strata that has been removed from above the Sisquoc Fm was made using data from the Airox Mine, 3 km southeast of the study area. Here the Sisquoc Fm is overlain by the Foxen, Careaga, and the Paso Robles Formations, which have a combined thickness of approximately 580 m (Farley and Wilson, 1983). The upper Sisquoc Fm in the study area is approximately 520 m thick. The combined thickness gives a total of approximately 1100 m of burial at the base of the study section. Based on these burial thicknesses, the top of the opal-CT transition zone formed at a burial depth of approximately 1000 m. This depth of burial agrees reasonably well with the depths of burial to the top of the opal-CT zone for various oil fields in the Santa Maria Valley (Fig. 1) calculated by Pisciotto (1981a). He determined the maximum overburden in: 1) the Orcutt Field to have been 723 m, 2) the Bradley Field area to have been 827 m, and 3) the Santa Maria Valley Field to have been 942 m.

A paleogeothermal gradient for the Casmalia Hills area can be calculated by assuming a burial depth of approximately 1000 m and a temperature of about 56°C for the opal-A:opal-CT boundary. Assuming a bottom water temperature of 4°C through the Pliocene (Savin and others, 1975), the difference in bottom-water temperature and the temperature of opal-A to opal-CT conversion at depth is approximately 52°C over the burial depth of 1000 m, which yields a geothermal gradient of approximately 5.2°C/100 m. This is in the range of present day geothermal gradients in the area (2.1 to 6.3°C/100m, Pisciotto, 1981a).

If the dolomite at 87.2 m in core T-20 formed at its deepest point of burial (approximately 660 m), the temperature should have been approximately 35°C (assuming a geothermal gradient of 5.2°C/100 m), a difference of 38°C from the 73°C determined from the oxygen-isotopic signature ($\delta^{18}O$=-6.14‰, PDB). To get a dolomite with this oxygen-isotopic composition at 35°C, the water's oxygen-isotopic composition should be approximately $\delta^{18}O$= -4.0 to -6.0‰, PDB.

A negative shift in $^{18}O/^{16}O$ values with increasing depth has been noted in pore waters from silicous sediments in DSDP cores (Lawrence et al., 1975; Anderson and Lawrence, 1976). This negative shift typically ranges from -0.5 to -1.0‰/100m. The shift is related to alteration of basalts, detrital clays, and volcanic ash (Lawrence and others, 1976). In the cores from this study basalts were not present and detrital clays have not undergone extensive diagenesis as indicated by XRD. However, volcanic ash beds do occur and may account for some of the depletion in the oxygen-isotopic values.

These results suggest that the pore waters were probably isotopically several per mil lighter than seawater (4 to 6‰) and that the temperature determinations for the formation of the diagenetic dolomites should be regarded as relative temperature values. Based on trends in the oxygen-isotopic data consistent with progressive burial shown by the ferroan dolomites (see below) these isotopic data are considered to be reliable indicators of relative burial depths.

Growth of Nodular Dolomites: The stable oxygen-isotopic ratios from closely-spaced samples in three ferroan dolomite nodules (108 m, Core V-19; 131 m, Core Q-14; and 27 m, Core V-19; Table 2), demonstrate that the dolomite nodules formed by accretion during burial. The $\delta^{18}O$ values demonstrate a systematic variation for the three dolomite nodules, with the heavier oxygen-isotopic ratios occurring at the centers of the nodules and becoming progressively lighter toward the margins. The $\delta^{18}O$ values for the V-19 dolomite nodule range from +4.5 at the center to -1.72‰, PDB, at the nodule margin (Fig. 16a), whereas the Q-14 dolomite nodule has $\delta^{18}O$ values ranging from +0.51 at the center to -1.35‰, PDB, at the margin (Fig. 16b). Two samples from the third nodule in core V-19 (Fig. 12c) suggest the same isotopic trends, the $\delta^{18}O$ values of +5.12 at the center and +2.95‰, PDB, at the margin indicating a strong shift toward isotopically lighter values at the margin.

This systematic shift in $\delta^{18}O$ values reflect accretionary growth during burial. The center of the nodule, with relatively heavy $\delta^{18}O$ values, began growing early in the diagenetic (and burial) history; with increasing temperature due to burial, the oxygen-isotopic ratio of the dolomite forming on the outside of the nodule became progressively lighter The difference in $\delta^{18}O$ values between the center and edge of the nodule at 108 m in Core V-19 reflect a maximum temperature difference of approximately 28°C. At a geothermal gradient of approximately 5.2°C/100 m, the $\delta^{18}O$ values indicates growth of dolomite over a burial range from 400 to 1000m. Based on our previous discussion of temperature and burial depth this represents a maximum range of burial since the actual temperatures probably were somewhat less.

Carbon Isotopes

The $\delta^{13}C$ values reflect the different processes that produced the CO_2 incorporated into the ferroan nodular dolomite and nonferroan disseminated dolomite. Claypool and Kaplan (1974) showed that extreme positive or extreme negative $\delta^{13}C$ values indicate that bacterial fermentation or abiotic breakdown of organic matter, respectively, is the dominant process producing CO_2. Intermediate values may indicate that both processes were operating, or that dissolution and precipitation of primary carbonate (sedimentary calcite) altered the carbon-isotopic pool.

Based on the carbon-isotopic values, the ferroan dolomite nodules formed in the upper zone of bacterial methanogenesis, and the nonferroan disseminated dolomite formed in the lower zone of bacterial methanogenesis, where isotopically light carbon CO_2, derived from abiotic processes, diluted the isotopically heavy carbon CO_2 pool produced by bacterial methanogenesis. This is consistent with the oxygen-isotopic analyses which indicate that the ferroan dolomite nodules formed at lower temperatures, and therefore shallower depths of burial compared to the nonferroan disseminated dolomite.

A much greater difference in $\delta^{13}C$ values exists among dolomite nodules than within the nodules, despite their formation at overlapping depths of burial, as indicated by their oxygen-isotopic compo-

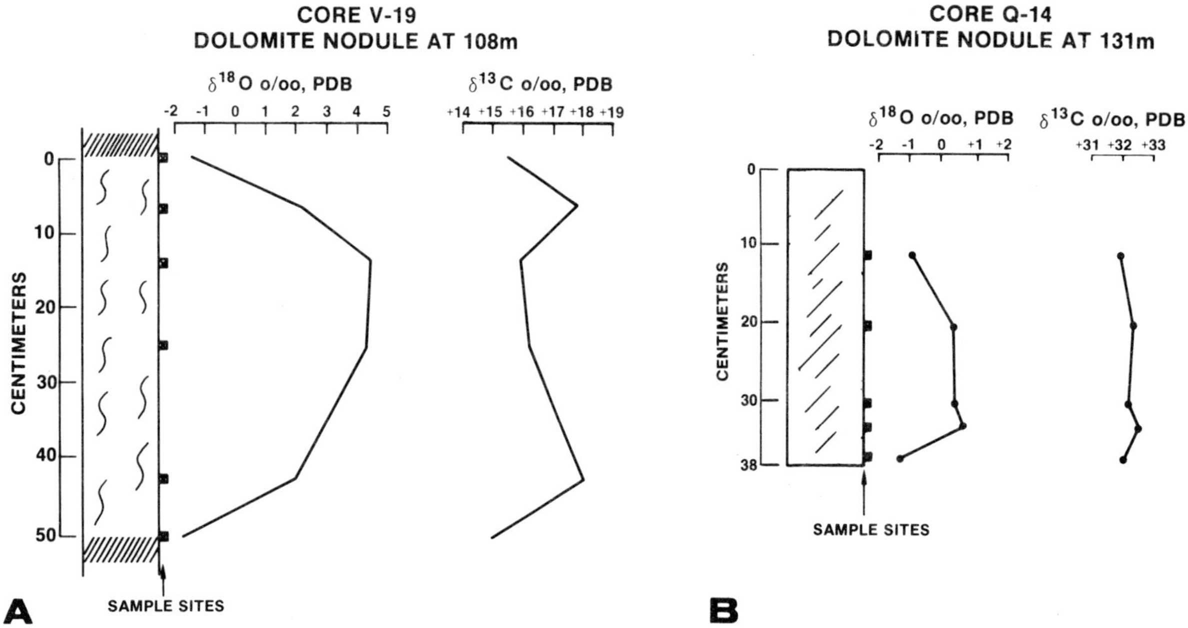

Figure 16. A. Plot of $\delta^{13}C$ and $\delta^{18}O$ values against sample localities from the dolomite nodule at 108m in core V-19. Note the distinct symmetry of both the $\delta^{18}O$ and $\delta^{13}C$ values which reflects the accretionary growth of the nodular dolomite. B. Plot of the $\delta^{13}C$ and $\delta^{18}O$ values against sample localities from the dolomite nodule at 131m in core Q-14. Only the lower well-cemented portion of the nodule was tested. The symmetry of the distribution of the $\delta^{18}O$ values is similar to that of the nodule at 108m in core V-19 (shown in part A). The $\delta^{13}C$ values are relatively constant over the sampled interval.

sition. For example, one nodule (core V-19, 108 m) has an average $\delta^{13}C$ of +16.4‰ (range +15.0 to +17.8‰, PDB) and a second nodule (core Q-14, 131 m) has an average $\delta^{13}C$ of +32.2‰ (range +32.0 to +32.5‰, PDB). This suggests that carbon-isotopic reservoirs from which the dolomite nodules formed, were isolated from one another. Isolation of the carbon-isotopic reservoirs is quite possible as the host lithologies are low-permeability diatomaceous mudstones.

Dolomite Habit, Nodular VS Disseminated Forms

Nodular dolomite formed during relatively shallow burial as indicated by the oxygen and carbon isotopic analysis, whereas the disseminated dolomite formed during deeper burial. Formation of nodular dolomite during early diagenesis and shallow burial suggests minimum compaction of sediments at the time of dolomite formation. Several isolated carbon-isotopic pools coexisted in the sedimentary column, as suggested by the carbon and oxygen-isotopic data. The CO_2 or bicarbonate within a pool probably migrated into more porous, diatom-rich terrigenous clastic-poor strata, where dolomite formed.

Later in the burial history, the host lithology became less permeable in response to progressive compaction. The CO_2 and bicarbonate were less mobile due to decreased porosity and permeability, and thus the bicarbonate was restricted to the immediate area in which it formed. As residual organic matter reacted and bicarbonate formed, the local pH rose and the interstitial fluid became supersaturated with respect to dolomite, which then precipitated. Thus the disseminated nature of the nonferroan dolomite may be a natural result of its late origin in very fine-grained, relatively tight rocks.

Summary of Dolomite Diagenesis

Dolomite occurs in the upper Sisquoc Fm in two principal forms: nodular ferroan dolomite and disseminate, nonferroan dolomite. A third form, dolomite pebble clasts,were probably derived from the ferroan dolomite nodules. The nodular, ferroan dolomite can be distinguished isotopically from the disseminated, nonferroan dolomite by the $\delta^{13}C$ values, which are generally greater than +15‰, PDB, and $\delta^{18}O$ values, which are generally heavier than -1.72‰, PDB. The dolomite formed early in the burial history of the sediment at relatively low temperature from carbon-isotopically heavy CO_2 produced by the isotopic fractionation of organic carbon by bacteria during methanogenesis. The sediment may have been less compacted, allowing CO_2 and bicarbonate to migrate within the sediments and enabling dolomite to form in relatively porous diatom-rich beds. Impermeable stratigraphic barriers limited the mobility of the bicarbonate within interstitial fluids and created isolated carbon-isotopic pools. Dolomite nodules with distinct carbon-isotopic composition formed from the pools by accretion during burial as inferred from the oxygen-isotopic composition.

Nonferroan, disseminated dolomite has $\delta^{13}C$ values less than +12‰, PDB, (most -4.5 to +5.0‰, PDB) and has $\delta^{18}O$ values more negative than -1.25‰, PDB. These carbon and oxygen-isotopic ratios indicate that the nonferroan dolomite formed later and at a greater burial depth than the ferroan dolomite. The carbon-isotopic ratios indicate that isotopically light-carbon bicarbonate, generated most probably by abiotic reactions degrading organic matter, were diluting the heavy carbon-isotopic pool generated by bacterial methanogenesis. The disseminated habit of the nonferroan dolomite may be a response to the relatively low porosity and permeability of the host lithology at the time abiotic reactions were contributing significant isotopically light bicarbonate to the interstitial fluids.

RELATIONSHIP OF DOLOMITE DEVELOPMENT TO LITHOSTRATIGRAPHIC UNITS

Dolomite types in the Casmalia Hills cores are facies specific. The nodular ferroan dolomites are most abundant in the bioturbated lower slope or basin facies (lithostratigraphic unit B) with rare occurrences in the laminated basinal or lower slope facies (lithostratigraphic unit A) and in interbedded mudstones of the allochthonous pebbly mudstone and conglomerate (lithostratigraphic units C and D). Ferroan dolomite clasts, derived from the ferroan dolomite nodules, are found in pebbly mudstones and conglomerates (units C and D). Disseminated nonferroan dolomite was noted in all lithostratigraphic units. A summary of the relationship of dolomite types with lithostratigraphic units is given below.

Nodular Dolomite

The majority of nodular dolomite occurs in bioturbated diatomaceous and opal-CT bearing mudstone (lithostratigraphic unit B) with minor occurrences in the laminated opal-CT bearing mudstone and claystone (lithostratigraphic unit A) and in mudstone and claystone interbedded with the pebbly mudstone and conglomerate (lithostratigraphic units C and D).

Differences in rock types betweeen facies give some clues to the origin of these nodular dolomites and the controls on their formation. Unit B, which contains the majority dolomite nodules (comprising up to approximately 4% of the unit) was deposited under dysaerobic conditions, with enough oxygen in the water column to permit a burrowing infauna. Lithostratigraphic unit A, which is very similar to unit B, but has only a few dolomite nodules, consists of laminated to thin opal-CT bearing claystone and minor mudstone. The difference in the laminated character suggests that the oxygen minimum zone was more intense during depositon of unit A. In addition, sedimentation rates appear to have been more rapid during deposition of unit B as indicated by an increased influx of siliciclastic silt and very fine sand. In unit A the combined effect of low oxygen levels in the overlying water column and slow sedimentation rate may have resulted in the zone of bacterial methanogenesis occurring very close to the sediment/water interface. Proximity of the zone to the sediment/water interface may have allowed diffusion or migration of dissolved carbon dioxide and bicarbonate to the sediment surface where it could escape into the overlying water column. The products of methanogenesis would be trapped only after sufficient burial had occurred to reduce the permeability of the sediment.

The increased sedimentation rate and slightly higher oxygen level in the overlying waters during deposition of unit B may account for the differences in dolomite occurrences in units A and B. Pisciotto and Mahoney (1981) have suggested that sedimentation rate may be the most important factor influencing the origin and isotopic composition of authigenic carbonates in organic-rich sediments, and Claypool and Kaplan (1974) have noted that significant methane is only present in thicker, more rapidly deposited deep sea sediments. The more rapid sedimentation in unit B along with differences in levels of dissolved oxygen in the water column may have caused a depression in the depth of the zone of methanogenesis below the sediment/water interface. This helped to isolate and preserve carbon dioxide and bicarbonate, the products of methanogenesis within the sedimentary column. Occasional diatom blooms or interruptions in delivery of siliciclastic sediment to the depositional site resulted in the deposition of porous, diatom-rich beds. Periodic slackening of diatom productivity and/or increases in rates of siliciclastic sedimentation resulted in deposition of low porosity, low permeability mudstone. The low permeability, muddying upward cycles identified by Farley and Wilson (1983), which are approximately 25 m thick in areas of the nearby Sisquoc Fm, may have stratified the sedimentary column with low-permeable zones that restricted bicarbonate diffusion between burial zones undergoing bacterial methanogenesis.

The bulk of the mudstone units in lithostratigraphic units C and D do not contain nodular ferroan dolomite. This may be a function of the thickness of the units or to disruption of the mudstone beds during deposition of the mass-flow units, which allowed the escape of bicarbonate before oversaturation of the pore water with respect to dolomite. Pebbly units are more permeable, as indicated by their higher degree of oil saturation, and may have acted as escape routes for the bicarbonate and carbon dioxide. The rare, thin ferroan dolomite nodules that do occur in the mudstones are inferred to have a similar origin to the nodular ferroan dolomite of unit B.

Dolomite Clasts

Dolomite clasts are found largely in pebbly mudstone and conglomerate, which are interbedded diatomaceous and opal-CT bearing mudstone and claystone (Lithostratigraphic units C and D). Oxygen and carbon isotopic values indicate that the dolomite clasts were derived from nodular dolomites similar to those in the interbedded mudstones and in the mudstones of unit B. The oxygen values from the clasts are heavy ($>+2.95$‰) suggesting a relatively shallow burial origin.

The occurrence of the clasts only in units C and D is a result of changing sedimentary processes through Sisquoc time. The occurrence of the pebbly mudstone and conglomerate suggests upslope erosion, probably by submarine slumping. These erosional events must have cut through overlying strata deeply enough to erode the burial dolomites.

Disseminated Dolomite

The nonferroan, disseminated dolomite was noted in all lithostratigraphic units in the Casmalia Hills cores. The nonferroan disseminated dolomite formed after considerable burial (several hundred meters), as shown by oxygen-isotopic ratios that indicate dolomite formation at elevated temperatures. With greater burial, the sediments became less porous and permeable, which may have inhibited migration of carbon dioxide and bicarbonate generated by late bacterial methanogenesis and abiotic reactions. The bicarbonate would raise the pH and may locally produce pore fluids supersaturated with respect to dolomite. The intermediate $\delta^{13}C$ values ofthe disseminated nonferroan dolomite ($\delta^{13}C=+4.53$ to -0.06‰,PDB), and the negative $\delta^{18}O$ values ($\delta^{18}O=-8.34$ to -1.25‰,PDB) indicate that the bicarbonate incorporated into the dolomite was produced by a mix of bacterial methanogenesis and abiotic reactions.

SUMMARY AND CONCLUSIONS

The type and style of formation of diagenetic dolomite in fine-grained, diatomaceous sediments deposited on a lower slope setting are dependent on sediment type, rate of deposition, depositional mechanism and oxygenation levels in the water mass overlying the sediment/water interface.

Slow rates of deposition under intense dysaerobic conditions, as typified by irregularly laminated claystone and mudstone of lithostratigraphic unit A, yield disseminated, nonferroan dolomite of intermediate carbon-isotopic composition. Earlier forming, isotopically heavy carbon-bicarbonate, may have been able to diffuse upward through the accumulating sediment and escape into the overlying water until compaction of the clay-rich sediment reduced the porosity and permeability. Subsequently, at deeper levels of burial, bicarbonate of intermediate carbon-isotopic composition formed as a result of a combination of ^{13}C-rich bicarbonate derived from bacterial methanogenesis, and isotopically light carbon-bicarbonate generated by abiotic reactions breaking down organic matter in the sediment. Initiation of abiotic processes may have caused an increase in the concentration of bicarbonate in the pore waters, which raised the pH, and caused oversaturation with respect to dolomite in the pore water. The low permeability of the clay-rich sediment may have prevented extensive diffusion of bicarbonate, causing localized supersaturation of the pore waters with respect to dolomite, resulting in the precipitation of the disseminated dolomite.

Faster rates of hemipelagic deposition of organic-rich sediment under less intensely dysaerobic conditions, as typified by bioturbated silt and sand-rich diatomaceous mudstone (unit B), yield nodular ferroan dolomite and minor disseminated nonferroan dolomite. The low level of dissolved oxygen in the water overlying the sediment/water interface preserved organic matter in the sediment as it was buried. An increased rate of deposition isolated the products of bacterial methanogenesis and prevented extensive diffusion of carbon dioxide or bicarbonate into the overlying water mass. Formation of bicarbonate early in the burial history of the sediment allowed diffusion of the bicarbonate to more porous, diatom-rich beds where precipitation of dolomite occurred.

Cyclic variations of the mud and diatom content of the sediment resulted in isolated carbon dioxide and bicarbonate pools of distinctive carbon-isotopic composition that yielded dolomite nodules of distinctive carbon-isotopic composition.

With increased burial and higher temperature, bacterial methanogenesis declined and thermal, abiotic processes increased the contribution of isotopically light-carbon bicarbonate and carbon dioxide to the pore water. Decreased permeability, probably due to compaction, may have prevented bicarbonate diffusion in the pore water to the nodular dolomite, stopping the growth of dolomite nodules. When bicarbonate locally raised pH and supersaturated the pore water with respect to dolomite, localized precipitation of the nonferroan disseminated dolomite.

Mass flow deposits, such as the pebbly mudstone and conglomerate of lithostratigraphic units C and D, typically contain ferroan dolomite clasts derived from nodular ferroan dolomites that isotopically reflect their previous burial history. Interbedded hemipelagic diatomaceous mudstone beds, inferred to be deposited under dysaerobic conditions rarely contain nodular dolomite. This may have been due to repeated interruptions of the hemipelagic units by debris flow or other mass flow units, which provided permeable pathways for the early forming bicarbonate to escape from the accumulating sediment. Reduction of permeability by compaction isolated late forming bicarbonate of intermediate isotopic composition that formed disseminated nonferroan dolomite.

Deposition of hemipelagic and mass-flow sediments under normal or near normal levels of dissolved oxygen are typified by units E and F. The higher levels of dissolved oxygen supported an indigeneous molluscan fauna that digested the organic matter deposited on the slope, and may have oxidized organic matter with free molecular oxygen. As a result, insufficient organic material reached the burial realm, and bacterial methanogenesis may have not have been an important process. Only disseminated, nonferroan dolomite developed which has the lightest carbon-isotopic composition ($\delta^{13}C$=-4.48‰, PDB) of the dolomites analyzed in this study.

ACKNOWLEDGEMENTS

Discussions with Dr. Caroline Isaacs and Andrew Soutar were useful and appreciated. We would like to express our thanks to Robert Arends for identifying diatoms and providing age dates, Greg Blake for analyzing the benthic foraminifera for paleobathymetric data and age dates, and to Dr. Cortez Hoskins for identifying molluscs and providing environmental information. Discussions with Dr. Bruce Bromley on the interpretation of the carbon-isotope data were most helpful. Sean Carey read the manuscript and provided editorial assistance. The manuscript benefited greatly from the reviews by Drs. Don Zenger, Miriam Kastner and Robert Garrison.

Special thanks goes to Union Oil Company, especially Art Lewis and Drs. Jim Friberg and Cortez Hoskins for permission to use the cores from the upper Sisquoc Fm, and for technical support. We also wish to thank Suzanne Henderson and Lorraine Carey for their invaluable assistance in the final preparation of this manuscript.

REFERENCES CITED

Anderson, T.F., and Lawrence,J.R., 1976, Stable isotope investigations of sediments, basalts, and authigenic phases from Leg 35 cores, in Hollister, C.D., Craddock, C., et al., eds., Initial Reports of the Deep Sea Drilling Project, v. xxxv, p. 497-505, U.S. Government Printing Office, Washington.

Byers, C.W., 1977, Biofacies Patterns in Euxinic Basins: A General Model, in Cook, H.E., and Enos, P., eds., Deep-Water Carbonate Environments: Soc. Econ. Paleontologists andMineralogists, Spec. Publ. No. 25, p. 5-17.

Claypool, G.E., and Kaplan, I.R., 1974, The Origin and Distribution of Methane in Marine Sediments, in Kaplan,I.R., ed, Natural Gases in Marine Sediments: New York(Plenum), p. 99-139.

Coplen, T.B., and Hanshaw, B.B., 1973, Ultrafiltration by a compacted clay membrane I. Oxygen and hydrogen isotopic fractionation: Geochimica et Cosmochimica Acta, v. 37, p. 2295-2310.

Farley, T., Wilson, M.L., 1983, Geology of the Airox Oil-saturated Diatomite Deposit, Santa Barbara County, California, in Isaacs, C.M., et al., eds., Petroleum generation and occurrence in the Miocene Monterey Formation, California: Spec. Publ. Soc. Econ. Paleontologists and Mineralogists, Pacific Sec., 228 p.

Field, M.E., 1981, Sediment Mass-transport on Basins: Controls and Patterns, in Douglas, R.G., et al, eds., Depositional Systems of Active Continental Margin Basins: Short Course Notes, Soc. Econ. Paleontologists and Mineralogists, Pacific Sec., Los Angeles, California, 165 p.

Field, M.E., and Clark, S.H.,Jr., 1979, Small-Scale Slumps and Slides and their Significance for Basin Slope Processes, Southern California Borderland, in Doyle, L.J. and Pilkey, O.H., eds., Geology of Continental Slopes: Soc. Econ. Paleontologists and Mineralogists Spec. Publ. No. 27, p. 223-230.

Friedman, I., and Murata, K.J., 1979, Origin of dolomite in Miocene Monterey Shale and related formations of the Temblor Range, California: Geochim. Cosmochim. Acta, v. 43, p. 1357-1365.

Garrison, R.E., Douglas, R.G., Pisciotto, K.E., Isaacs, C.M., and Ingle, J.C. eds., 1981, The Monterey Formation and Related Siliceous Rocks of California: Spec. Publ. Soc. Econ. Paleontologists and Mineralogists, Pacific Section, Los Angeles, California, 327 p.

Govean, F.M., and Garrison, R.E., 1981, Significance of Laminated and Massive Diatomites in the Upper Part of the Monterey Formation, California, in Garrison, R.E. and others, eds, The Monterey Formation and Related Siliceous Rocks of California: Spec. Publ. Soc. Econ. Paleontologists and Mineralogists, Pacific Sec., Los Angeles, California, p. 181-198.

Henderson, N.C., Jr, 1983, Relationship of diagenetic dolomite to depositional facies in diatomaceous strata, upper Sisquoc Formation, Casmalia Hills, Santa Barbara County, California: Masters thesis, University of Southern California, 200 p.

Ingle, J.C., 1980, Cenozoic paleobathymetry and depositional history of selected sequences within the Southern California Continental Borderland: Cushman Foundation Spec. Publ. No. 19, p. 163-195.

Irwin, H., 1980, Early diagenetic carbonate precipitation and pore fluid migration in the Kimmeridge Clay of Dorset, England: Sedimentology, v. 27, p. 577-591.

Irwin, H., Curtis, C., and Coleman, M., 1977, Isotopic evidence for source of diagenetic carbonates formed during burial of organic-rich sediments, Nature, v. 269, p. 209-213.

Isaacs, C.M., 1980, Diagenesis in the Monterey Formation examined laterally along the coast near Santa Barbara, California: Stanford Univ. unpubl. Ph.d. thesis, 329 p.

-------, 1981a, Lithostratigraphy of the Monterey Formation, Goleta to Point Conception, Santa Barbara Coast, California,

-------, 1981b, Outline of diagenesis in the Monterey Fm examined laterally along the Santa Barbara Coast, California: in Isaacs, C.M., ed., Guide to the Monterey Formation in the California Coastal Area, Ventura to San Luis Obispo: Am. Assoc. Petroleum Geologists Pacific Sec. Publ. No. 52, p. 25-38.

-------, 1982, Influence of rock composition on kinetics of silica phase changes in the Monterey Formation, Santa Barbara area, California: Geology, v. 10, p. 304-308.

Kelts, K., and McKenzie, J.A., 1982, Diagenetic Dolomite Formation in Quaternary Anoxic Diatomaceous Muds of Deep Sea Drilling Project Leg 64, in Blakeslee, J., Platt, L.W., et al., ed., Init. Reports, Deep Sea Drilling Project, Leg 64: Washington (U.S. Govt. Printing Office), p. 553-569.

Land, L.S. 1980, The Isotopic and Trace Element Geochemistry of Dolomite: The State of the Art, in Zenger,D.H., Dunham,J.B., Ethington,R.L., Concepts and Models of Dolomitization: Soc. Econ. Paleontologists and Mineralogists, Spec. Publ. No.. 28, p. 87-110. v. 28, p. 87-110.

_______, 1983, The application of stable isotopes to studies of the origin of dolomite and to problems of diagenesis of clastic sediments: in Stables isotopes in sedimentary geology; S.E.P.M. Short Course No. 10, p. 44-1--4-22.

Lawrence, J.R., Gieskes, J.M., and Broecker, W.S., 1975, Oxygen isotope and cation composition of DSDP pore waters and the alteration of layer II basalts: Earth Planet. Sci. Lett., v. 27, p. 1-10.

Lowe, D.R., 1976, Grain flow and grain flow deposits: Jour. Sed. Petrology, v. 46, p. 188-199.

McCrea, J.M., 1950, On the isotopic chemistry of carbonates and paleotemperature scale: Jour. Chem. Phys., v.18, p. 849-857.

Murata, K.J., Friedman, I. and Madsen, B.M., 1969, Isotopic composition of diagenetic carbonates in marine Miocene formations of California and Oregon: U. S. Geological Survey Prof. Paper 614B.

Murata, K.J. and Larson, R.R., 1975, Diagenesis of Miocene siliceous shales, Temblor Range, California: U. S. Geological Survey J. Res., v. 3, p. 553-566.

Murata, K.J. and Nakata, J.K., 1974, Cristobalite stage in the diagenesis of diatomaceous shale: Science, v. 184, p. 567-568.

Pisciotto, K.A., 1981a, Diagenetic trends in the siliceous facies of the Monterey shale in the Santa Maria region, California: Sedimentology, v. 28, p. 547-571.

---------, 1981b, Review of Secondary Carbonates in

the Monterey Formation, California, in Garrison, R.R. et al., eds., The Monterey Formation and Related Siliceous Rocks of California: Spec. Publ. Soc. Econ. Paleontologists and Mineralogists, Pacific Sec., p. 273-283.

Pisciotto, K.A., and Mahoney, J.J., 1981, Isotopic Survey of Diagenetic Carbonates, Deep Sea Drilling Project Leg 63, in Yeats, R.S., Haq, B.U., et al.ed., Init. Reports Deep Sea Drilling Project, Leg 63: Washington (U. S. Govt. Printing Office) p. 595-609.

Porter, W.W.,III, 1932, Lower Pliocene in Santa Maria District, California: Am. Assoc. Petroleum Geologists Bull., v. 16 No. 2, p. 135-143.

Potter, P.E., and Pettijohn, F.J., 1977, Paleocurrents and Basin Analysis: New York, Springer-Verlag, 425 p.

Potter, P.E., Maynard, J.B., and Pryor, W.A., 1980, Sedimentology of Shale: New York, Springer-Verlag, 306 p.

Savin, S.M., Douglas, R.G., Stehli, F.G., 1975, Tertiary marine paleotemperatures: Geological Society of America, v. 86, p.1499-1510.

Sheppard, S. M. F., and Schwarcz, H. P., 1970, Fractionation of carbon and oxygen isotopes and magnesium between metamorphic calcite and dolomite: Contrib. Mineralogy and Petrology, v. 26, p. 161-198.

Woodring, W.P., and Bramlette, M.N., 1950, Geology and paleontology of the Santa Maria District, California: U. S. Geological Survey Prof. Paper # 222, 185 p.

Woodring, W.P., Bramlette, M.N., and Lohman, K.E., 1943, Stratigraphy paleontology of the Santa Maria District, California: Am. Assoc. Petroleum Geologists Bull., v. 27 # 10, p. 1335-1360.

STRATIGRAPHY AND DOLOSTONE OCCURRENCE IN THE MIOCENE MONTEREY FORMATION, SANTA MARIA BASIN AREA, CALIFORNIA

John S. Compton and Raymond Siever
Department of Geological Sciences
Harvard University
Cambridge, MA 02138

ABSTRACT

The detailed stratigraphy and mode of occurrence of dolostone in the dominantly siliceous Sisquoc and Monterey Formations of the Santa Maria basin are guides to the timing and nature of the diagenetic processes by which dolomite was precipitated. The Monterey Formation may be divided into four members; from oldest to youngest they are: (1) lower calcareous-siliceous, (2) phosphatic, (3) upper calcareous-siliceous, and (4) siliceous. Both the siliceous (uppermost) member of the Monterey Formation and the overlying Sisquoc Formation consist of thickly laminated clayey diatomite and porcelanite, but the Sisquoc Formation is generally richer in detrital clay and silt, more thickly laminated, and more massive (bioturbated). Porcelanite occurs interbedded with diatomite or as bedded intervals several meters thick. Dolostone and chert are minor rock types present as isolated nodules or nodular beds. These appear to have formed by pore-filling and replacement processes at a stage later than burial compaction. The diatomite and porcelanite beds are essentially carbonate-free.

The upper calcareous-siliceous member has rhythmically interbedded siliceous mudstone and porcelanite, black and white banded, frequently intensely folded chert beds, and a marked increase in dolostone content. Dolostone occurs as dolomitized interbedded siliceous mudstone and porcelanite, as homogeneous beds, nodular beds, or more rarely as isolated early nodules. Dolostone beds average 40 to 80 cm in thickness and have a variable (30 to 95 wt %) but homogeneous dolomite content. Dolostone beds are more or less regularly spaced 1 to 20 m apart. Rocks between dolostone beds contain little or no carbonate.

The phosphatic member consists of phosphatic, organic-rich mudstone, porcelanite and dolostone. The dolostone occurs as beds, nodular beds, and as isolated early nodules. Dolostone beds average 10 to 30 cm in thickness, and have a homogeneously distributed dolomite content of 50 to 90 wt %. Most beds appear to have formed early. The lower calcareous-siliceous member has a gradational depositional contact with the underlying Tranquillon Volcanics, and consists of siliceous mudstone, porcelanite, dolostone, and volcanic tuff beds.

The timing of dolomitization relative to burial compaction can be determined for most dolostone nodules and nodular beds, but not for most dolostone beds. The area studied in the Santa Maria basin differs from stratigraphically equivalent rocks in the Santa Barbara coastal area (Isaacs, 1981a, 1983) by having more dolostone and a far lower abundance of calcareous-rich rock types.

INTRODUCTION

General aspects of the paleontology, stratigraphy and structure of the Santa Maria basin area were described about 40 years ago, partly in relation to its large hydrocarbon reserves (Canfield, 1939; Woodring and others, 1943; Bramlette, 1946; Woodring and Bramlette, 1950; Dibblee, 1950). Several recent studies have covered specific aspects of the stratigraphy in relation to diagenesis of the Monterey Formation in the Santa Maria basin (Pisciotto, 1981), the Santa Barbara coastal region (Isaacs, 1980, 1981a and b), and in the Pismo Syncline (Surdam and Stanley, 1981). Several well-exposed sections of the Monterey Formation in coastal and inland areas of the Santa Maria basin have been described and measured by Grivetti (1982). Actively eroding beach cliffs provide fresh exposures with excellent vertical and lateral stratigraphic control. Field work for this study included detailed stratigraphy of five areas in the Santa Maria basin: Point Pedernales, Lion's Head, Rodeo Canyon, Sweeny Road and Jalama Beach. Additional less detailed work was done at Mussel Rock and Lompoc diatomite quarry (Figure 1).

The Monterey Formation has been divided into four members (Isaacs, 1981a). From oldest to youngest they are the (1) lower calcareous-siliceous, (2) phosphatic, (3) upper calcareous-siliceous, and (4) siliceous members (see Table I). The Sisquoc Formation overlies the Monterey Formation and is lithologically similar to the siliceous member, often making it difficult to determine unambiguously the contact between the two formations in the field. In sections where the formational contact cannot be distinguished clearly, the Sisquoc Formation and siliceous member will be discussed together. The lower calcareous-siliceous member will not be discussed since it has only limited, poor exposure in the Point Pedernales and Rodeo Canyon areas. In those two areas the Monterey Formation has a depositional basal contact with the Tranquillon Volcanic breccia. At Lion's Head there is a faulted contact with the Coast Range ophiolite. The basal contact of the Monterey in the other areas is covered.

The assignment of a specific rock name to Monterey rocks in the field is often difficult because most rocks are fine-grained, complex assemblages with complete gradations in composition (Isaacs, 1983). For example, with decreasing silica and increasing detrital content the rock type grades from chert to porcelanite to siliceous mudstone to mudstone. More accurate classification of rock types requires detailed petrographic and geochemical study.

in Garrison, Robert E.; Kastner, Miriam and Zenger, Donald H., eds., 1984, **Dolomites of The Monterey Formation and Other Organic-Rich Units:** Pacific Section S.E.P.M., Vol. 41, p. 141-153.

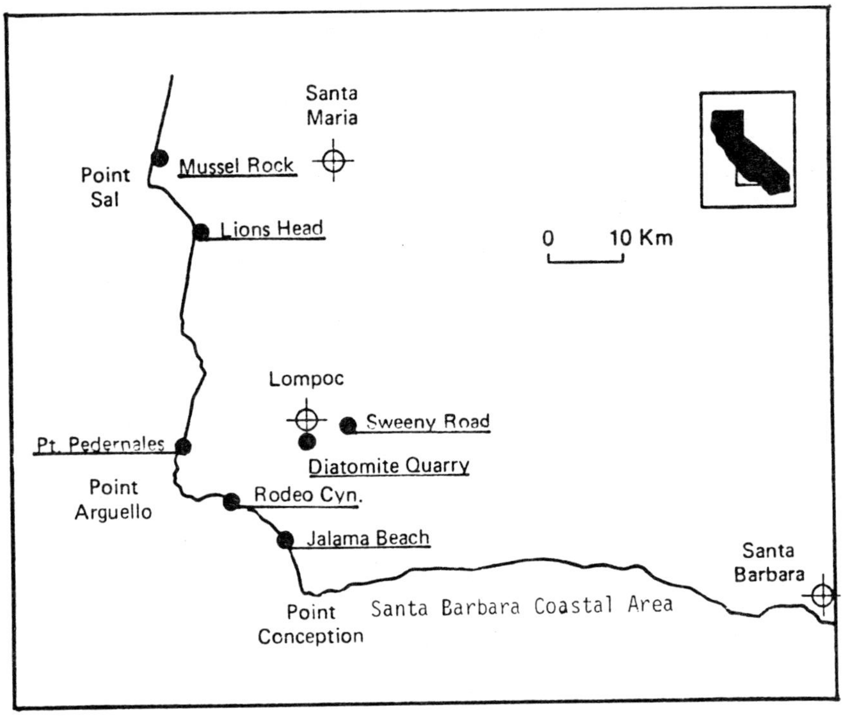

Figure 1. Location map of the areas studied, south-central coastal California.

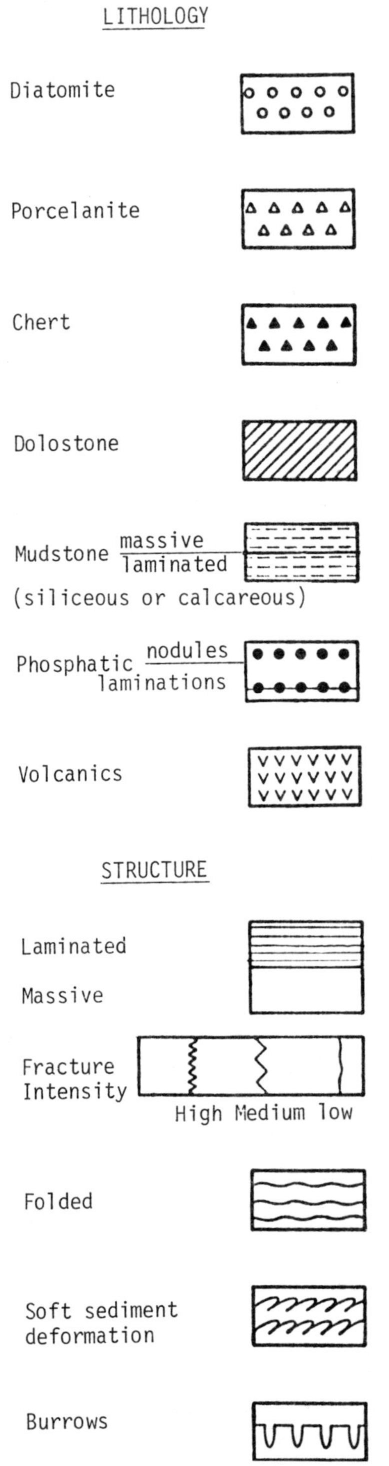

Figure 2. Key to lithologic and textural symbols used in the stratigraphic columns.

METHODS

Representative stratigraphic intervals of all the members present in each area were measured in detail. These intervals, here designated as locations, ranged from meters to tens of meters in thickness. The sections were measured with a resolution of one centimeter, described, photographed, and sampled. Figure 2 shows the lithologic and textural symbols used in the stratigraphic columns. This detailed stratigraphy provides the basis for characterizing the occurrence of dolostone throughout the different members of the Monterey Formation. It also allows the comparison of dolostone occurrence from different areas of variable lithology and diagenetic grade.

Rock types were identified in the field using the definitions of Isaacs (1981b). Dolostone, originally proposed by Shrock (1948) as a rock type largely composed of dolomite, was defined by Pettijohn (1975, p. 360) as a rock containing > 50% carbonate of which more than one-half is dolomite. For dolostone the following field definition was used: an aphanitic to sucrosic rock which, when powdered, reacts with dilute (4%) hydrochloric acid. On fresh surfaces the color of dolostone ranges from dark to light gray or brown, weathering to a distinctive yellow or white color at outcrops. Dolostone is denser (2.4 to 2.7 g/cc bulk rock density) and usually more resistant to erosion than other rock types. Typically very hard, the dolostones ring when hit with a hammer. Most of the rocks that satisfied the above field definition contain > 50 vol % dolomite, although several have between 25 and 50 vol % carbonate occurring as dolomite and are technically dolomitic porcelanites or siliceous mudstones.

The three modes of dolostone occurrence observed are defined as nodules, nodular beds, and dolostone beds: (1) Nodules are somewhat flattened and isolated pods aligned along bedding planes. (2) Nodular beds consist of the partial or complete coalesence of individual nodules aligned along a stratigraphic horizon to form wavy bedding. (3) Dolostone beds are more or less uniform in thickness, laterally continuous and have planar lithologic contacts (stratiform); they are not necessarily homogeneous.

SILICEOUS MEMBER OF THE MONTEREY FORMATION AND OVERLYING SISQUOC FORMATION

The siliceous member of the Monterey Formation and the overlying Sisquoc Formation are present throughout the area and have an aggregate thickness in excess of 300 to 600 meters. These units consist of thickly laminated clayey diatomite and porcelanite with rare chert and dolostone nodules or nodular beds. The Sisquoc Formation is distinguished from the Monterey Formation by a generally higher detrital clay-silt fraction, lower silica and carbonate content, thicker laminations, and more massive (bioturbated) intervals. All except the last are presumably responses to higher terrigenous sedimentation rates. The Sisquoc Formation and siliceous member of the Monterey are both low in phosphate and relatively low in organic matter (0.5 to 4 wt %). A phosphatic nodule conglomerate in several areas (Sweeny Road, Mussel Rock, and the Lompoc diatomite quarry) may represent a significant period of nondeposition. The stratigraphic position of this conglomerate approximately coincides with the transition from the Monterey to the more detrital-rich diatomite of the Sisquoc Formation.

Figures 3a and 3b show the typical lithology of the Sisquoc Formation. It is a fairly homogeneous, continuous, thickly laminated (1 to 3 mm thick couplets), opal-A, clayey diatomite. The laminations are the result of compositional differences, the darker green-brown colored laminae containing more clay and organic matter. Laminations generally are laterally continuous. The thickness of individual laminae ranges between 0.45 and 2.0 mm, and their vertical distribution appears to be random. Some angular unconformities, ranging from < 1 to 6 degrees inclination of upper to lower bed, are present within the laminated sediment. Soft sediment deformation and local microfaulting are common. Massive clayey diatomites with burrow marks at the contact between laminated and massive intervals occur sporadically throughout the section. The Sisquoc Formation at Jalama Beach shows all the sedimentary structures and textures of other areas, but is of higher diagenetic grade. At Jalama Beach the silica of diatoms has recrystallized to opal-CT. Elsewhere, the Sisquoc diatoms are opal-A.

The siliceous member of the Monterey Formation (Figures 4a and 4b) is similar to the Sisquoc, but contains more porcelanite and chert. The laminated clayey diatomite and bedded porcelanite alternate in two ways. On a larger scale they occur in alternating zones several meters thick. On a smaller scale, they are found as alternating beds 4 to 10 cm thick. The transformation of opal-A to opal-CT, seen in the field as the change from diatomite to porcelanite or chert, occurs over an interval 30 to 40 m thick. The stratigraphically highest occurrence of opal-CT is in isolated chert nodules. Chert nodules preserve diatomite textures, including laminations, which are continuous through the chert nodules and show no evidence of differential compaction.

Both the Sisquoc Formation and the siliceous member of the Monterey contain little carbonate. The carbonate present occurs mostly as dolostone nodules and nodular beds that constitute about < 1 to 2 vol % and 1 to 4 vol %, respectively, of each section. The nodules range in size from 4 to 60 cm vertically and 10 to 200 cm horizontally. They are symmetrical spheroids to elongated ellipsoids. Some are isolated

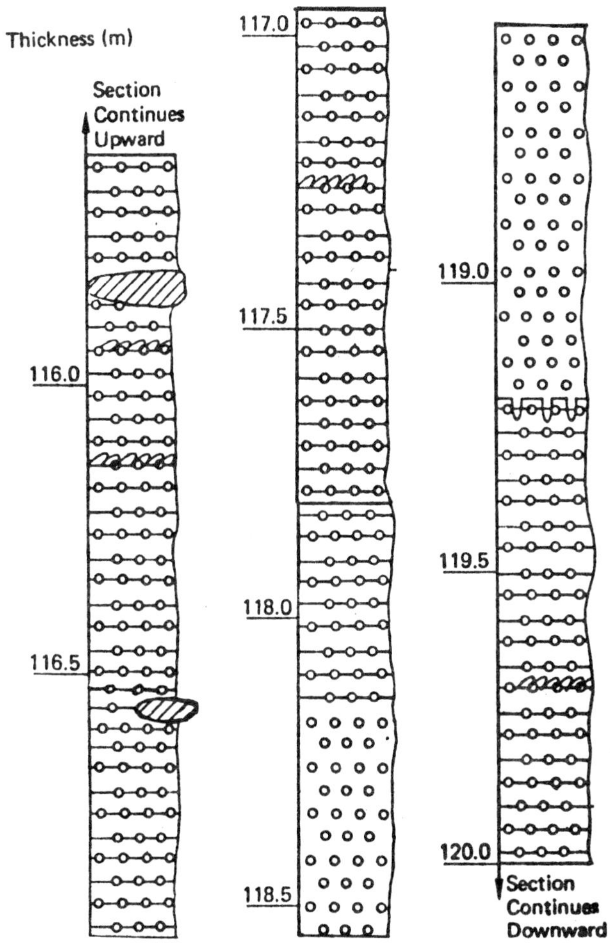

Figure 3a. Detailed measured section within the Sisquoc Formation, Pt. Pedernales area.

Figure 3b. Thickly laminated clayey diatomite typical of the Sisquoc Formation. Isolated dolostone nodule at center. Laminations are of variable thickness and randomly distributed, but laterally continuous except where microfaulted.

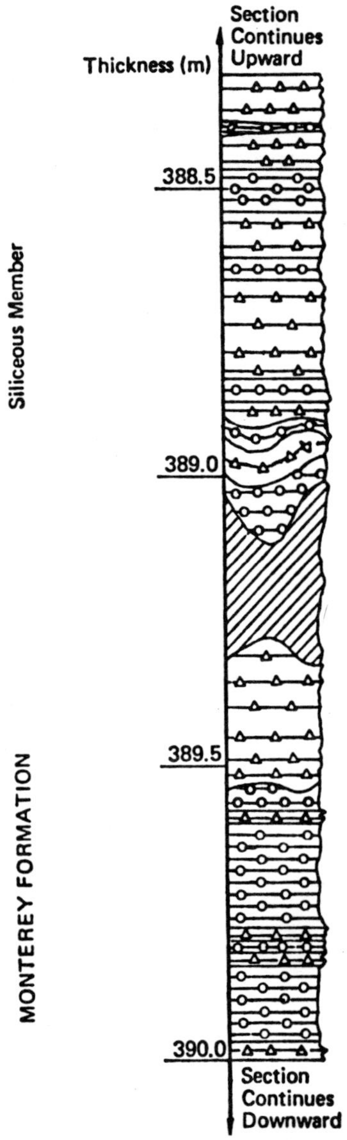

Figure 4a. Detailed measured section within the siliceous member, Monterey Formation, Sweeny Road area.

Figure 4b. Interbedded diatomite and porcelanite of the siliceous member, Monterey Formation in the Sweeny Road area. Nodular dolostone occurring along the same stratigraphic horizon with thin connecting dolostone beds outlined.

(Figure 3b); others are restricted to a common stratigraphic horizon. In some places nodules within the same horizon are connected by a thin dolostone layer (Figure 4b). Elsewhere, nodules have coalesced to form a nodular bed. The dolostone preserves all the sedimentary structures and textures of the host diatomite and, like the chert nodules, their laminations show no evidence of differential compaction. The diatomite and porcelanite beds between dolostone nodules contain little detectable carbonate. The dolostone contains 85 to 95 wt % dolomite, some of it consisting of dolomitized foraminiferal tests. Insoluble residues consist of unreplaced siliceous microfossils, organic matter, and detrital clay and silt.

UPPER CALCAREOUS-SILICEOUS MEMBER OF THE MONTEREY FORMATION

The upper calcareous-siliceous member of the Monterey is marked in all areas by an increase in carbonate content compared to the overlying siliceous member. It is also more variable in lithology and bed thickness than the overlying member (Figure 5). Figure 6 shows a detailed stratigraphic interval at Point Pedernales typical of the upper calcareous-siliceous member. The member is characterized by black and white banded chert beds. About one meter thick, many have intensely folded black and white bands. The member also contains more or less regularly spaced dolostone beds and rhythmically alternating beds of siliceous mudstone and porcelanite (Figures 7a and 7b). The member is present throughout the area, ranging in thickness from 305 to 670 m. There is a fairly sharp transition to the overlying siliceous member except at Point Pedernales, where the transition is gradational, occurring over many tens of meters. The upper calcareous-siliceous member at Point Pedernales has a higher detrital content than elsewhere. At Lion's Head it is more phosphatic and at Jalama Beach more silicic.

The interbedded siliceous mudstone and porcelanite occur most commonly as finely laminated, conformable, continuous beds with sharp contacts. They range from 4 to 12 cm in thickness. The two lithologies also occur as highly disrupted and locally faulted pinch-and-swell beds with siliceous mudstone laminations showing differential compaction relative to porcelanite laminations (Figure 7c). The porcelanite is hard, brittle, and highly fractured. The siliceous mudstone is soft and plastically deformed. There is little detectable carbonate between dolostone beds. Phosphatic laminations and nodules are rare except within certain intervals and in the Lion's Head area.

The amount of dolostone in the upper calcareous-siliceous member of the Monterey Formation is dramatically higher than in the overlying siliceous member or the Sisquoc Formation. Dolostone constitutes between 12 and 20 vol % of the rock present. The dolostone is far more variable in dolomite content than in the siliceous member, ranging from 30 to 95 wt %. Dolostone beds range from 20 to 200 cm in thickness, averaging 40 to 80 cm. They are more or less regularly spaced between 1 to 20 meters apart. In this member dolomitization is rarely confined to a specific lithology. More typically it involves an entire stratigraphic interval, 1 to 2 m in thickness, of alternating siliceous mudstone and porcelanite beds. All of the original sedimentary textures of the interbedded

Figure 5. White weathering dolostone beds and dark interbedded siliceous mudstone and porcelanite, upper calcareous-siliceous member, Point Pedernales area.

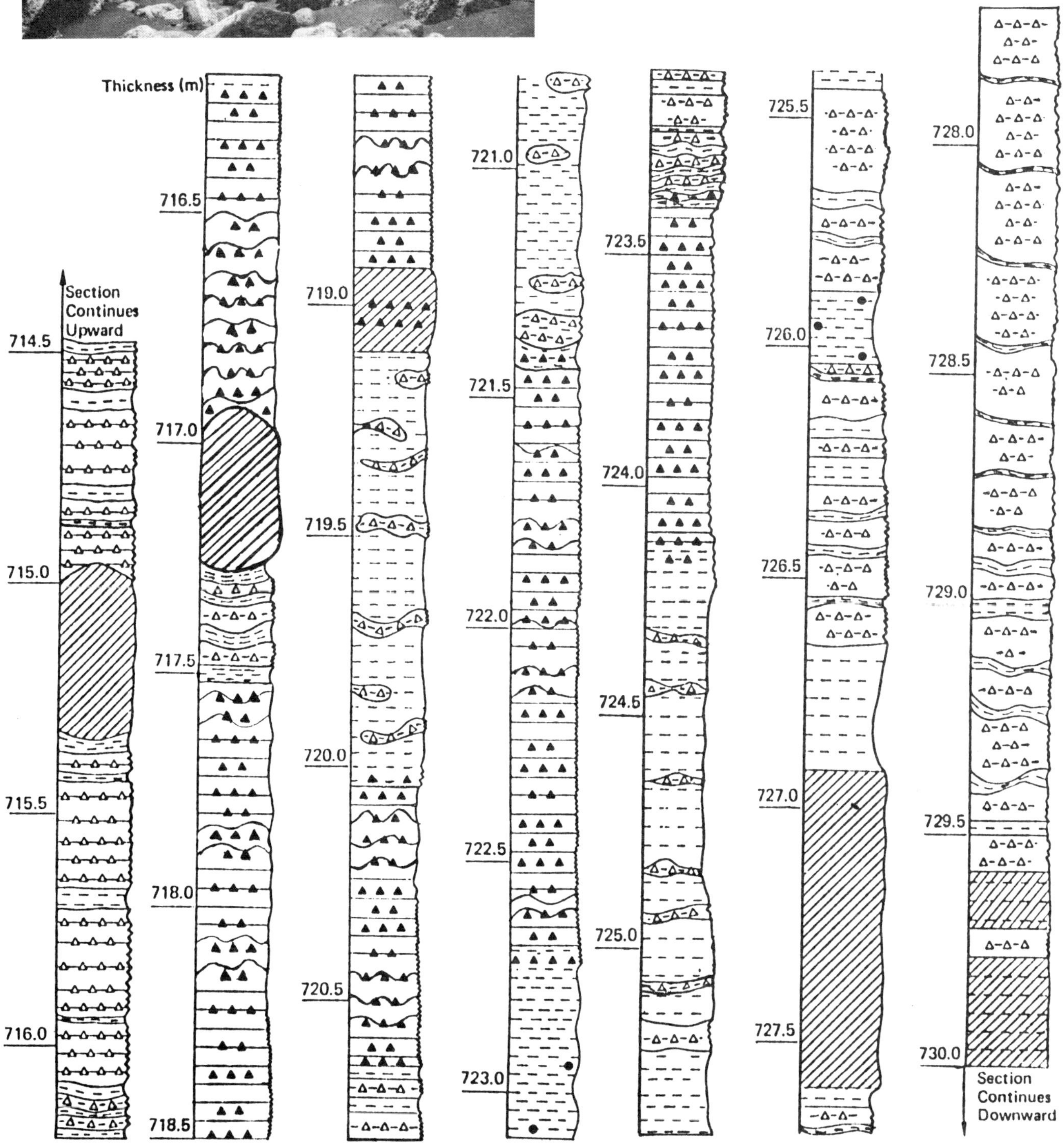

Figure 6. Detailed measured section in the upper calcareous-siliceous member of the Monterey Formation, Point Pedernales area.

Figure 7a. Dolostone bed (light colored, center) preserving texture of siliceous mudstone and porcelanite beds. Interbedded siliceous mudstone and porcelanite directly to the right of dolostone bed. Black and white banded chert to far right and left. Point Pedernales, Figure 6 (716.50-718.60m). Scale is 10 cm.

Figure 7b. Close-up of black and white banded chert bed to the right in gradational contact with adjacent dark siliceous mudstone on the left. Point Pedernales, Figure 6 (721.25-721.85m). Scale is 10 cm.

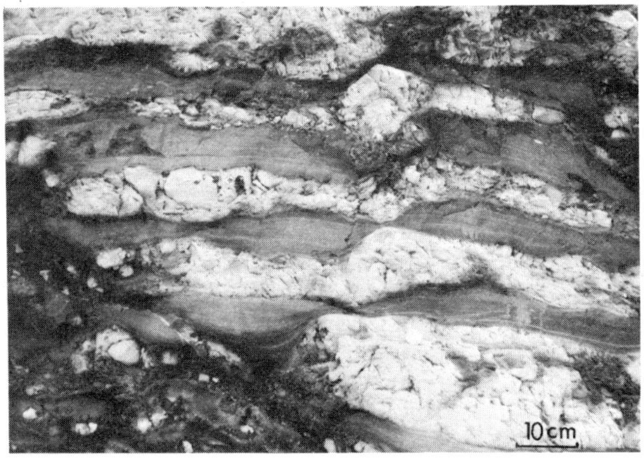

Figure 7c. Interbedded light-colored, highly fractured porcelanite and dark, plastically deformed siliceous mudstone. Both are finely laminated. Note faulted area in lower left corner. Upper calcareous-siliceous member, Pt. Pedernales.

Figure 7d. Close-up of dolostone bed in figure 5 showing preserved interbedded siliceous mudstone and porcelanite textures, and poorly defined lithologic contacts of dolostone bed.

Figure 7e. Highly irregular partial dolomitization (light-colored rock) of interbedded siliceous mudstone and porcelanite (dark-colored, upper left and lower right corners). Lithologies are laminated and in places the porcelanite occurs as discontinuous pinch-and-swell type beds.

Figure 7f. Light-colored, 2m thick dolostone bed and dolostone nodules. Nodules occur within the same stratigraphic horizon. Freshly exposed dark interbedded siliceous mudstone and porcelanite weather quickly to a white-gray color, Pt. Pedernales, upper calcareous-siliceous member.

siliceous mudstone and porcelanite are preserved (Figures 7d and 7e). The concentration of dolomite in such cases is not uniform throughout the dolostone bed, but varies among the lithologies present. Siliceous mudstone beds usually contain significantly more dolomite than porcelanite beds. Some dolostone beds contain internal laminations or thin beds of chert. Dolostone beds with a high silica content tend to be more resistant to erosion. Beds of dolomitized interbedded siliceous mudstone and porcelanite may be poorly defined by gradational contacts with adjacent beds. Homogeneous dolostone beds are common; large dolostone nodules and nodular beds are less common (Figure 7f).

PHOSPHATIC MEMBER OF THE MONTEREY FORMATION

The phosphatic member of the Monterey is exposed in all areas except Sweeny Road, where it is covered. The phosphatic member is characterized by beds rich in phosphate and organic matter. The member is highly variable in composition and occurrence among the different areas studied. The lithologies present throughout the entire section at Lion's Head contain abundant phosphatic matter; even the upper calcareous-siliceous member is highly phosphatic. In contrast, the phosphatic member in the Point Pedernales area has less phosphatic matter and a correspondingly higher silica and carbonate content. Phosphate rich stratigraphic intervals occur sporadically throughout the section at Point Pedernales, from the base up to the lower part of the upper calcareous-siliceous member. The transition between the upper calcareous-siliceous and the phosphatic member is gradual in most areas, and is characterized by a downward increase in phosphate rich beds and a highly variable decrease in silica and carbonate rich beds. One kind of phosphate occurs as discrete white particles ranging from 0.1 to 1.0 mm in diameter, randomly dispersed throughout the rock. Another variety is found as pure white isolated nodules or series of nodules, 1 to 2 cm in thickness, aligned along a stratigraphic horizon. Phosphate also occurs as randomly (0.2 to 5 cm apart) or evenly (2 to 3 cm apart) spaced laminations composed of concentrated phosphatic particles. The laminations are occasionally microfaulted.

In the phosphatic member dolostone occurs as beds, nodular beds, and as isolated nodules. Dolostone beds are typically 10 to 30 cm thick. Nodules are spherical to ellipsoidal and 10 to 50 cm across. Dolostone makes up 10 to 18 volume % of the member. Most dolostone beds replace the phosphatic mudstone lithology and have a homogeneously distributed dolomite content ranging from 50 to 90 wt %. The phosphatic laminations preserved in dolostone nodules are always about double the spacing of those in the laterally equivalent beds, indicating significant differential compaction. Where an interval has sufficient vertical homogeneity to allow comparison, the laminations in dolostone beds are either equally or more widely spaced than the phosphatic mudstone above and below, again indicating differential compaction. Both dolostone beds and nodules tend to be associated with stratigraphic intervals that are richer in phosphate and organic matter.

Figure 8a shows a segment of the phosphatic member at Lion's Head, where the major lithology present is a fairly homogeneous calcareous or siliceous phosphatic mudstone, rich in clay and organic matter. This mudstone's dolomitized phosphatic mudstone equivalent is shown in Figures 8b and 8c. Dolostone beds occur more or less regularly at 0.5 to 3 m intervals; the beds tend to be thicker where separated by thicker intervals. The phosphatic member at Lion's Head is calcareous; a few white laminae are composed entirely of foraminiferal tests. The dolostone is far more resistant to erosion than the surrounding phosphatic mudstone. The dolostone is highly fractured, with most fractures oriented perpendicular to bedding; many fractures are carbonate filled. At Lion's Head there are rare intervals several meters thick of massive mudstone and dolostone.

An interval of disturbed dolostone and phosphatic mudstone is also exposed in the Lion's Head area. In this interval blocks of a dolostone bed have parallel and undistorted laminations, whereas the surrounding mudstone has highly contorted laminations. The simplest interpretation of this structure is slumping of the still soft mudstone after at least initial lithification of the dolomite bed. This origin implies early dolomitization for these beds.

Figure 9a shows a typical part of the phosphatic member at Point Pedernales, where the major lithologies are phosphatic mudstone, porcelanite and dolostone. There is far less phosphatic matter and it occurs much more sporadically than at Lion's Head. The silica content is much higher at Point Pedernales and the dolostone occurs mostly as homogeneous to partly phosphatic beds (Figures 9b, c and d) or as large isolated nodules (Figure 10). The interbedded mudstone and porcelanite within this interval contain no detectable carbonate. There are discrete chert laminations or chert blebs within the center of many porcelanite beds (Figure 9c).

SUMMARY AND DISCUSSION

Stratigraphic Occurrence of Dolostone and Timing of Dolomitization

The varied yet distinctive modes of stratigraphic occurrence of dolostone in the Sisquoc and Monterey Formations are summarized in Table I. The presence or absence of differential compaction in dolostone occurring within laminated lithologies can be used to determine the timing of lithification relative to other diagenetic processes. Early dolostone is indurated by dolomite pore-filling and replacement processes that took place before appreciable compaction. Early <u>dolostone nodules</u> have laminations spaced about twice as far apart as adjacent beds, depending on the extent to which the original sediment was able to compact. Laminations above and below early <u>dolostone nodules</u> are usually deformed around the nodule. Early <u>dolostone beds</u> have laminations spaced about twice as far apart as beds above and below. Only by assuming that laminations are evenly spaced throughout the section, which is rarely the case, could we deduce with confidence differential compaction. Some have deduced that early dolostone may start to form within meters of the sediment-seawater interface and continue to a burial depth of 200 to 500 m (Pisciotto and Mahoney, 1981). Late dolostone forms after compaction is largely completed, i.e. at the depth beyond which the porosity changes little with depth, and some late dolostone may form during or after uplift.

In many places dolomitization will include both

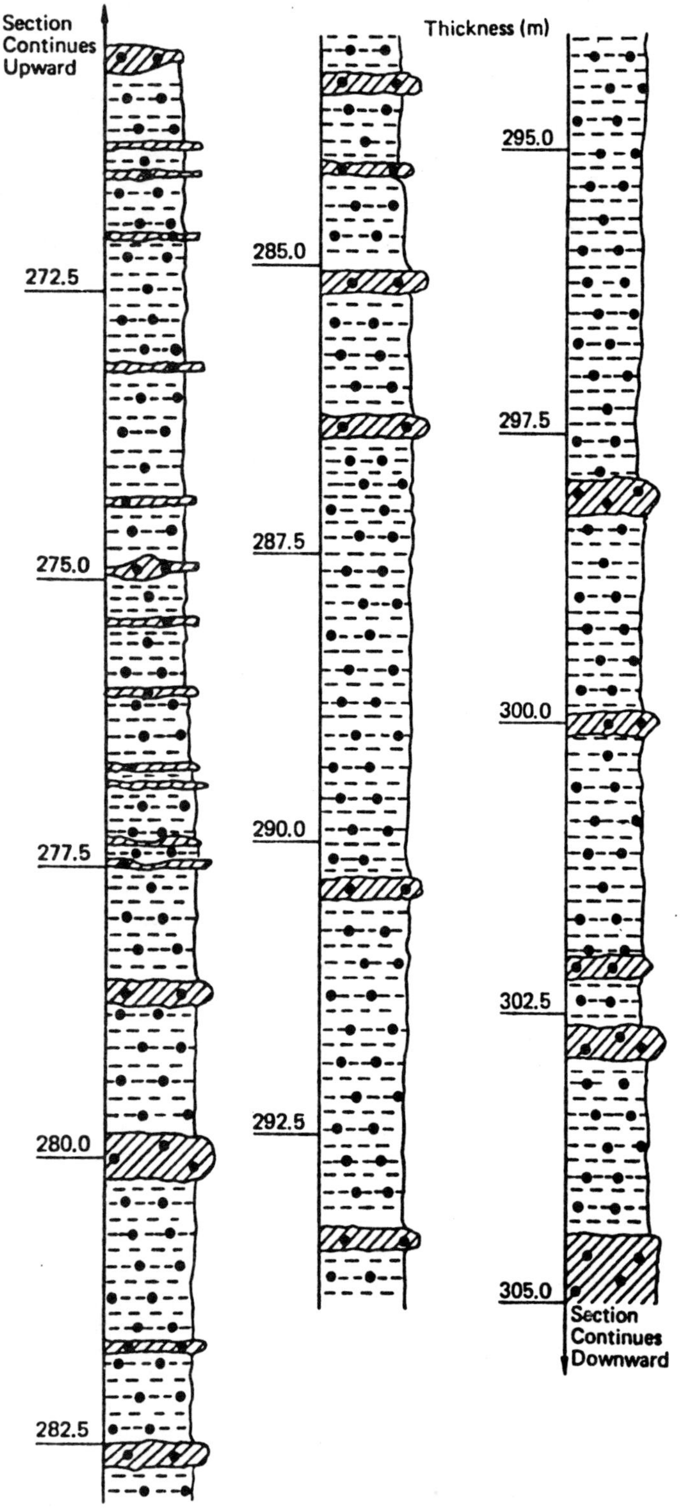

Figure 8a. Detailed measured section within the phosphatic member of the Monterey Formation in the Lion's Head area.

Figure 8b. Partial overview of Figure 8a, Lion's Head area. More or less evenly spaced nodular to stratiform dolostone beds within phosphatic mudstone.

. .gure 8c. Close-up of Figure 8b just to the right of hammer head. Continuous phosphatic laminations and nodules in mudstone in sharp contact with dolostone bed. Dolostone has finer and more randomly spaced laminations, is more resistant to erosion, and has numerous fractures perpendicular to bedding, many of which are carbonate filled.

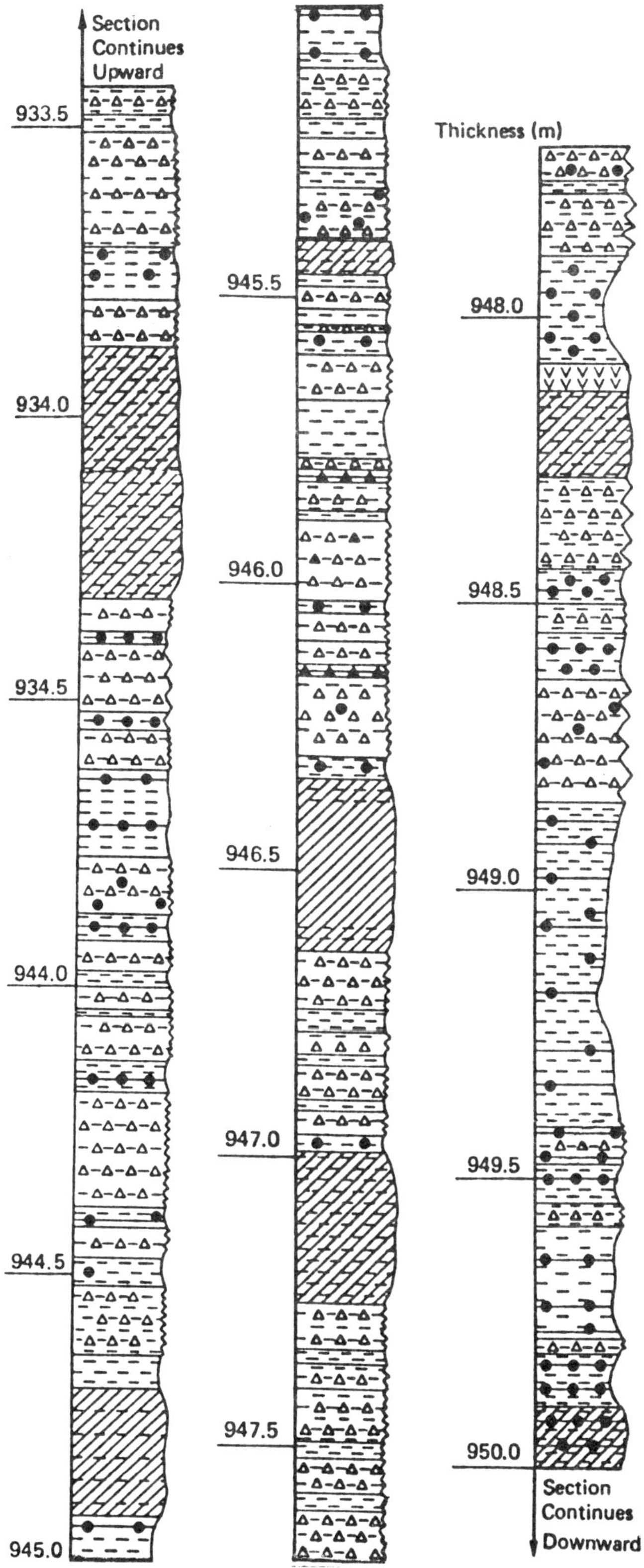

Figure 9a. Detailed measured section within the phosphatic member of the Monterey Formation in the Point Pedernales area.

Figure 9b. Overview of Figure 9a within the phosphatic member, Point Pedernales area (933.40-950.00m). Continuous dolostone beds (labeled D), and interbedded siliceous mudstone and porcelanite that are phosphate rich in places.

Figure 9c. Close-up of Figure 9b (third dolostone from the left) showing massive dolostone bed with several thick laminations confined to left side. Interbedded dark phosphatic mudstone and light-colored porcelanite, black chert laminations in porcelanite to the right of 10 cm scale. 946.02-947.10m.

Figure 9d. Close-up of fourth dolostone bed from left in Figure 9b (947.10-947.50m) showing massive clayey dolostone at far left and interbedded dark phosphatic mudstone and light-colored porcelanite. Scale is 10 cm.

a central nodule with laminations spaced more widely apart than at the periphery, and the adjacent and surrounding beds. Dolomitization of the compactionally distorted laminations around the nodule give evidence for progressive dolomitization during burial (Figure 10).

Dolostone has two principle modes of occurrence: nodular and bedded. Nodular dolostone occurs in all members; it is the only type of dolostone in the Sisquoc Formation and siliceous member of the Monterey. Nodules usually have gradational contacts, with highly dolomitic centers becoming less dolomitic outward into the host lithology. There is evidence of differential compaction between the laminations of most nodules and adjacent beds in all parts of the Monterey Formation except for the uppermost part of the siliceous member. This suggests that the nodules in the lower parts of the section formed early. However, most nodules in the Sisquoc Formation and the higher parts of the siliceous member of the Monterey show no evidence of differential compaction, suggesting that they formed later. But the above argument assumes an extent of compaction sufficient for differential compaction to be observed. Given porosities as high as 67% in the surrounding thickly laminated host diatomite, the Sisquoc Formation and siliceous member of the Monterey may not have undergone enough compaction after dolomitization to show that the nodules formed early. From field evidence alone we cannot resolve the choice between late origin or insufficient general compaction to show early origin.

Many nodules occur along a single stratigraphic horizon, as observed by Bramlette (1946) and Isaacs (1981a). Some nodular beds appear to have formed by the coalescence of numerous nodules. Observations of dolomite-filled diatom frustules, relict diatom ghosts in dolomite, and general preservation of sedimentary structures of the host lithology in dolostone thin sections suggest that nodules formed by both pore-filling and replacement processes. In dolomitization of porous diatomite pore-filling is probably the dominant process (Murata and others, 1969).

The other common occurrence of dolostone is as beds that contain between 25 to 95 wt % dolomite. The timing of dolostone bed formation is not obvious from the field evidence. In the upper calcareous-siliceous member beds of dolomitized interbedded siliceous mudstone and porcelanite have a variable dolomite content that seems to depend on the original composition and porosity of the host sediment. Murata and others (1969) studied dolostones from many localities in California, and found that the dolomite content, which ranged between 40 and 85 wt %, was generally dependent on the presumed original porosity of the host sediment, increasing with greater porosity. If the present dolomite content largely depends on the porosity of the host sediment at the time of dolomitization, and if dolomitization occurred before the silica phase transformation, then the dolomite content of the porcelanite should be as high or higher than that of the siliceous mudstone. This assumes that a higher silica content in the form of a diatom-rich, clay-poor, high porosity host sediment would favor its more rapid transformation from opal-A to opal-CT. And yet siliceous mudstone usually has a significantly higher dolomite content than porcelanite, and contains many dolomite-filled diatom frustules. In addition, if pore-filling is a major mechanism of dolomitization, then

Figure 10. Large isolated dolostone nodule, with thick laminations and phosphatic matter, phosphatic member, Point Pedernales area. Note dolomitization of compactionally distorted laminations above and to the right of hammer.

dolomitization would tend to create a more closed (less permeable) rock, making it difficult for the opal-A to opal-CT solution-dependent reaction to occur. This is supported by the presence of metastable opal-A diatoms in the insoluble residue of early formed dolostones from opal-CT and quartz grade diagenetic zones.

These dolostone beds show no evidence of differential compaction, and texturally resemble the carbonate-free interbedded siliceous mudstone and porcelanite. Fracturing of the several rock types in the upper calcareous-siliceous member gives clues to their origin. Porcelanite is highly fractured with curved, cross-cutting fractures. Dolostone has fewer, more widely spaced fractures that run more or less perpendicular to bedding. Siliceous mudstone is broken by hairline fractures. Dolostone beds not only have their characteristic perpendicular fractures, but also retain the fracture patterns, as well as lamination textures, of the replaced porcelanite and siliceous mudstone. Those fractures observed in thin sections that cross cut opal-CT-dolomite and quartz-dolomite laminations are dolomite filled. Some quartz particles include secondary dolomite rhombohedra, suggesting some dolomite precipitation followed quartz precipitation. It is unclear whether the fracture-filling dolomite is syngenetic with the finely crystalline matrix dolomite (10 to 50 microns) that is intergrown with cryptocrystalline opal-CT and microcrystalline quartz.

The above evidence is not conclusive in deciding the relative timing of the silica phase transformation and dolomitization, but does suggest the following sequence of events: (1) opal-A to opal-CT phase transformation, forming interbedded siliceous mudstone and porcelanite; (2) fracturing of the porcelanite; and (3) dolomitization.

The spacing of phosphatic laminations within numerous dolostone beds is greater than in the beds above and below, suggesting that dolomitization of the phosphatic mudstone was early with respect to compaction in the phosphatic member. However, phosphatic laminations are usually too randomly spaced within the stratigraphic interval of interest to be conclusive. The timing of dolomitization is

Table I.

SUMMARY OF LITHOSTRATIGRAPHY AND DOLOSTONE OCCURRENCE, SANTA MARIA BASIN AREA

	Thickness (m)	Lithology	Vol % Dolostone	Dolostone Occurrence
Sisquoc Formation	0 - >400	Clayey diatomite or porcelanite, thickly laminated or massive.	<1 - 2	Late(?) isolated, round to elongate ellipsoidal nodules 10-80cm across, gradational and sharp contacts.
Monterey Formation	300 - 1000			
Members				
Siliceous	0 - 350	Medium to thickly laminated or massive diatomite, and 4-10cm or 1-4m thick interbeds of porcelanite. Chert nodules, nodular beds minor.	1 - 4	Late(?) 4-30cm thick elongate nodules that occur isolated, in same horizon, or as 20-40cm thick nodular beds.
Upper Calcareous-Siliceous	50 - 600	Rhythmically interbedded siliceous mudstone and porcelanite, continuous and discontinuous 4-12cm thick beds. 1m black and white banded, often intensely folded chert beds. Rare phosphatic mudstone. Phosphate rich at Lion's Head.	12 - 20	1m thick dolomitized interbedded siliceous mudstone and porcelanite. 20-80cm thick stratiform beds, homogeneous and laminated. Occasional large early isolated ellipsoid nodule or nodular bed.
Phosphatic	0 - 200	Organic-rich phosphatic mudstone. Porcelanite common at Point Pedernales; calcareous at Lion's Head.	10 - 18	Early and late(?) dolomitized phosphatic mudstone beds or as 20-80cm thick early nodules. Associated with sediment intervals high in phosphate.
Lower Calcareous-Siliceous	0 - 150	Porcelanite, siliceous mudstone (phosphatic in places), and tuffaceous beds.	20 - 30(?)	Dolomitized phosphatic mudstone beds.

even less clear for homogeneous, nonphosphatic dolostone beds that occur in both the upper calcareous-siliceous and phosphatic members.

Other than in dolostone beds, carbonate is either absent or a minor constituent of other rock types except in the Lion's Head area where calcareous rocks are common. The carbonate occurs dispersed throughout the rock as biogenic, low-Mg calcite foraminiferal and coccolith tests, recrystallized calcite, or as discrete dolomite rhombs. Carbonate veins and carbonate filled vugs are common within calcareous rich intervals. There are rare zones of extensively brecciated dolostone at Point Pedernales and Jalama Beach. In the breccias angular rock fragments of varied lithology are carbonate- and silica-cemented, with evidence of several brecciation and cementation episodes.

Comparison with the Santa Barbara Coastal Area

The lithostratigraphy of the areas studied in the Santa Maria basin is generally similar to, but has distinct differences from that in the Santa Barbara coastal area (Isaacs, 1981a). Isaacs (1983) has recently revised member names so that the phosphatic and transition members referred to below are referred to as the carbonaceous marl and transitional marl-siliceous members. The average thicknesses of all the members except the lower calcareous-siliceous member are far greater in the Santa Maria basin area where the Monterey Formation is 500 to 800 m thick, as opposed to about 400 m thick in the Santa Barbara coastal area. All the lithologies of the Santa Maria area tend to be more thickly laminated with the average thickness of lamination couplets ranging between 0.1 and 3.0 mm versus 0.02 to 0.2 mm for the Santa Barbara coastal area.

Other than the difference in lamination thickness, the lithologies of the Sisquoc Formation and siliceous member of the Monterey are similar in the two areas. Rock types of the upper calcareous-siliceous member of the Monterey in both areas have highly variable compositions and significantly greater carbonate contents. However, the carbonate distribution is different. In the Santa Maria basin almost all the carbonate is concentrated in dolostone beds and nodules, with most interbedded lithologies devoid of or very low in carbonate. In the Santa Barbara coastal area all rock types within the upper calcareous-siliceous member are calcareous or dolomitic, and most contain between 5 and 30% carbonate. Black and white banded chert beds, about a meter thick and often intensely folded, are common within the upper calcareous-siliceous member of the Santa Maria basin area. The same kinds of beds occur only locally and are uncommon in the Santa Barbara coastal area.

The phosphatic member of the Santa Maria basin and its equivalent phosphatic and transition members of the Santa Barbara coastal area are similar, with a downward gradational decrease in siliceous rock types and an increase in fairly homogeneous, organic-rich phosphatic mudstone. But this member in the two areas differs in its carbonate content and distribution. In the Santa Maria basin area, dolostone beds and early nodules are more common in the phosphatic member, constituting 10 to 18 vol % of the section. Most interbedded rock types have a low carbonate content, except in the Lion's Head area. In contrast, the phosphatic and transition members of the Santa Barbara coastal area are less dolomitic and far more calcareous, with most rock types present containing 25 to 75% calcite.

The lower calcareous-siliceous member appears to be similar in both areas, except that it includes more dolostone and tuffaceous beds and is more poorly exposed in the Santa Maria basin area. The underlying formation in the Santa Barbara coastal area is the Rincon Shale as opposed to the Tranquillon Volcanics, although a volcanic unit that correlates with the Tranquillon Volcanics (Dibblee, 1950) does occur locally between the Rincon and Monterey Formations near Santa Barbara.

Origin of Dolostone

The controls on the site of initial dolomitization and continued growth of a nodule or bed are not completely understood but appear related to the composition and physical properties of the host lithology as well as the interstitial water chemistry. The original carbonate content and composition of a dolostone-bearing horizon is difficult to determine because most of the carbonate has been dolomitized. Some dolomitized foraminifera, presumably originally low-Mg calcite, have been observed in nodules within the Sisquoc Formation and Monterey siliceous member. In areas where calcareous lithologies occur, low-Mg calcite foraminiferal and coccolith tests are commonly preserved, suggesting that a major source of carbonate for diagenetic carbonate is biogenic calcite. Most nodules and beds do not appear to contain any phases or textural elements that may have acted as nucleation sites for dolomitization, other than foraminiferal tests. Recent detailed isotopic and mineralogical studies show that early nodules can grow continuously outward from the center, whereas an early dolostone bed was found to be homogeneous, with no systematic geochemical trends (Kushnir and Kastner, this volume).

Fifteen dolostone samples from the Point Pedernales area contain 0 to 12.7 wt % total organic matter* (mean of 3.3 wt %), 0.1 to 3.4 wt % pyrite (mean of 0.95 wt %), and 6 to 26 wt % clay (mean of 14 wt %), along with detrital quartz and feldspar. There is a positive correlation between the organic matter, pyrite and clay contents as would be expected from the general relations known for marine sediments. The Monterey Formation is rich in organic matter with a mean of 6 to 13 wt % (Isaacs, 1983). For dolomitization, the organic matter is an important source of carbon that is oxidized to carbonate, and pyrite is evidence of a reducing environment in which sulfate, a dolomite crystallization inhibitor, has been reduced to sulfide (Baker and Kastner, 1981). Detrital clay is one potential source of Mg for dolomitization through cation exchange with the interstitial water.

In addition to dolomite formed by the replacement of biogenic calcite, primary dolomite can form directly from solution, with Ca and Mg in interstitial waters derived from seawater and carbonate from the oxidation of organic matter. If sulfate is removed by sulfate reducing bacteria, and the sediment contains sufficient calcite and organic matter, then dolomitization will depend only on a source of Mg.

Sufficient Mg can be obtained by diffusion of Mg from the overlying seawater to account for the dolomite present if dolomitization occurs at burial depths of less than 500 m. Continued growth of a dolostone nodule or bed depends on the continual supply of Ca, Mg and carbonate. A more complete discussion on the controls of dolomitization in relation to the stratigraphic occurrence of dolostone, taking into account the effects of Mg diffusion, sedimentation rates, etc. is developed in detail elsewhere (Compton and Siever, 1984).

From observations on the occurrence of various rock types present in the Monterey Formation, it is clear that in addition to original differences in sediment composition, diagenetic processes involving a solution-precipitation mechanism have caused significant redistribution of the original sediment. The biogenic constituents of the sediment: carbonate, silica, and some organic matter are considered as mobile; whereas the detrital clay frameworks, silt, and more refractory organic matter are considered immobile. In clays aluminum is not easily mobilized, but the silica and cations may be. Bramlette (1946) proposed that carbonate was leached from adjacent beds and recrystallized in concentrated nodules or beds. He suggested that the relative abundance of carbonate within a horizon determined whether or not nodular or bedded dolostone forms. Compared with the Santa Barbara coastal area, the Santa Maria basin area has more nodular and bedded dolostone and the lithologies occurring between dolostone beds are far less calcareous. Other than in the phosphatic member, the total carbonate abundance between the two areas does not appear to be significantly different. Based on the difference in their carbonate mineralogy and occurrence, the two areas probably had different diagenetic histories. The common occurrence within the Santa Maria basin area of meter thick chert beds as well as thin beds, nodules, and laminations of chert within porcelanites and dolostones, suggests that silica redistribution was also significant and in many places closely associated with carbonate redistribution.

ACKNOWLEDGEMENTS

We gratefully acknowledge financial support provided for this work by NSF Grant #EAR-82-12261, and the Chevron Oil Field Research Company. Part of this work was completed by J. Compton while working as a summer employee at Chevron Oil Field Research Company, and he would like to give special thanks to the following individuals and groups at Chevron Oil Field Research Company for their help and suggestions: A. B. Carpenter, T. C. MacKinnon, S. L. Mitchell, W. E. Seixas, the Biostratigraphy and Office Services groups, and the summer employment program. A. B. Carpenter reviewed the manuscript.

*total organic matter = (total carbon- carbonate carbon) x 1.45

REFERENCES CITED

Baker, P.A. and Kastner, M., 1981, Constraints on the formation of sedimentary dolomites: Science, 213, p. 214-216.

Bramlette, M.N., 1946, The Monterey Formation of California and the origin of its siliceous rocks: U.S. Geol. Survey Prof. Paper 212, 57 p.

Canfield, C.R., 1939, Subsurface stratigraphy of Santa Maria Valley oil field and adjacent parts of Santa Maria Valley, California: Bull. Am. Ass. Petrol. Geol., 23, p. 45-81.

Compton, J.S. and Siever, R., 1984, Magnesium mass balance calculations for the Monterey Formation, Santa Maria basin, California (abstr.): Program and Abstracts, 59th Annual Meeting; Pacific Sections AAPG, SEG and SEPM, San Diego, April 1984. Also, ms. in preparation.

Dibblee, T.W., Jr., 1950, Geology of southwestern Santa Barbara County, California: California Div. of Mines and Geol. Bull. 150, 95 p.

Grivetti, M.C., 1982, Aspects of stratigraphy, diagenesis, and deformation in the Monterey Formation near Santa Maria-Lompoc, California: Unpublished Master's Thesis, Univ. of California Santa Barbara, 155 p.

Isaacs, C.M., 1980, Diagenesis in the Monterey Formation examined laterally along the coast near Santa Barbara, California: Stanford University, unpublished Ph.D. thesis, 329 p.

_____, 1981a, Field characterization of rocks in the Monterey Formation along the coast near Santa Barbara, California. In: C.M. Isaacs (Editor), Guide to the California coastal area, Ventura to San Luis Obispo: Amer. Ass. Petrol. Geol. Pacific Section 52, p. 39-53.

_____, 1981b, Lithostratigraphy of the Monterey Formation, Goleta to Point Conception, Santa Barbara coast, California. In: C.M. Isaacs (Editor), Guide to the Monterey Formation in the California coastal area, Ventura to San Luis Obispo: Amer. Ass. Petrol. Geol. Pacific Section 52, p. 9-23.

_____, 1983, Compositional variation and sequence in the Miocene Monterey Formation, Santa Barbara coastal area, California. In: Larue, D.K. and Steel, R.J. (Editors), Cenozoic marine sedimentation, Pacific margin, U.S.A., p. 117-132.

Murata, K.J., Friedman, I. and Madsen, B.M., 1969, Isotopic composition of diagenetic carbonates in marine Miocene formations of California and Oregon: Geol. Survey Prof. Paper 614-B, 24 p.

Pettijohn, F.J., 1975, Sedimentary rocks (third edition): Harper and Row, New York, 628 p.

Pisciotto, K.A., 1981, Diagenetic trends in the siliceous facies of the Monterey Shale in the Santa Maria region, California: Sedimentology, 28, p. 547-571.

Pisciotto, K.A. and Mahonev, J.J., 1981, Isotopic survey of diagenetic carbonates, DSDP Leg 63. In: Yeats, R.S., Haq, B.U., et al. (Editors), Initial Reports of the Deep Sea Drilling Project 63, U.S. Govt. Printing Office, Washington, D.C., p. 595-609.

Shrock, R.R., 1948, A classification of sedimentary rocks: J. of Geology, 56, p. 118-129.

Surdam, R.C. and Stanley, K.O., 1981, Stratigraphic and sedimentologic framework of the Monterey Formation, Pismo Syncline, California. In: C.M. Isaacs (Editor), Guide to the Monterey Formation in the California Coastal Area, Ventura to San Luis Obispo, Amer. Ass. Petr. Geol., Pacific Section, 52, p. 83-91.

Woodring, W.P., Bramlette, M.N. and Lohman, K.D., 1943, Stratigraphy and paleontology of the Santa Maria district, California: Bull. Amer. Ass. Petrol. Geol., 27, p. 1335-1360.

Woodring, W.P. and Bramlette, M.N., 1950, Geology and paleontology of the Santa Maria district, California: Prof. Paper U.S. Geol. Surv. 222, 185 p.

the Monterey Formation, California, in Garrison, R.R. et al., eds., The Monterey Formation and Related Siliceous Rocks of California: Spec. Publ. Soc. Econ. Paleontologists and Mineralogists, Pacific Sec., p. 273-283.

Pisciotto, K.A., and Mahoney, J.J., 19[illegible], Isotopic Survey of Diagenetic Carbonates, Deep Sea Drilling Project Leg 63, in Yeats, R.S., Haq, B.U., et al.ed., Init. Reports Deep Sea Drilling Project, Leg 63: Washington (U. S. Govt. Printing Office) p. 595-609.

Porter, W.W.,III, 1932, Lower Pliocene in Santa Maria District, California: Am. Assoc. Petroleum Geologists Bull., v. 16 No. 2, p. 135-143.

Potter, P.E., and Pettijohn, F.J., 1977, Paleocurrents and Basin Analysis: New York, Springer-Verlag, 425 p.

Potter, P.E., Maynard, J.B., and Pryor, W.A., 1980, Sedimentology of Shale: New York, Springer-Verlag, 306 p.

Savin, S.M., Douglas, R.G., Stehli, F.G., 1975, Tertiary marine paleotemperatures: Geological Society of America, v. 86, p.1499-1510.

Sheppard, S. M. F., and Schwarcz, H. P., 1970, Fractionation of carbon and oxygen isotopes and magnesium between metamorphic calcite and dolomite: Contrib. Mineralogy and Petrology, v. 26, p. 161-198.

Woodring, W.P., and Bramlette, M.N., 1950, Geology and paleontology of the Santa Maria District, California: U. S. Geological Survey Prof. Paper # 222, 185 p.

Woodring, W.P., Bramlette, M.N., and Lohman, K.E., 1943, Stratigraphy paleontology of the Santa Maria District, California: Am. Assoc. Petroleum Geologists Bull., v. 27 # 10, p. 1335-1360.

DISSEMINATED DOLOMITE IN THE MONTEREY FORMATION, SANTA MARIA AND SANTA BARBARA AREAS, CALIFORNIA

Caroline M. Isaacs
U.S. Geological Survey
Menlo Park, CA 94025

ABSTRACT

Though mainly studied in dolostones (>75% dolomite), the mineral dolomite is more abundant in the Monterey Formation of the Santa Maria and Santa Barbara areas within other rock types. Estimates based on data from both surface and subsurface Monterey samples from these areas suggest that (1) only about 2-15% of the dolomitic rocks are dolostones, and (2) the dolomite in dolostones represents only about 10-35% of the total dolomite, whereas the dolomite disseminated in other rock types represents as much as 65-90% of the total dolomite.

Minor amounts of disseminated dolomite are widely present in calcareous strata, and all zones where disseminated dolomite is widespread and abundant are sedimentologically similar to and stratigraphically correlative with calcareous members. Field relations indicate that disseminated dolomite was formed from and replaced, in a broad sense, biogenous calcite locally. Though more common in sequences where silica is diagenetic, disseminated dolomite is not consistently more abundant in rocks exposed to higher maximum burial temperatures (as inferred from silica diagenesis) or to greater depth (as determined from superposition). For example, many zones 50-100 ft thick in which disseminated dolomite is ubiquitous are underlain by thick (500+ ft) calcareous sequences with little disseminated dolomite. Moreover, all the carbonate minerals are dolomite in some sequences of opal CT strata both onshore (as at Cojo and Jalama Canyons) and offshore (as in the South Elwood Oil Field), whereas all the carbonates are calcite in some sequences of quartz strata in the same areas.

The carbon-isotope ratios of disseminated dolomite in 10 samples range from -8.2 to +3.8 permil $\delta^{13}C$; they are close to values in comparatively unaltered and partially recrystallized biogenic calcite (-12.1 to +1.5) and much more restricted in range than values (-15 to +21) reported in the literature from dolomite in Monterey dolostones. This difference suggests that organic-geochemical reactions may have been much less involved in the formation of disseminated dolomite than in the formation of dolomite in dolostone. The oxygen-isotope ratios of the disseminated dolomite samples range from -5.2 to +0.5 permil $\delta^{18}O$, indicating that this type of dolomite formed in the higher part of the temperature range inferred for dolomite in dolostones and also at somewhat higher temperatures.

Formation of disseminated dolomite did not generally result in tight cementation but did somewhat increase rock cohesion. In most rocks, only slight reduction in porosity and slight increase in dry bulk-density apparently accompanied the formation of disseminated dolomite, indicating that pore-filling was not a major associated process.

INTRODUCTION

The mineral dolomite in the Miocene Monterey Formation is most commonly studied as a constituent of dolostone, a rock containing a predominance of dolomite generally concluded to have formed by primary precipitation in association with carbon dioxide and methane released during maturation reactions of organic matter (Murata and others, 1969, 1972; Pisciotto, 1981b). Detailed analyses of numerous sequences of the Monterey have shown, however, that dolomite is also common in rock types other than dolostone as a constituent in minor or moderate abundance (Isaacs, 1980, 1981a; Surdam and Stanley, 1981). Detailed field work in the Santa Barbara coastal area (Fig. 1) indicates that such dolomitic rocks had calcareous precursors and that the dolomite disseminated in such rocks derived from and broadly replaced the original calcite.

The contrast in occurrence and inferred processes of formation of these two types of dolomite suggests that more detailed study of disseminated dolomite may be important to an understanding of dolomite in the Monterey. Because the total number of beds containing dolomite in rock types other than dolostone far outweighs the number of dolostones in the Monterey of the Santa Barbara and Santa Maria areas, the study of the disseminated dolomite in these areas may elucidate major mechanisms of dolomitization not important in the formation of the dolomite in dolostone. Inasmuch as the total amount of dolomite disseminated in other rock types outweighs the amount in dolostones in the Monterey of the Santa Barbara and Santa Maria areas, determining the abundance of disseminated dolomite is also critical in the evaluation of magnesium sources for dolomite formation in these areas. Moreover, the formation of disseminated dolomite is important to stratigraphic correlation in offshore areas inasmuch as "dolomite" (dolostone) is a principal lithologic category in mud-logs. Because of their hardness and cohesiveness, marls that have been dolomitized would usually be categorized in mud-logs separately from unaltered calcareous marls, which are quite chalky and comparatively friable. Both in the onshore and offshore, however, dolomitized sequences can be adjacent to depositionally equivalent undolomitized sequences in puzzling patterns which make "dolomite" an unreliable correlation tool.

The main objective of this paper is to focus attention on a geologically significant form of dolomite that has received little direct study in geochemical publications on dolomite in the Monterey (Murata and others, 1969, 1972; Pisciotto, 1981b). The paper describes the geologic occurrence of disseminated dolomite in the Monterey of

in Garrison, Robert E.; Kastner, Miriam and Zenger, Donald H., eds., 1984, **Dolomites of The Monterey Formation and Other Organic-Rich Units:** Pacific Section S.E.P.M., Vol. 41, p. 155-170.

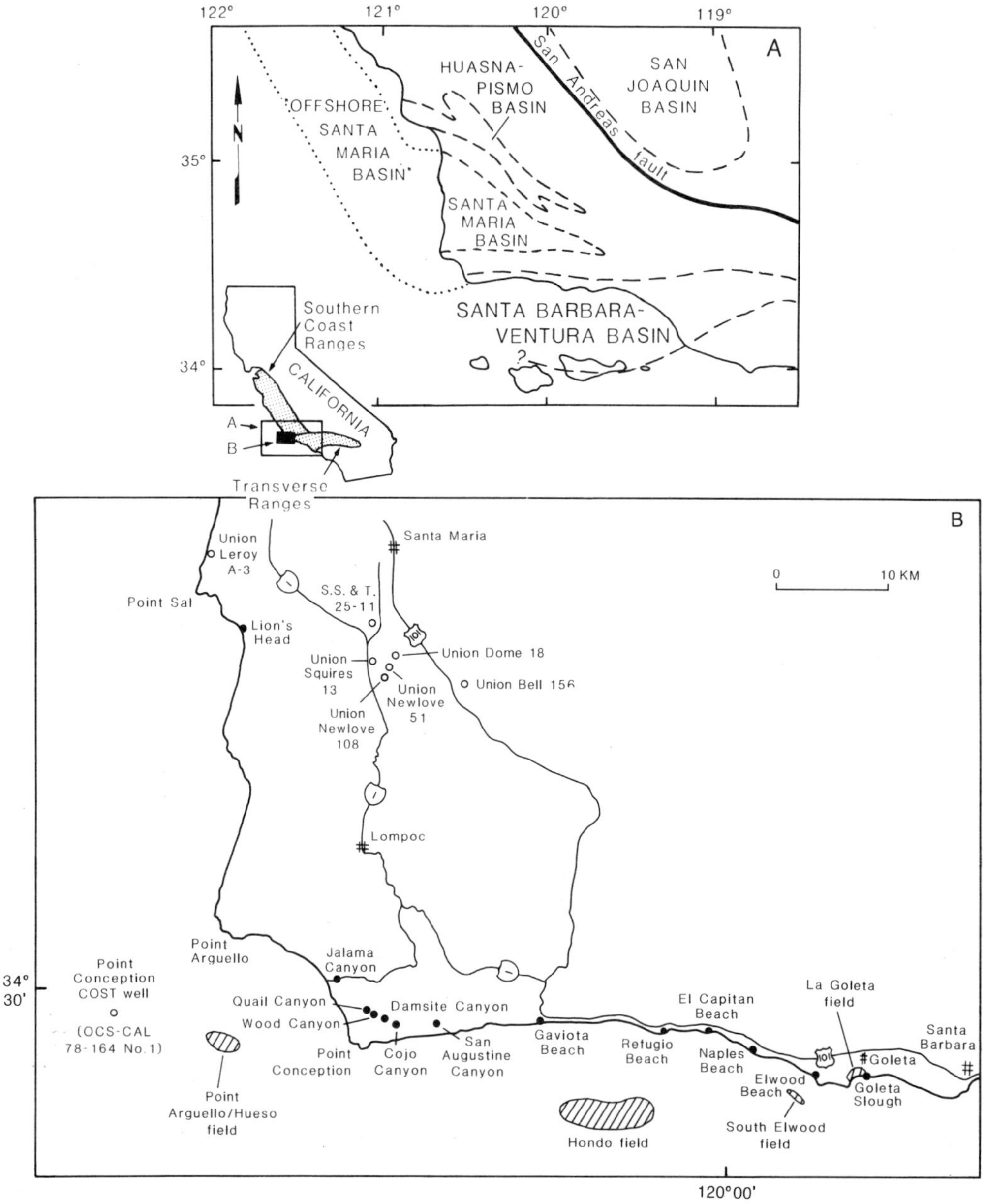

Figure 1. Maps of the Santa Barbara and Santa Maria areas, California.

the Santa Maria and Santa Barbara areas, estimates the relative proportion of dolomite in dolostones and in other rock types, describes the relation between silica diagenesis and formation of disseminated dolomite, and summarizes the analytical data available on isotopic composition, density, and porosity in this type of dolomite.

TERMS

Dolostone, a term which is variously used to indicate a rock having a minimum dolomite content anywhere in the range of 50-90% (Bates and Jackson, 1980), is here restricted to rocks with 75% or more dolomite by weight. This definition generally encompasses the highly cemented dolomite nodules and beds classed as "diagenetic carbonates" in the Monterey Formation and excludes rocks that are recognizable in the field as dark siliceous mudstones, porcelanites, cherts, etc. containing dolomite. Note, however, that many nodules and concretionary layers (particularly in strata otherwise free of carbonate minerals) contain only 40-60% dolomite; these rocks are called impure dolomite nodules and concretionary layers in this paper.

DATA AND METHODS

Compositional data used in this report are based on major elemental abundances combined with X-ray diffraction analyses of Monterey samples from the Santa Maria and Santa Barbara areas (Fig. 1) as follows: (1) 477 surface samples from the Santa Barbara coastal area (Goleta to Point Conception), 321 samples listed in Isaacs (1980) and 156 additional samples; (2) 70 surface samples from the northwestern Ventura basin (Goleta east to Ojai) listed in Keller (1982, 1984); (3) 37 composite and 39 individual cutting samples from the Point Conception COST well listed in Isaacs and others (1983); (4) 187 core samples from the Union Newlove 51, Union Newlove 108, Union Squires 13, Union Bell

156, Union Leroy A-3, and S.S.&T. 25-11 wells in the Santa Maria onshore basin; (5) 32 core samples (including some from the Rincon Shale) from the G.P. More 1, Marathon Oakley 1, and S.O. Chase & Bryce 1 wells in the La Goleta gas field; and (6) 28 core samples from the Arco Ames 3242-11, -13, and -16 wells in the South Elwood oil field.

Abundances of biogenic and diagenetic silica and of terrigenous detritus in samples were calculated from analyzed SiO_2 and Al_2O_3 by ratios described in Isaacs and others (1983). Abundances of other major sediment components in samples that are listed in Isaacs (1980) and Keller (1982, 1984) were determined by a combination of partial chemical analyses and X-ray diffraction. For this report, abundances of dolomite and calcite for other samples have been calculated from MgO and CaO abundances adjusted for amounts in aluminosilicate minerals and apatite as described in Isaacs and others (1983); samples with small amounts of carbonate minerals have also been checked by X-ray diffraction. Because the calculation assumes a stoichiometric composition for dolomite, whereas dolomite in most dolostones contains excess CaO (Murata and others, 1972; Pisciotto, 1981b), the method probably slightly underestimates dolomite and slightly overestimates calcite.

Techniques used to determine major elemental abundances included atomic absorption, wet chemical analysis, and X-ray fluoresence; for samples other than those listed in Isaacs (1980), Keller (1982, 1984) and Isaacs and others (1983), all elements were determined by X-ray fluoresence methods described in Taggart and Wahlberg (1980a, b) and Taggart and others (1981). Techniques used to determine organic carbon included (1) wet oxidation (direct measurement) and (2) difference between total carbon (measured by dry combustion with a LECO* WR12 or CR12 apparatus) and carbonate carbon (the acid-soluble fraction measured after digestion with perchloric acid by a gasometric procedure described by Rader and Grimaldi in 1961 or by a coulemetric procedure described by Huffman in 1977).

Porosity values were calculated on a dry basis from dry bulk-density values (determined with a mercury pycnometer) and dry grain-density values (determined with a helium pycnometer), as described in Isaacs (1981b); saturated bulk-density values were calculated from dry bulk-density and porosity values, assuming that pores were filled with water at 1.0 g/cm^3.

All carbon- and oxygen-isotope ratios were determined by Global Geochemistry by standard procedures. Isotopic values were determined on gas collected during reaction of the sample with phosphoric acid at 25°C, between 6 and 8 hours of reaction time for calcite and between 24 and 96 hours of reaction time for dolomite. Samples were pretreated with chlorox to remove organic matter.

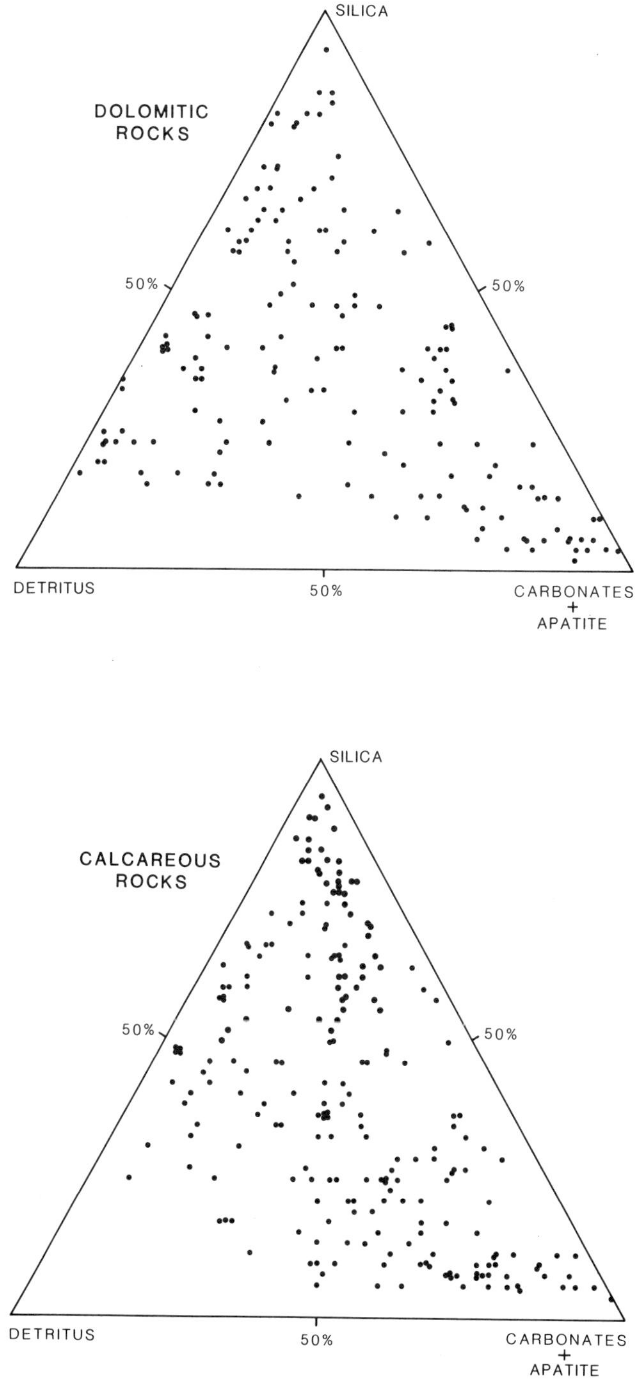

Figure 2. (A) Representative compositions of dolomitic rocks from the Santa Maria and Santa Barbara areas. Includes some calcite-bearing rocks that have >50% of total carbonate minerals as dolomite. (B) Representative compositions of calcareous rocks from the Santa Maria and Santa Barbara areas. Includes some dolomite-bearing rocks that have >50% of total carbonate minerals as calcite.

* Use of trade names in this report is for descriptive purposes only and does not imply endorsement by the U.S. Geological Survey.

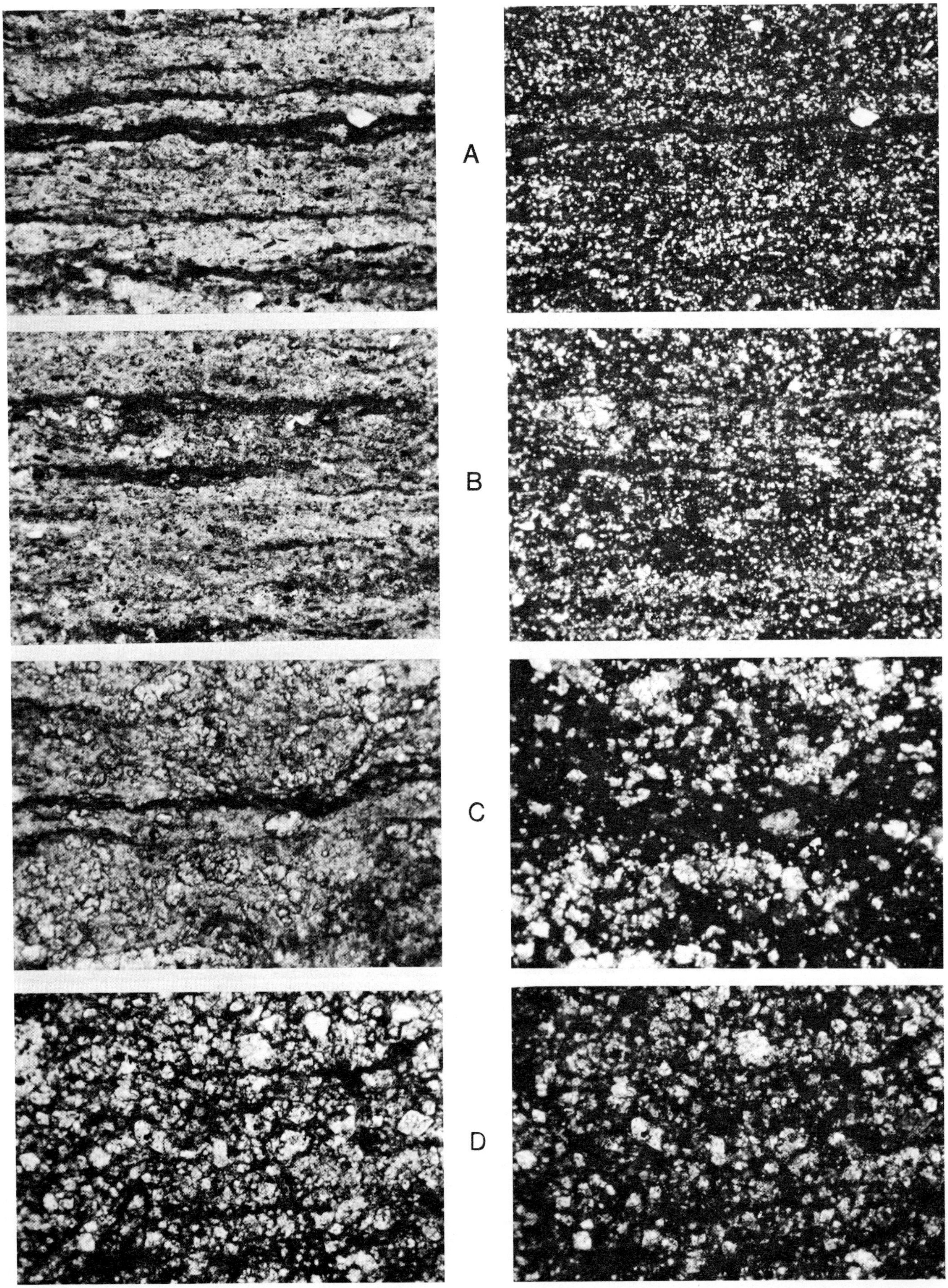
A
B
C
D

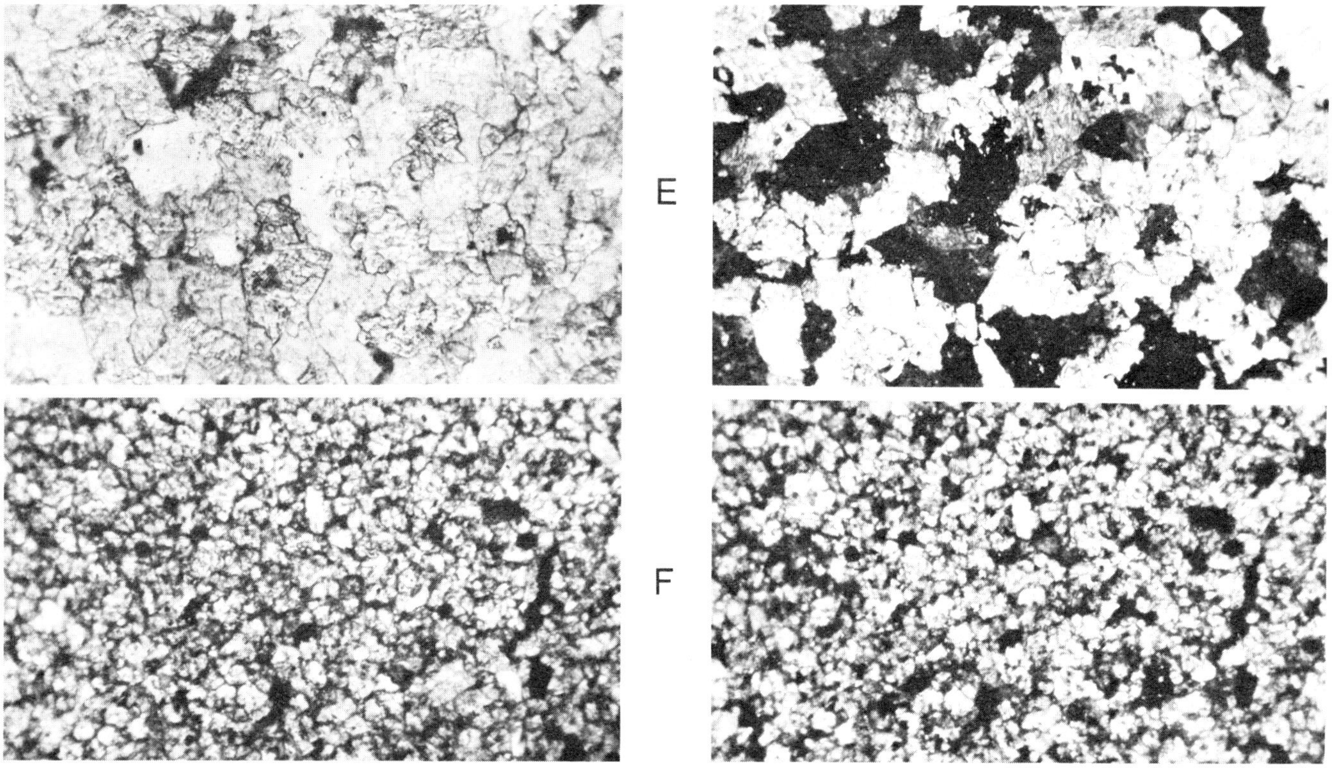

Figure 3. Photomicrographs of dolomite-bearing beds. Note the contrast between the textures of rocks with disseminated dolomite (facing page) and textures in dolostone and in an impure dolomitic nodule (above). Views are 9 mm across. Left view, plane polarized light; right view, crossed nicols. (A) Dolomitic opal-CT porcelanite (having about 6% dolomite by weight) from the Ames 3242-13 well. (B) Dolomitic quartzose porcelanite (having about 19% dolomite by weight) from the Ames 3242-13 well. (C) Dolomitic opal-CT-bearing siliceous shale (having about 47% dolomite) from Gaviota Beach. (D) Dolomitic quartz-bearing siliceous shale (having about 37% dolomite by weight) from Damsite Canyon. (E) Dolostone from Wood Canyon (having about 93% dolomite by weight). (F) Impure dolomitic nodules from Gaviota Beach (having about 55% dolomite by weight).

Figure 4. Outcrops of strata having widespread and abundant disseminated dolomite (top and bottom left) contrasted with outcrop of a dolostone nodule in clay-rich siliceous mudstone (above). Damsite Canyon (top left), Jalama Canyon (bottom left), Gaviota Beach (above).

GEOLOGIC OCCURRENCE AND ORIGIN

In the Monterey Formation of the Santa Barbara and Santa Maria areas as well as in the Pismo syncline to the north (Fig. 1), dolomitic rocks have a wide range of composition similar to the range in calcareous rocks (Fig. 2). Dolomitic rock types are also similar to calcareous rock types and include dolomitic chert, dolomitic porcelanite, dolomitic siliceous shale/mudstone, and dolomitic marl as well as dolostone (Fig. 3). In the field, dolomitic rocks are generally a little more cohesive than calcareous rocks of otherwise similar composition and tend to weather to an orange or ochre tone similar to that of dolostones (Isaacs, 1981a). As a result, thick-bedded dolomitic porcelanites and dolomitic siliceous mudstones are sometimes mistaken for dolostones (Fig. 4). In thin section, disseminated dolomite is generally dispersed and, in contrast to the dolomite in dolostone, does not form a tight interlocking framework (Fig. 3).

Considerable field evidence indicates that the dolomite dispersed in these rocks is derived directly from calcite. In the Santa Barbara coastal area, sequences of the Monterey Formation that were originally similar have been subjected to varying levels of diagenetic alteration due to a progressive increase of temperature and overburden from east to west over a distance of 55 km; diagenetically altered rocks can thus be compared with precursors at varying stages of diagenesis (Isaacs, 1980, 1982). In this area, the Monterey Formation has been divided into a series of 5 members (Fig. 5) distinguished by a combination of rock composition, layering characteristics, and both small-scale (1-3 ft) and large-scale (100-300 ft) variations reflecting paleoproductivity and bottom-water changes that were widespread and apparently synchronous throughout the area (Isaacs, 1983). Broadly, the Monterey is divided into an upper (clayey-siliceous) member in which carbonate minerals are restricted to nodules and concretionary layers and an underlying series of calcareous members (or carbonate-bearing members) in which carbonate minerals are ubiquitous (abundance >5% in virtually all beds; average about 30%).

Where strata in the lower carbonate-bearing members are least diagenetically altered (as in the diatomaceous sequences at Elwood Beach and Goleta Slough), dispersed carbonates are entirely calcite, in coccolith and foraminiferal debris. At some other localities, disseminated dolomite is commonly present within individual beds in the calcareous members. In zones where many beds have minor (<10%) amounts of dolomite, dolomite tends to be somewhat more common in calcite-rich rocks, which also have less-than-average silica abundance and somewhat more-than-average clay abundance (Table 1).

Detailed field work in the Santa Barbara coastal area has shown that some sequences of rocks 50-100 ft thick--which are readily identified by composition, layering, and stratigraphic position as correlative with one of the calcareous members--contain no calcite in any beds but have an approximately similar dolomite content. In outcrop, these dolomitic rocks are practically indistinguishable from their stratigraphically equivalent calcareous counterparts, except that dolomitic zones are somewhat more resistant to erosion, and both dolomitic and calcareous sequences have similar layering and characteristic bed cycles (Fig. 3). Detailed examination of the 3 least thermally mature sequences having zones of ubiquitous dolomite (Gato Canyon just east of Cojo Canyon, Cojo Canyon, and Damsite Canyon) and comparison with stratigraphically equivalent calcareous strata in adjacent canyons showed a striking similarity in all sedimentary details (Isaacs, 1980). Compositions of dolomitic rocks are also generally similar to compositions of calcareous rocks (Fig. 2). These relations together indicate that dolomite locally replaced calcite, in a broad sense, within individual beds and zones.

Dolomitized zones are very localized in the Santa Barbara coastal area, but some general relations are apparent. Stratigraphically, zones of widespread and abundant disseminated dolomite are most common within the upper Miocene upper calcareous-siliceous member (as at Damsite, Cojo, and Quail Canyons) and are also present at the base of the upper lower Miocene lower calcareous-siliceous member (as at San Augustine and Damsite Canyons). By contrast, organic-rich phosphate-bearing marl sequences (in the middle Miocene carbonaceous marl member) are not dolomitized in the Santa Barbara coastal area, although individual beds of organic-rich marl are the rock type most commonly dolomitized in other lithostratigraphic members. Disseminated dolomite in this area is thus most abundant in silica-poor beds within silica-rich calcareous members. However, strata are more generally dolomitized in the Point Conception COST well (Isaacs and others, 1983), at the Shell Beach section in the Pismo basin (Surdam and Stanley, 1981), and in parts of the Lion's Head section (Pisciotto, 1981a; Redwine, 1981).

ABUNDANCE OF DISSEMINATED DOLOMITE

The overall abundance of dolomite (and total carbonate) in an extensively dolomitized sequence of the Monterey Formation can be estimated from the composition of cuttings from the deep stratigraphic test well OCS-Cal 78-164 No. 1, here informally called the Point Conception COST well. Cuttings from this well were collected at 30-foot intervals, and analyses from every third interval in the Miocene sequence were published by Isaacs and others (1983). Monterey strata (drilled depths of 5845-9355 feet) average 19% dolomite and 3% calcite, totalling 22% carbonate minerals by weight. The average abundance varies among stratigraphic intervals (Table 2). For example, the uppermost interval (5845-6655 ft), correlative with the part of the Monterey in onshore sequences having carbonates restricted to nodules and concretionary layers, averages 8% dolomite and 0.6% calcite. The interval below 6835 feet, correlative with the part of the Monterey in onshore sequences having dispersed calcite and locally disseminated dolomite, averages 23% dolomite and 4% calcite, for a total of 27% carbonate minerals.

The relative proportions of dolomite in dolostones and in other rock types were approximated from compositional data on the following samples: (1) individual cuttings from the Point Conception COST well; (2) cores from the onshore Santa Maria basin; and (3) surface samples from the Santa Barbara coastal area. All of these data have biases. For example, because of analytical constraints in samples size, individual cuttings from the Point Conception COST well represent only the

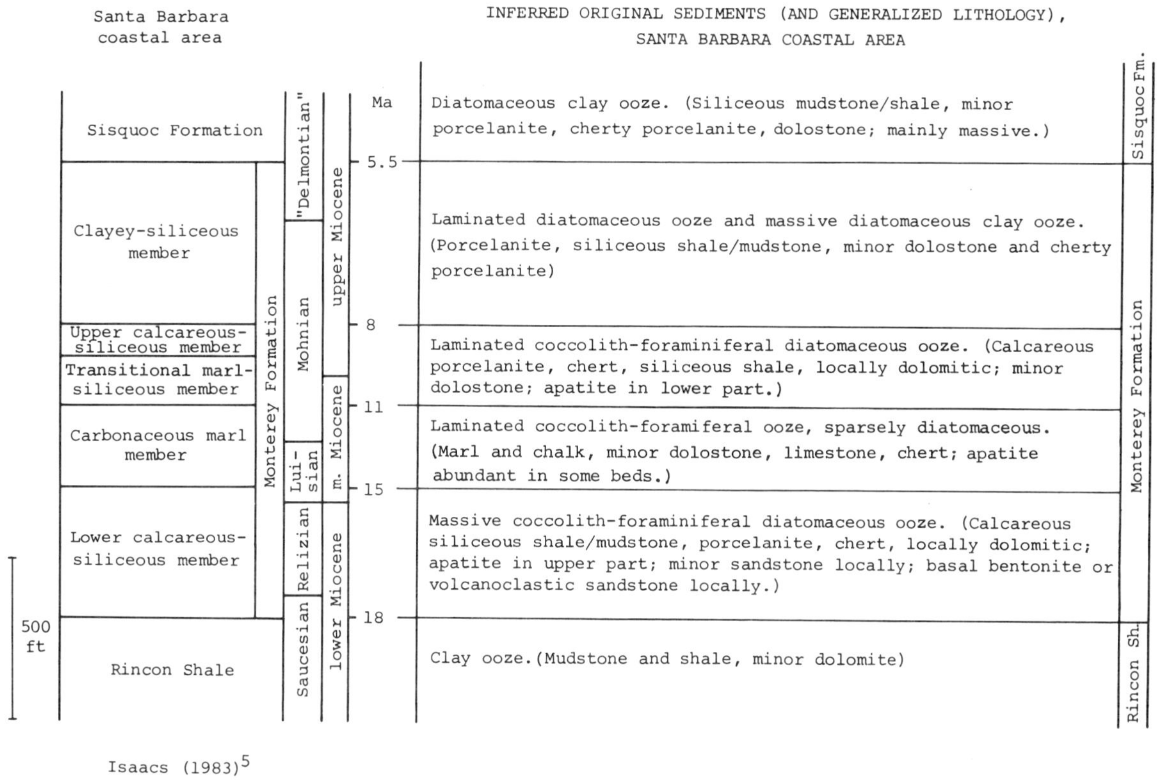

Figure 5. Lithostratigraphic members of the Monterey Formation in the Santa Barbara coastal area showing the inferred composition of the original sediments and present lithology in sequences where silica is diagenetic.

Table 1. Composition of rocks in zones where minor amounts of disseminated dolomite are common. Data are from the lower calcareous-siliceous member of the Monterey Formation near Goleta (surface samples from Naples and El Capitan Beaches and subsurface samples from the La Goleta gas field).

Sample number	Sample composition, weight %				
	Silica	Detritus	Calcite	Dolomite	Apatite
1	81	9	10	0	0
2	71	18	11	0	0
3	53	29	18	0	0
4	52	27	20	0	0
5	59	17	24	0	0
6	60	17	23	1	0
7	77	9	12	2	0
8	64	16	18	2	0
9	57	21	20	2	0
10	53	24	22	2	0
11	34	26	38	2	0
12	22	27	49	2	0
13	11	75	8	3	3
14	53	20	23	4	0
15	53	21	22	5	0
16	11	36	47	5	0
17	31	23	38	7	0
18	10	24	59	7	0
19	24	27	42	7	0
20	27	37	27	8	1
21	14	54	22	8	2
22	8	48	36	9	0
23	12	21	24	42	1
24	16	21	14	49	0

larger (>0.4 g) pieces recovered during drilling. Core samples analyzed are representative of cores available but may be biased by differential core recovery and by prior handling and division of samples. However, these data provide approximations of the proportion of dolomite present in dolostones. Dolostones (>75% dolomite) represent 2-14% of dolomitic beds, and dolomite in dolostones ranges from about 10% to about 35% of total dolomite present in the sections (Table 3). Even highly dolomitic rocks (>50% dolomite) represent only about 5-25% of all dolomitic rocks, and dolomite in highly dolomitic rocks represents only about 35-55% of total dolomite (Table 3).

Another approach to estimating the proportion of total dolomite represented by disseminated dolomite utilizes the analytical data from composite cuttings of the Point Conception COST well. In the uppermost part of the onshore Monterey (arenaceous zone or clayey-siliceous member), carbonate minerals are generally restricted to discrete nodules, and strata contain virtually no disseminated carbonates. Assuming that the dolomite in the correlative interval (5845-6655 ft) in the Point Conception COST well is similarly restricted to nodules and concretionary layers, the total abundance of dolomite in this interval (8%; Table 2) may approximate the average abundance of dolomite in dolostone. If dolomite in dolostone is equally abundant in the underlying intervals which have abundant dispersed carbonates, then dolomite in dolostone represents about 35% of the total dolomite in these intervals (8% in dolostone out of 23% total dolomite, Table 2).

In summary, dolomitic rock types other than dolostones (>75% dolomite) probably represent the vast majority (>80%) of dolomitic rocks, and disseminated dolomite is estimated to represent 65-90% of the total dolomite (by weight) in the Monterey of the Santa Barbara and Santa Maria areas. These abundances demonstrate that disseminated dolomite is geologically significant in these areas and should be taken into account in any evaluation of the formation mechanisms and magnesium sources of dolomite in the Monterey.

DISTRIBUTION COMPARED TO DIAGENETIC TEMPERATURE AND DEPTH

In a general sense, disseminated dolomite tends to be more abundant in strata that are thermally more mature.* In the Santa Barbara coastal area, for example, those zones in which disseminated dolomite is widespread and abundant and calcite is virtually absent occur mainly in strata where opal-CT is "ordered" or where silica is predominantly diagenetic quartz.

On the other hand, the occurrence and abundance of disseminated dolomite is not closely related to maximum diagenetic temperature or depth as inferred from either superposition or silica diagenetic grade. For example, many zones 50-100 ft thick where disseminated dolomite is ubiquitous and calcite is absent are underlain by sequences as much as 800 ft thick in which calcite is abundant and disseminated dolomite is virtually absent (e.g., Damsite and Cojo Canyons; Table 4). Even strata that are age-equivalent and sedimentologically similar, such as sequences of the upper part of the upper calcareous-siliceous member near Point Conception, are locally dolomitized in a pattern unrelated to that of silica diagenetic grade (Isaacs, 1980). Furthermore, disseminated dolomite is present (although not the dominant carbonate mineral) in individual beds within some thermally immature sections, such as the transitional zone of interbedded opal-A to opal-CT rocks at Naples Beach, and is abundant and ubiquitous in some zones where opal-CT is "disordered", as at Jalama Canyon.

A similarly complex relation exists between maximum diagenetic temperature and the abundance of disseminated dolomite in the subsurface (Table 4). In the Point Conception COST well (Table 2), for example, over 85% of carbonate minerals are dolomite in most Monterey strata (5845-9355 ft), but only about 65% of the carbonates are dolomite in the comparatively deep interval 8545-8725 ft. In the offshore South Elwood field, strata analyzed from the Ames 3242-11 well are entirely quartzose and contain abundant calcite but little dolomite. By contrast, in cores from the Ames 3242-13 well, in strata exposed to lower maximum diagenetic temperatures as inferred from the presence of both quartz- and opal-CT-bearing beds, carbonates are almost entirely dolomite.

Differences between the distribution of dolomite in dolostones and the distribution of disseminated dolomite suggest that disseminated dolomite generally formed at higher temperatures than many dolostones. For example, dilation-brecciated dolostones (Roehl, 1981) are present in the calcareous diatomaceous strata of the least diagenetically mature section along the Santa Barbara coast (the Goleta Slough section where the entire upper Mohnian section is diatomaceous) (Keller, 1982, 1984; for photograph, see Belfield and others, 1983, p. 216). Dolostone nodules are also present in diatomaceous sequences (as at Elwood Beach) of the clayey-siliceous member of the Monterey (the member which has virtually no dispersed carbonates). Disseminated dolomite, by contrast, has not been recognized in sequences that are entirely diatomaceous. The least thermally mature sequence in which disseminated dolomite has been detected is at Naples Beach, where strata consist mainly of interbedded opal-A and opal-CT beds. Here, disseminated dolomite is present in minor amounts in a few individual diatomaceous beds and represents 1-49% (median 2%) of analyzed samples in opal-CT-bearing rocks just underlying the transition interval of interbedded opal-A and opal-CT rocks (Isaacs, 1980, p. 295-296).

In summary, comparison of the field distribution of disseminated dolomite with that of dolostone indicates that disseminated dolomite formed

---

* Thermal maturity is inferred from silica diagenesis according to the following sequence of increasing diagenetic grade: (1) diatomaceous (opal-A) strata; (2) interbedded opal-A and opal-CT strata; (3) opal-CT strata with "disordered" opal-CT (opal-CT having comparatively high d-spacings); (4) opal-CT strata with "disordered" opal-CT (opal-CT having comparatively low d-spacings); (5) interbedded opal-CT and quartz strata; and (6) quartz strata (Murata and Larson, 1975; Pisciotto, 1981a; Isaacs, 1982). No diagenetic relation between silica diagenesis and carbonate diagenesis is implied by comparison of silica diagenetic grade with the abundance of dolomite; silica diagenetic grade is used here merely as an indication of thermal maturity.

Table 2. Abundance of dolomite and calcite and percentage of dolomite among carbonate minerals in strata from the Point Conception COST well as correlated with the onshore Monterey Formation. Based on chemical analysis and X-ray diffraction of composite cuttings collected at 30-ft intervals (every third sample was analyzed, representing one-third of total strata). Depths represent drilled depths from the derrick floor, with the seafloor at 1465 ft. Summarized from Isaacs and others (1983).

Preliminary correlative onshore member	Depth interval, ft	Dolomite, mean wt %	Calcite, mean wt %	Percentage of Dolomite among carbonate minerals
Clayey-siliceous member	5845-6655	8	0.6	93
Upper calcareous-siliceous & transitional marl-siliceous members	6835-7375	20	3	87
Carbonaceous marl member	7555-7915	34	2	94
Lower calcareous-siliceous member	8005-9355	20	4	82
	(8545-8725)	15	8	66
All intervals	5845-9355	19	3	87
Intervals below 6835 feet.	6835-9355	23	4	86

Table 3. Abundances of dolomite in the Monterey Formation at selected locations in the Santa Maria and Santa Barbara areas. Samples included here are restricted to those having at least 2% dolomite. For sources of data, see text.

Location	Type of Sample	Number of Samples	Percentage of samples having various abundances of dolomite				Percentage of total dolomite in samples having various abundances of dolomite			
			Samples having 2-25% dolomite	Samples having 25-50% dolomite	Samples having 50-75% dolomite	Samples having >75% dolomite	Samples having 2-25% dolomite	Samples having 25-50% dolomite	Samples having 50-75% dolomite	Samples having >75% dolomite
Point Conception COST well	Individual cuttings	29	89	6	3	3	35	22	19	24
Union Newlove 51 well	Cores	47	74	15	9	2	34	33	25	8
Union Bell 156 well	Cores	21	43	38	5	14	12	42	10	35
Santa Barbara coastal area*	Surface samples	71	59	18	10	13	20	23	20	37

* Excludes samples from the clayey-siliceous member.

at higher temperatures than many dolostones. Comparison of disseminated dolomite distribution with maximum diagenetic temperature (as inferred from silica diagenesis) suggests that disseminated dolomite does not begin to form at temperatures lower than that of the opal-A to opal-CT transformation, approximately 45°-50°C (Murata and others, 1977; Keller, 1982, 1984). In strata exposed to higher temperatures, however, disseminated dolomite has an apparently erratic distribution, being locally abundant in patterns not closely related to maximum diagenetic temperature even within single stratigraphic horizons.

ISOTOPIC COMPOSITION*

Carbon- and oxygen-isotopes were determined on a preliminary set of samples in order to compare the original isotopic compositions of biogenic calcite, partially recrystallized dispersed calcite, and disseminated dolomite in the Monterey Formation of the Santa Barbara coastal area (Tables 5 and 6). Also analyzed were 1 sample from Lion's Head and a few samples from the Union Newlove 51 and Union Dome 18 wells in the onshore Santa Maria basin and from the Ames 3242-11 well in the South Elwood oil field (Table 6). The results, together with preliminary conclusions, are presented here with the caution that data are sparse. Also note that some CO_3 in the apatite in samples analyzed here may affect values as reported (M. Kastner, written communication, 1984).

Most biogenic calcite in the Santa Barbara coastal area was originally in the form of coccoliths, generally composed of low-magnesium calcite which is geochemically stable at temperatures near the sediment-water interface. Because diatom frustules support a rock framework that is extremely porous and resistant to compression (Hamilton, 1976), solution and reprecipitation of calcite in

* Isotope ratios are reported in conventional notation: $\delta^{13}C$ or $\delta^{18}O$ permil = [R(sample)/R(standard) - 1]x10^3, where R = $^{13}C/^{12}C$ or $^{18}O/^{16}O$ respectively, and the standard is the Peedee belemnite (PDB).

Table 4. Abundance of calcite and disseminated dolomite and the percentage of disseminated dolomite among carbonate minerals in selected samples analyzed from the Monterey Formation. Based on chemical analysis and X-ray diffraction of individual samples.

Location	Depth interval, ft	Number of samples	Disseminated dolomite, mean wt %	Calcite, mean wt %	Percentage of dolomite among carbonate minerals
	Subsurface cores				
UNION NEWLOVE 51 WELL					
Cherty zone	1913-2154	11	3	9	28
Bentonitic-brown zone	2154-2515	13	14	7	66
Dark brown zone	2515-2841	31	20	4	82
Oil sand zone, plus siltstone, and shell zones	2841-4011	10	9	5	65
All zones	1913-4011	65	14	6	71
UNION SQUIRES 13 WELL	2642-3589	19	20	4	84
UNION NEWLOVE 108 WELL	3417-3744	14	19	0	98
UNION BELL 156 WELL	4954-5780	72	15	1	94
UNION LEROY A-3 WELL	4975-5263	9	23	33	42
S.S.&T. 25-11 WELL	8011-9155	8	23	18	57
	Surface samples				
DAMSITE CANYON					
Upper calcareous-siliceous member (upper part)	485-570*	11	33	0	100
Upper calcareous-siliceous member (lower part) plus transitional marl-siliceous member	570-600*	9	19	16	54
Carbonaceous marl member	804-846*	2	0	48	0
Lower calcareous-siliceous member	951-1158*	26	4	36	10

* Measured from the top of the Monterey Formation.

Table 5. Oxygen- and carbon-isotope ratios of calcite from diatomaceous and "immature" opal-CT sequences of the Monterey Formation. Except as marked (*), samples are from the upper calcareous-siliceous and transitional marl-siliceous members, Santa Barbara coastal area. (Si = biogenic and diagenetic silica, Det = terrigenous detritus, Ca = calcite, Do = dolomite, Ap = apatite, Org = organic matter; values are normalized to an organic-matter-free basis.)

Location / Sample number	$\delta^{13}C$, permil	$\delta^{18}O$, permil	Silica phase	Sample composition, wt % Si	Det	Ca	Do	Ap	Org
Goleta Slough									
UGO-2-3	-0.9	+1.1	Opal-A	12	34	45	0	8	-
UGO-2B-1	-2.6	+0.6	Opal-A	20	55	22	0	3	9.4
UGO-2B-11	-0.4	+0.1	Opal-A	62	25	12	0	1	5.2
Elwood Beach									
ELW-1-5	+0.4	+0.8	Opal-CT	76	9	15	0	0	3.1
ELW-1-6	+0.7	+0.1	Opal-A	37	27	32	0	4	6.8
Naples Beach									
NAP-13-2	+0.1	+0.6	Opal-CT	84	10	6	0	0	1.6
NAP-13-18	-1.6	-0.8	Opal-A	29	64	5	0	2	7.2
NAP-13-24	-3.1	+0.3	Opal-A	39	43	15	0	2	6.4
NAP-11-5	-0.9	+0.2	Opal-A	6	46	39	0	9	19.1
NAP-11-17	-1.1	+0.2	Opal-A	17	33	45	0	4	14.4
San Augustine Canyon									
AUG-7B-11	+0.6	+0.2	Opal-CT	23	30	44	0	2	9.6
AUG-7B-16	-1.6	+0.2	Opal-CT	70	14	16	0	0	4.5
AUG-6B-15	-1.6	-0.7	Opal-CT	5	21	66	0	7	12.2

diatomaceous rocks are not particularly promoted by overburden pressure, which is the major cause of calcite recrystallization in chalks (Scholle, 1977). The original isotopic composition of carbonates in the Monterey of the Santa Barbara coastal area can thus be approximated by the isotopic composition of calcite in sequences where rocks are diatomaceous. Isotopic values in diatomaceous (opal-A) rocks range from -3.1 to +0.7 permil $\delta^{13}C$ and from -0.8 to +1.1 permil $\delta^{18}O$ (Table 5). Because the isotopic composition of calcite is quite similar in both diatomaceous and "immature" opal-CT sequences (Table 5), recrystallization of calcite was apparently not strongly promoted by the compaction associated with the formation of opal-CT.

Isotopic compositions of calcite changed markedly, however, with increased diagenesis. Values in "mature" opal-CT sequences (those in which diagenetic quartz has begun to form in some beds) and in diagenetic quartz sequences are -6.6 to 0.0 permil $\delta^{18}O$ and -12.1 to +1.5 permil $\delta^{13}C$ (Table 6, Fig. 6B). Because only part of the calcite may have recrystallized, these isotopic values probably represent the isotopic composition of mixtures of original and recrystallized calcite rather than just recrystallized calcite (Scholle, 1977). The trend resulting from recrystallization, however, is clearly toward lighter oxygen- and lighter carbon-isotopes.

Experimental evidence indicates that dolomite forms from calcite by a dissolution-precipitation reaction (Katz and Matthews, 1977), so the isotopic composition of disseminated dolomite should represent equilibrium with pore water during crystallization. Isotopic compositions of the 5 samples of disseminated dolomite from the Santa Barbara coastal area (Table 6) are close to the range in biogenic and partially recrystallized calcite in the same area (Tables 5 and 6). Carbon isotopic values for

Table 6. Oxygen- and carbon-isotope ratios of carbonate minerals from "mature"-opal-CT and diagenetic-quartz sequences of the Monterey Formation. Values in parentheses are not included in Figure 6 as X-ray diffraction does not indicate any calcite present. For explanation of composition (righthand columns), see Table 5.

Location	Calcite		Dolomite		Sample composition, wt %					
Sample number	$\delta^{13}C$	$\delta^{18}O$	$\delta^{13}C$	$\delta^{18}O$	Si	Det	Ca	Do	Ap	Org
Surface samples										
Damsite Canyon										
DAM-7D-1	(-8.3	+0.5)	-8.2	-0.2	27	31	0	40	2	6.6
DAM-7D-2*	(+5.9	+1.3)	+7.1	+2.1	7	1	0	93	0	0.2
Wood Canyon										
WOOD-9-9I	-0.3	-4.4	-1.5	-5.2	80	12	5	2	0	4.6
WOOD-9-9II	-1.1	-0.8	-1.5	+0.5	83	12	3	2	0	4.8
WOOD-5-4	-2.3	0.0			32	12	55	0	0	14.3
WOOD-4-6	-7.5	-2.4	-	-	5	24	70	0	0	5.2
WOOD-4-3	-3.5	-0.9	-	-	8	28	60	0	3	12.8
WOOD-2B-6	-1.6	-3.4	-	-	61	9	29	0	0	6.8
WOOD-2-4Ch	-2.6	-5.2	-	-	90	5	4	0	1	-
Quail Canyon										
BIX-6-4*	(-1.4	-2.0)	-1.8	-0.5	4	8	0	88	0	-
BIX-6-5	-0.1	-4.2	-	-	88	8	4	0	0	2.8
BIX-6B-13	(-2.1	-2.7)	-2.8	-3.0	92	4	0	4	0	2.0
Lion's Head										
LYN-1-2	-0.6	-2.4	-1.0	-1.0	39	52	6	3	1	5.7
Subsurface samples										
Union Newlove 51 well										
1946A	-9.2	-4.6	-	-	72	17	10	0	0	-
1968A(2)	-12.1	-6.6	-	-	53	31	9	0	0	4.9
2705A(4)	(-3.3	-4.4)	-2.9	-2.6	7	34	0	58	0	2.8
2722A(2)*	(-1.8	-3.2)	-1.2	-2.0	2	8	0	89	0	1.8
2722A(6)	(-3.9	-3.6)	-3.9	-2.4	39	40	0	21	0	5.8
Ames 3242-11 well, South Elwood field										
4953	-0.3	-1.4	-	-	38	25	38	0	0	8.9
5033	-0.5	-3.7	-	-	84	3	12	0	0	1.9
5055	-0.8	-6.3	-	-	87	4	9	0	0	-
Union Dome 18 well										
10317A(5)	+1.2	-0.6	+2.6	+0.5	59	37	1	3	0	2.6
10522A(2)	-1.2	-5.6	+2.2	-1.5	58	36	6	0	0	-
10593A	-1.1	-5.4	+2.5	-1.0	56	38	6	0	1	3.4
10953A(3)	+1.5	-5.9	+3.8	-3.2	65	21	2	11	1	3.5

* Nodular and bedded dolostone, not included on Figure 6.

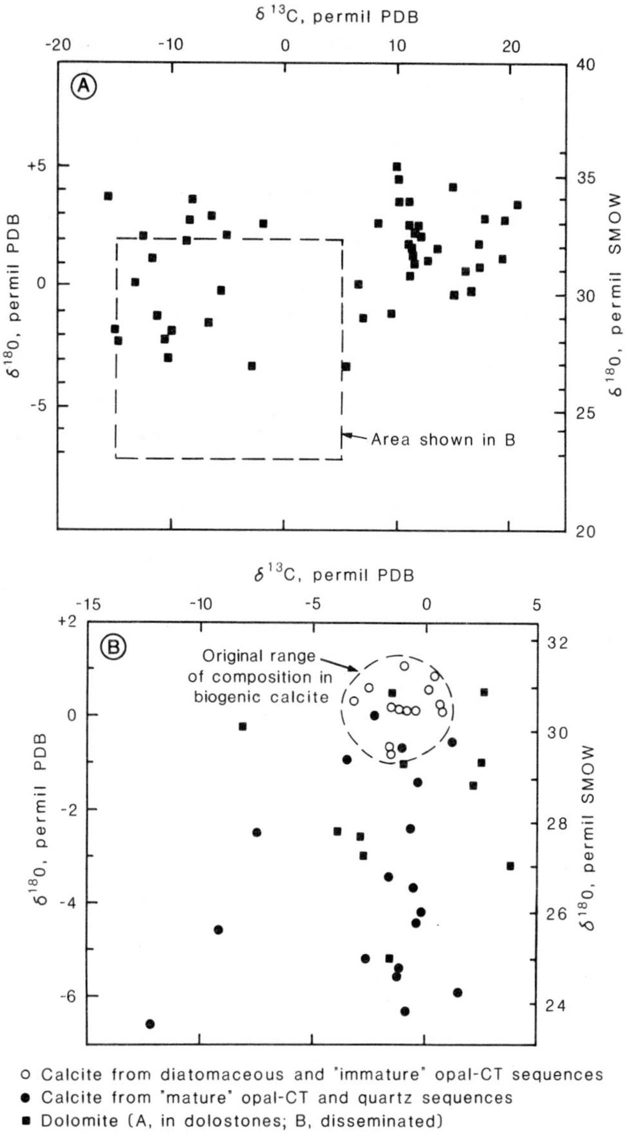

Figure 6. Isotopic composition of carbonate minerals. (A) Dolostones from the Monterey Formation analyzed by Murata and others (1969). (B) Calcite and disseminated dolomite in samples listed in Tables 5 and 6.

disseminated dolomite in all 10 analyzed samples (-8.2 to +3.8 permil $\delta^{13}C$) cluster in the center of the range of values generally reported for dolostones in the Monterey (e.g., Fig. 6A). Dolostones analyzed by Murata and others (1969), for example, were divided into light-carbon and heavy-carbon groups, with the wide range of values (all samples, -25 to +25 permil $\delta^{13}C$; Monterey samples, -15 to +21 permil $\delta^{13}C$) attributed to derivation from carbon released during organic-geochemical reactions; only 2 of their 50 dolomite samples fall in the range between -5 to +5 permil $\delta^{13}C$. Nine out of 10 samples of disseminated dolomite analyzed here, however, have carbon-isotope ratios in this intermediate range (Table 6, Fig. 6B). This contrast suggests that organic-geochemical reactions may not have been closely involved in formation of disseminated dolomite, a relation consistent with its derivation from biogenic calcite.

Samples of disseminated dolomite analyzed here have oxygen-isotope ratios (-5.2 to +0.5 permil $\delta^{18}O$) either lighter or on the light side of the range (-3.3 to +4.9 permil $\delta^{18}O$) in dolostones from the Monterey Formation reported by Murata and others (1969) (Table 6, Fig. 6). Assuming that the isotopic composition of pore water was 0.0 permil $\delta^{18}O$ as suggested by Pisciotto (1981b), oxygen-isotope ratios of disseminated dolomite* approximately correspond to temperatures in the range 35-70°C. If pore water was enriched in ^{18}O as observed in some California oil fields (Murata and others, 1972 and references therein), oxygen-isotope ratios would correspond to higher temperatures (e.g., 50-90°C for +3.0 permil $\delta^{18}O$). Whatever isotopic composition of pore water is assumed, oxygen isotopic ratios suggest that disseminated dolomite generally formed within the higher part of the temperature range inferred for dolomite formation in dolostones (corresponding to $\delta^{18}O$ of -3.3 to +0.5 permil) and also at somewhat higher temperatures (corresponding to $\delta^{18}O$ of -5.2 to -3.3 permil).

Because of the sparsity of data, general conclusions drawn from the isotopic composition of disseminated dolomite would be premature. However, the contrast in range of isotopic composition between dolomite in dolostones and dolomite in other rock types suggests that the diagenetic histories of the two types of dolomite may differ. In particular, carbon-isotope ratios indicate that disseminated dolomite may have formed largely from carbonate derived from calcite, with little involvement in the organic-geochemical reactions that are important in dolostone formation. Oxygen-isotope ratios, on the other hand, indicate that disseminated dolomite may generally have formed at higher temperatures than the dolomite in many dolostones.

POROSITY AND DENSITY CHARACTERISTICS

Data on the density and porosity characteristics of dolomitic rocks are sparse (Table 7), but some general relations are apparent. For example, matrix porosities for dolostone nodules and beds are relatively low (<10%) compared with those for other dolomitic rocks, in which porosities range as high as 37%. Low porosity values together with compaction around nodules--even the impure ones listed in Table 7--indicates that the formation of dolostone nodules involved pore-filling cementation, and the nodules are generally highly cohesive.

An important question about disseminated dolomite is whether or not its formation also involved pore-filling cementation. If so, bulk-density values should be markedly higher and porosity values markedly lower in dolomitic rocks than in precursor calcareous rocks. Because the sparse data available includes rocks of varying diagenetic maturity, the slightly higher average density and slightly lower average porosity of dolomitic rocks (Table 7) are probably not generally significant. For example, a comparison of dolomitic and calcareous siliceous-shale/marl samples that are compositionally similar and from the same lithostratigraphic member at equal diagenetic maturity suggests that formation of disseminated dolomite

* Adjusted for acid fractionation relative to calcite by subtracting 0.8 permil from reported values (Pisciotto, 1981b and references therein).

Table 7. Mass properties of dolomitic and calcareous rock types from the Monterey Formation, Santa Barbara Coastal area. In this table, "marl" is restricted to rocks having 25-75% carbonates and <30% silica. Data are from surface samples, summarized from Isaacs (1980).

Lithology	Number of samples	Saturated bulk density, g/cm^3	Dry bulk density, g/cm^3	Dry grain density, g/cm^3	Dry porosity, %
OPAL-CT ROCKS					
Calcareous cherty	15	1.9-2.1	1.7-2.0	2.0-2.3	2-26
Dolomitic cherty	2	2.0	1.8	2.2	19-20
Calcareous porcelaneous/grainy	21	1.8-2.0	1.4-1.6	2.2-2.6	26-40
Dolomitic porcelaneous/grainy	3	2.0-2.1	1.6-1.8	2.4	26-33
Calcareous marl	28	1.6-2.0	1.2-1.7	2.0-2.5	27-47
Dolomitic marl	2	2.0-2.1	1.6-1.8	2.4-2.6	27-37
QUARTZ ROCKS					
Calcareous cherty	5	2.3-2.5	2.2-2.5	2.3-2.5	0-12
Dolomitic cherty	2	2.5-2.6	2.5	2.5-2.6	1
Calcareous porcelaneous/grainy	6	2.2-2.3	2.0-2.1	2.3-2.5	11-17
Dolomitic porcelaneous/grainy	2	2.2-2.5	2.0-2.5	2.4-2.6	5-16
Calcareous marl	16	1.9-2.3	1.5-2.0	1.9-2.7	5-33
Dolomitic marl	7	2.1-2.5	1.9-2.5	2.3-2.6	4-19
OTHER DOLOMITIC ROCKS					
Impure dolomitic nodules (40-60% dolomite)	4	2.2-2.6	1.9-2.5	2.2-2.6	13-24
Dolostone beds and nodules (>75% dolomite)	8	2.5-2.8	2.4-2.8	2.6-2.8	1-9

Table 8. Density and porosity data for compositionally similar calcareous and dolomitic rocks at equal diagenetic maturity from the same lithostratigraphic member of the Monterey Formation. Note the similarity in the mean density and porosity values of the sets. Data are from the upper calcareous-siliceous member at Gaviota Beach and San Augustine Canyon (from Isaacs, 1980).

Sample number	Dry bulk density, g/cm^3	Dry grain density, g/cm^3	Dry porosity, %	Sample composition, wt %					
				Opal-CT	Det	Ca	Do	Ap	Org
CALCAREOUS ROCKS									
1	1.62	2.29	29	44	21	32	0	2	8.5
2	1.68	2.33	28	23	30	44	0	2	9.6
3	1.71	2.37	28	24	26	46	0	3	9.1
4	1.62	2.28	29	50	23	25	0	2	7.9
Mean	1.66	2.32	28						
DOLOMITIC ROCKS									
1	1.70	2.39	29	33	54	3	10	0	6.2
2	1.65	2.44	33	31	35	2	32	0	5.1
3	1.80	2.42	26	38	12	0	50	0	4.3
4	1.78	2.44	27	22	35	0	43	0	5.9
Mean	1.73	2.42	29						

caused less than 0.1 g/cm^3 increase in bulk density and no porosity decrease (Table 8). Among quartzose marl beds, on the other hand, dolomitic beds appear to be significantly denser and less porous than calcareous beds (Table 7). Calcareous marl has a comparatively weak (and uncemented) rock framework resembling that of chalk and seems to respond to increasing overburden pressure, in contrast to more rigid diagenetic-silica-rich rocks that bear calcite, with continuous incremental compaction (Isaacs, 1981b). Rather than indicating cementation, then, the higher density in beds of dolomitic marl may indicate additional compaction permitted by dissolution of calcite during dolomitization.

Although not conclusively, therefore, porosity data seem to rule out extensive pore-filling and (simple) cementation as major processes associated with formation of disseminated dolomite. This tentative conclusion contrasts processes of dolomite formation in dolostone with processes of disseminated dolomite formation and suggests that further study of both dolomite types would be valuable for understanding the processes involved in dolomite formation in the Monterey Formation.

SUMMARY

In the Monterey Formation of the Santa Maria and Santa Barbara areas, the vast majority of dolomitic rocks are not dolostones (>75% dolomite) but are rather other rock types bearing disseminated dolomite. Weighted estimates of dolomitic rocks in these areas indicate that dolomite in dolostones represents 10-35% of total dolomite and that dolomite in other rock types represents 65-90% of total dolomite. These estimates demonstrate the importance of disseminated dolomite in these areas, both in terms of mechanisms of dolomitization and in terms of sources of magnesium.

The distribution of disseminated dolomite is not closely related to increasing diagenetic temperature (inferred from silica diagenesis) or to increasing depth (determined by superposition). Though generally more common in strata that are more thermally mature in the Santa Barbara coastal area, zones in which disseminated dolomite is widespread and abundant are in many places underlain by thick calcareous sequences where disseminated dolomite is virtually absent. Similar relations are present in the subsurface in both the Santa Maria and Santa Barbara areas.

Dolomite in dolostone differs in a number of ways from dolomite formed in other rock types. (1) In contrast to dolomite in nodules and concretions, many of which formed in strata having little or no carbonate minerals, disseminated dolomite generally formed in calcareous beds and was probably derived, by a dissolution-precipitation reaction, from preexisting carbonate. (2) In contrast to carbon-isotopic values of dolomite in dolostone, which have an extremely wide range of values indicating involvement in organic-geochemical reactions, carbon-isotopic values of disseminated dolomite analyzed in this study have a restricted range more similar to the range in biogenic and partially recrystallized dispersed calcite. (3) In contrast to formation of dolomite in nodules and concretions, only slight increase in density and little or no porosity reduction accompanied formation of disseminated dolomite; this relation suggests that pore-filling was not particularly important in formation of disseminated dolomite.

The many contrasts between dolomite in dolostone and dolomite disseminated in other rock types indicate that different processes may habe been important in their formation and suggest that more extensive study of disseminated dolomite would be extremely valuable to an understanding of dolomite in the Monterey Formation.

ACKNOWLEDGMENTS

I particularly thank Peter A. Scholle for many valuable discussions of chalks during an early stage of this study and Toby Scott for invaluable organization of analytical data. I also thank Joseph Taggart and Larry Jackson of the U.S. Geological Survey Denver for overseeing most of the chemical analyses.

William Bazeley, Larry Beyer, William Dahleen, and Kenneth Pisciotto generously provided samples used in this study. Mary Lou Callas, Robert Compton, Robert Garrison, Miriam Kastner, Margaret Keller, James O'Neill, Kenneth Pisciotto, and Donald Zenger provided useful discussions and/or editorial suggestions, and Brigit Fulop and Lillian Wood provided technical assistance.

REFERENCES CITED

Bates, R. L., and Jackson, J. A., eds., 1980, Glossary of Geology: Falls Church, Virginia, American Geological Institute, 751 p.

Belfield, W. C., Helwig, J., La Pointe, P., and Dahleen, W. K., 1983, South Ellwood oilfield, Santa Barbara Channel, California, a Monterey Formation fractured reservoir, in Isaacs, C. M., and Garrison, R. E., eds., Petroleum generation and occurrence in the Miocene Monterey Formation, California: Soc. Econ. Paleontologists Mineralogists Pacific Sec. Spec. Pub., p. 213-221.

Hamilton, E. L., 1976, Variations of density and porosity with depth in deep-sea sediments: Jour. Sedimentary Petrology, v. 46, p. 280-300.

Huffman, E. W. D., Jr., 1977, Performance of a new automatic carbon dioxide coulometer: Microchemical Jour., v. 22, p. 567-573.

Isaacs, C. M., 1980, Diagenesis in the Monterey Formation examined laterally along the coast near Santa Barbara, California: Stanford, Calif., Stanford Univ. unpubl. Ph.D. dissertation, 329 p.

______, 1981a, Field characterization of rocks in the Monterey Formation along the coast near Santa Barbara, California, in Isaacs, C. M., ed., Guide to the Monterey Formation in the California coastal area, Ventura to San Luis Obispo: Am. Assoc. Petroleum Geologists Pacific Sec., v. 52, p. 39-53.

______, 1981b, Porosity reduction during diagenesis of the Monterey Formation, Santa Barbara coastal area, California, in Garrison, R. E.., and Douglas, R. G., eds., The Monterey Formation and related siliceous rocks of California: Soc. Econ. Paleontologists Mineralogists Pacific Sec. Spec. Pub., p. 257-271.

______, 1982, Influence of rock composition on kinetics of silica phase changes in the Monterey Formation, Santa Barbara area, California: Geology, v. 10, p. 304-308.

______, 1983, Compositional variation and sequence in the Miocene Monterey Formation, Santa Barbara coastal area, California, in Larue, D. K., and Steel, R. J., eds., Cenozoic marine sedimentation, Pacific Margin, U.S.A.: Soc. Econ. Paleontologists Mineralogists Pacific Sec. Spec. Pub., p. 117-132.

Isaacs, C. M., Keller, M. A., Gennai, V. A., Stewart, K. C., and Taggart, J. E., Jr., 1983, Preliminary evaluation of Miocene lithostratigraphy in the Point Conception COST well OCS-Cal 78-164 No. 1, off southern California, in Isaacs, C. M., and Garrison, R. E., eds., Petroleum generation and occurrence in the Miocene Monterey Formation, California: Soc. Econ. Paleontologists Mineralogists Pacific Sec. Spec. Pub., p. 99-110.

Katz, A., and Matthews, A., 1977, The dolomitization of $CaCO_3$: an experimental study at 252-295°C: Geochimica Cosmochimica Acta, v. 41, p. 297-308.

Keller, M. A., 1982, Lithostratigraphy and diagenesis of the Monterey Formation near Ojai, California: California State Univ. Los Angeles, unpublished Ms. thesis, 111 p.

______, 1984, Silica diagenesis and lithostratigraphy of the Monterey Formation of the northwestern Ventura basin, California, including biostratigraphy, kerogen maturity, chemical analyses, and a preliminary temperature zonation of the opal-CT zone: U.S. Geol. Survey Open-file Rep., 111 p. (in press).

Murata, K. J., Friedman, I., and Cremer, M., 1972, Geochemistry of diagenetic dolomites in Miocene marine formations of California: U.S. Geol. Survey Professional Paper 724-C, 12 p.

Murata, K. J., Friedman, I., and Gleason, J. D., 1977, Oxygen isotope relations between diagenetic silica minerals in Monterey shale, Temblor Range, California: Am. Jour. Science, v. 277, p. 259-272.

Murata, K. J., Friedman, I., and Madsen, B. M., 1969, Isotopic composition of diagenetic carbonates in marine Miocene formations of California and Oregon: U.S. Geol. Survey Professional Paper 614-B, 24 p.

Pisciotto, K. A., 1981a, Notes on Monterey rocks near Santa Maria, California, in Isaacs, C. M., ed., Guide to the Monterey formation in the California coastal area, Ventura to San Luis Obispo: Am. Assoc. Petroleum Geologists Pacific Sec., v. 52, p. 73-81.

______, 1981b, Review of secondary carbonates in the Monterey Formation, California, in Garrison, R. E., and Douglas, R. G., eds., The Monterey Formation and related siliceous rocks of California: Soc. Econ. Paleontologists Mineralogists Pacific Sec. Spec. Pub., p. 273-283.

Rader, L. F., and Grimaldi, F. S., 1961, Chemical analyses for selected minor elements in Pierre Shale: U.S. Geol. Survey Professional Paper 391-A, 45 p.

Redwine, L. E., 1981, Hypothesis combining dilation, natural hydraulic fracturing, and dolomitization to explain petroleum reservoirs in Monterey Shale, Santa Maria area, California, in Garrison, R. E., and Douglas, R. G., eds., The Monterey Formation and related siliceous rocks of California: Soc. Econ. Paleontologists Mineralogists Pacific Sec. Spec. Pub., p. 221-248.

Roehl, P. O., 1981, Dilation brecciation - a proposed mechanism of fracturing, petroleum expulsion and dolomitization in the Monterey Formation, California, in Garrison, R. E., and Douglas, R. G., eds., The Monterey Formation and related siliceous rocks of California: Soc. Econ. Paleontologists Mineralogists Pacific Sec. Spec. Pub., p. 285-315.

Scholle, P. A., 1977, Chalk diagenesis and its relation to petroleum exploration: oil from chalks, a modern miracle?: Am. Assoc. Petroleum Geologists Bull., v. 61, p. 982-1009.

Surdam, R. C., and Stanley, K. O., 1981, Stratigraphic and sedimentologic framework of the Monterey Formation, Pismo syncline, California, in Isaacs, C. M., ed., Guide to the Monterey Formation in the California coastal area, Ventura to San Luis Obispo: Am. Assoc. Petroleum Geologists Pacific Sec., v. 52, p. 83-91.

Taggart, J. E., Jr., Lichte, F. E., and Wahlberg, J. S., 1981, Methods of analysis of samples using X-ray fluorescence and induction-coupled plasma spectroscopy, in Lipman, P. W., ed., The 1980 eruptions of Mount St. Helens, Washington: U.S. Geol. Survey Professional Paper 1250, p. 683-687.

Taggart, J. E., Jr., and Wahlberg, J. S., 1980a, A new in-muffle automatic fluxer design for casting glass discs for X-ray fluorescence analysis [abs.]: Federation of Analytical Chemists and Spectroscopy Societies conference, Sept. 19, 1980, Philadelphia,. Pa.

______, 1980b, New mold design for casting fused samples: Advances in X-ray analysis, v. 23, p. 257-261.

TWO FORMS OF DOLOMITE OCCURRENCES IN THE MONTEREY FORMATION, CALIFORNIA: CONCRETIONS AND LAYERS -- A COMPARATIVE MINERALOGICAL, GEOCHEMICAL, AND ISOTOPIC STUDY

J. Kushnir
Scripps Institution of Oceanography
La Jolla, California 92093

M. Kastner
Scripps Institution of Oceanography
La Jolla, California 92093

ABSTRACT

Concretions and layers of early diagenetic dolomite (and protodolomite) are common in the Monterey Formation, California. A comparative mineralogical, geochemical, and isotopic study of a dolomite concretion (1.0 x 0.3 m) from the Sandholdt Member at Arroyo Seco and of a dolomitic layer (0.4 m thick) in the opal-CT zone from the easternmost proximal side of Santa Maria Basin records important differences in their diagenetic histories.

In the concretion $\delta^{13}C$ and $\delta^{18}O$ values of the dolomite decrease systematically from the center outward, in a peneconcentric pattern. The $\delta^{13}C$ values range from 10.7°/₀₀ in the center to 2.1°/₀₀ (PDB) at the periphery, indicating formation in the anaerobic carbonate reduction zone. The lighter values toward the edges reflect either formation in a "closed" system or the presence of isotopically-light CO_2 derived from thermocatalytic decarboxylation of organic matter, as burial depth and temperature increased. The $\delta^{18}O$ values vary from 0.3°/₀₀ at the center to -5.1°/₀₀ (PDB) at the periphery, indicating continuous dolomite formation with increasing temperature, from about 25° to 60°C.

These changes in isotopic composition are accompanied by systematic changes in the mineralogy of the concretion and the chemistry of the dolomite crystals. The latter account for up to 89 weight percent in the center, decreasing to 16% at the periphery; insoluble residue (quartz, some feldspar, pyrite, and clay minerals) increases simultaneously from 9 to 65%, calcite from 2 to 19%. The dolomite is rich in Ca and Fe; $CaCO_3$ increases outward from 55 to 58 mole %, and $FeCO_3$ from 5 to 15 mole %.

In the dolomitic layer, however, the composition of the dolomite crystals stays nearly constant at about 56 mole percent $CaCO_3$; $\delta^{13}C$ and $\delta^{18}O$ values of the dolomite vary slightly but do not exhibit any obvious trend. The $\delta^{13}C$ values range from 14.8 to 13.4°/₀₀ (PDB), showing that this type of dolomite also formed in the anaerobic carbonate reduction zone; the $\delta^{18}O$ values, -0.9 to -2.1°/₀₀ (PDB) show that dolomitization was rather rapid and took place between 30° and 40°C at a depth of only a few hundreds of meters below the sediment/water interface. Diagenetic quartz formed later, deeper, and at higher temperatures, between approximately 60° and 70°C, than the dolomite.

The amount of dolomite varies irregularly between 5 and 55 weight percent; calcite was not detected. The insoluble residue consists of various mixtures of quartz, opal-CT ($d_{(101)} \leq 4.06$ Å), opal-A, feldspars, and traces of clay minerals. There is an approximate inverse relationship between percent quartz and $d_{(101)}$ spacing of opal-CT.

Whereas pyrite is ubiquitous in the concretion, it is present in small quantities only in the dolomitic layer. The dolomite crystals in the latter contain a maximum of 3 mole percent $FeCO_3$, demonstrating that in this type of depositional environment the amount of Fe in the diagenetic phases seems largely to be controlled by the chemistry of the primary sediment.

On the basis of textural relationships and mass balance calculations, both carbonate replacement and void-filling dolomite constitute the concretion and layer.

INTRODUCTION

The widespread occurrence of dolomitite concretions and layers in the Monterey Formation, California was already emphasized by Bramlette (1946) in his classical work on this formation. On the basis of the relations of the concretions to the bedding in the enclosing rocks, he concluded that they are diagenetic, formed during the early stages of compaction and lithification. He further suggested that the dolomitic layers formed much in the same way as the concretions. The origin of these dolomitites was, however, unclear to him; he could not observe nuclei of carbonate or of other materials in these concretions. Since then, numerous extensive textural, mineralogical, geochemical, and isotopic investigations of the diagenetic dolomites and protodolomites in the Monterey Formation and in other Neogene and Quaternary continental margin and small ocean basin organic-rich sediments indicate that the majority started to form early, before important compaction, and continued to grow during burial -- during progressive microbial and abiogenic decomposition of organic matter (Spotts and Silverman, 1966; Russell _et al_., 1967; Hathaway and Degens, 1969; Murata _et al_., 1969; Deuser, 1970; Supko _et al_., 1974; Friedman and Murata, 1979; Hein _et al_., 1979; Kulm _et al_., 1981; Pisciotto, 1981; Pisciotto and Mahoney, 1981; Kelts and McKenzie, 1982; and references therein). Most are Ca-rich, between 52 and 57 mole percent (%) $CaCO_3$, and are imperfectly ordered. According to the definition of Graf and Goldsmith (1956), they are protodolomites. Many are also Fe-rich, up to 16 mole % $FeCO_3$. In these sediments, in addition to layers (mostly thin) and concretions, dolomite and protodolomite commonly occur as euhedral disseminated rhombs.

The extensive geochemical studies of diagenetic carbonate concretions, and to a lesser extent of layers (for example, Degens _et al_., 1962; Weber _et al_., 1964; Hodgson, 1966; Hoefs, 1970; Curtis _et al_., 1972; Sass and Kolodny, 1972; Irwin _et al_., 1977; Hudson, 1978; Coleman and Raiswell, 1981; Kelts and McKenzie, 1982; and references therein) greatly advanced the understanding of their mode of growth. Calcite concretions commonly closely associated with well preserved fossils in their center may have formed by

in Garrison, Robert E.; Kastner, Miriam and Zenger, Donald H., eds., 1984, **Dolomites of The Monterey Formation and Other Organic-Rich Units:** Pacific Section S.E.P.M., Vol. 41, p. 171-183.

bacterial decay of the soft parts of the original fossil, as suggested by the light $\delta^{13}C$ values of these calcites, and as shown experimentally by Berner (1968). Except for these calcite concretions, the problem of why diagenetic carbonate concretions grow is not well understood.

Recent experimental work by Baker and Kastner (1981) and Kastner and Baker (1982) showed that the presence of sulfate ions is a major inhibitor of dolomitization. This result provides the necessary explanation for the origin of dolomite in organic-rich sediments, such as the Monterey Formation, California, or other Neogene and Quaternary continental margin sediments.

Detailed textural, petrological, geochemical, and isotopic studies of diagenetic concretions and layers are invaluable. Because both may start to grow before compaction and continue to grow during burial, they preserve some of the original sedimentary textures and record the post-depositional physico-chemical changes. Hence, a comparative detailed study of the diagenetic histories of (1) a protodolomite concretion from the Sandholdt Member, Monterey Formation, Salinas Basin at Arroyo Seco, and (2) of a protodolomitic layer from the Middle Member, Monterey Formation, from a road cut along State highway 166 on the easternmost proximal side of Santa Maria Basin, is hereby presented.

METHODS OF STUDY

The concretion and layer were subdivided systematically into 91 and 28 samples, respectively; 3 samples outside the concretion and 1 sample outside the layer were also studied. Microtextures were observed in thin sections, stained for calcite with Alizarin Red-S and for Fe-rich dolomite with potassium ferricyanide (Friedman, 1959). Mineralogical analyses of bulk samples and of insoluble residues were carried out by X-ray diffraction (XRD) using a Philips APD 3600 instrument, at a scanning rate of 1°2θ/min. (Royse et al., 1971; Goldsmith and Graf, 1958); a few samples were also analyzed by the electron microprobe using a Cameca Camebax instrument, a 3μm diameter beam, at 15 kV acceleration voltage and 0.015 mA sample current; corrections were made for atomic number, absorption, and fluorescence. These analyses were performed on polished carbon-coated thin sections. Carbonate standards were obtained from the Smithsonian Institution (Jarosewich and MacIntyre, 1983). In addition to Mg, Ca, and Fe, the samples were also analyzed for Si and S.

Opal-CT $d_{(101)}$ spacing was determined by XRD at a scanning rate of 1/4°2θ/minute.

Weight percent quartz (plus feldspars, which constitute only a small fraction of the insoluble residues of the samples studied) was obtained following the removal of clay minerals by sodium bisulfate fusion (Syers et al., 1968) and dissolution of the opal phases (opal-A and -CT) with sodium hydroxide (Pisciotto, 1978).

Prior to oxygen isotope analyses of diagenetic quartz, by the BrF_5 method described by Clayton and Mayeda (1963), the feldspars were removed by treatment with fluorosilicic acid (Syers et al., 1968). The pure quartz fraction was analyzed for its cathodeluminescence properties, which allow distinction between diagenetic versus detrital quartz from igneous or metamorphic rocks (e.g. Sippel, 1968; Zinkernagel, 1978).

Oxygen and carbon isotopic ratios were measured on a 6 cm VG Micromass 602 C mass spectrometer by analysis of CO_2 gas released at 50°C from the dolomites by reaction with 100% phosphoric acid (McCrea, 1950). The isotopic analyses were corrected according to Craig (1957), and the results are reported as per mil (‰) relative to the PDB standard for the dolomites and relative to the SMOW standard (Craig, 1961) for the quartz. The standard deviation is ± 0.15 for $\delta^{18}O$ and ± 0.20 for $\delta^{13}C$ values, because of the presence of some organic C impurities.

Prior to isotopic analyses of the dolomites, the small amounts of calcite present were dissolved with the acetic acid-Na acetate buffer (pH ≃ 4.7). Complete removal of calcite was certified by XRD. The oxygen and carbon isotopic composition of the protodolomites were not affected by this mild dissolution treatment, as verified by subjecting pure protodolomites to the same chemical treatment; the isotopic compositions of the untreated and acid pre-treated dolomites were identical.

Temperatures of formation were calculated using the experimentally-determined fractionation factors (1) for dolomite by Northrop and Clayton (1966) and Matthews and Katz (1977), and (2) for protodolomite by Fritz and Smith (1970) and using a constant fractionation of 3.1‰ between dolomite and calcite, given by Friedman and O'Neil (1977). For all the equations $\alpha\ CO_2\text{-}H_2O$ at 25°C of 1.0412 was used (Friedman and O'Neil, 1977). It was assumed that the protodolomites equilibrated with a fluid phase, having $\delta^{18}O = 0.0$‰ relative to SMOW. The results are given in Tables 1 and 2.

The estimated temperatures discussed in the text are those calculated according to Fritz and Smith (1970) for the following main reasons: (1) their experiments were conducted between 25° and 79°C, the temperature range of formation of most dolomites in the Monterey Formation; and (2) the concretion is composed of Fe-rich protodolomite and the layer of protodolomite, not of dolomite.

Temperatures of dolomite formation calculated from the experimental protodolomite-water fractionation expression (Fritz and Smith, 1970) are intermediate, between higher temperatures calculated by Northrop and Clayton (1966) and somewhat lower temperatures calculated by Matthews and Katz (1977).

FIELD OCCURRENCE

Dolomite Concretion

The dolomite concretion, shown in Figure 1, is from the Sandholdt Member of the Monterey Formation, late Early Miocene (Durham, 1974), in the northern part of the Santa Lucia Range, Salinas Basin, near Arroyo Center, along the Arroyo Seco. The concretion is an oblate spheroid, approximately 0.3 meters wide and 1 m long, with the long axis parallel to bedding.

The concretion is laminated, hard, and weathers to yellow-brown. It is surrounded by a laminated gray calcareous-siliceous shale, with total organic carbon of 2.5 to 7.0 % (Mertz and Garrison, 1983), deposited in an anaerobic environment, most probably on a slope in the oxygen minimum zone. According to Graham (1976), the Sandholdt Member in this area was deposited at 500 to 2500 m water depth. Sedimentation rates were high, approximately 400 m/10^6 yrs (Mertz and Garrison, 1983).

Fig. 1. The dolomite concretion at Arroyo Seco, 0.3 m by 1 m diameters. The layers directly above and below the concretion are wrapped around it due to compaction.

The laminae in the enclosing host rock can be followed into the concretion, where they are about two to three times thicker, indicating early formation prior to extensive compaction. The layers directly below but mainly above the concretion wrap around it due to subsequent compaction.

The left half (the eastern side) of the concretion was subdivided into 91 samples, according to the scheme shown in Figure 2. The surrounding shale was also studied. Sample V1A lies directly above the concretion, Sample V1N directly below, and Sample H1L is about 1 meter to the left, laterally continuous with the concretion (Fig. 2)

Dolomitic Layer

The dolomitic layer is from the opal-CT zone, Middle Member of the Monterey Formation (Woodring and Bramlette, 1950; Pisciotto and Garrison, 1981), Middle Miocene, in the easternmost proximal side of the Santa Maria Basin, at a road cut along State highway 166, between Twitchell Reservoir and the intersection with Tepusquet Road (hereafter referred to as Road Cut 166).

The layer is about 0.4 meters thick, laminated, hard, and weathers to light yellow-beige. The laminated siliceous silty shale above and below is white and light gray.

These sediments as well must have been deposited on a slope in the oxygen minimum zone. Sedimentation rates must have been high, probably similar to those at Arroyo Seco.

The layer was subdivided into 28 samples, listed in Table 2. Sample RC2000 is from the siliceous shale above the dolomitic layer.

MINERALOGY, PETROLOGY, AND CHEMISTRY

Dolomite Concretion

The concretion is composed mainly of dolomite with lesser amounts of quartz, calcite and pyrite, and very small amounts of clay minerals (mostly kaolinite and some mixed-layer smectite/illite) and feldspars. Weight percent insoluble residue, dolomite and calcite, are given in Table 1. The quartz and mixed-layer smectite/illite are detrital, calcite is primarily biogenic, small amounts of calcite and all the pyrite and dolomite are authigenic, and the origin of the kaolinite has not been resolved; it may be authigenic.

The mineralogy of the concretion and the chemistry of the dolomite change systematically from the

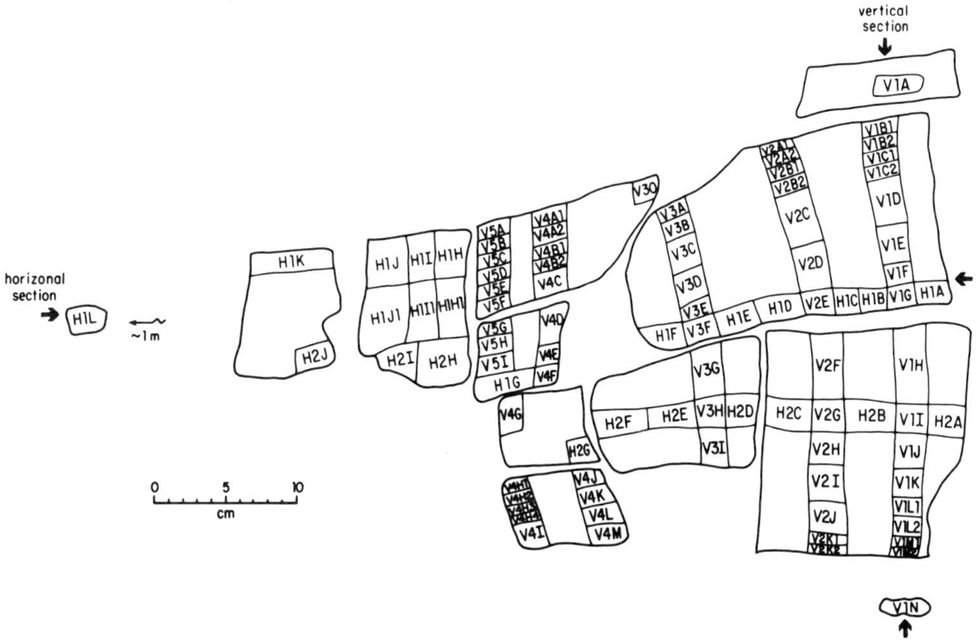

Fig. 2. Sampling scheme of the left half of the dolomite concretion shown in Fig. 1. Major cross-sections are marked by arrows. Samples V1A, V1N, and H1L are outside the concretion boundaries.

Table 1

Mineralogy, Carbon and Oxygen Isotopes of Arroyo Seco Dolomite Concretion**

Sample	Insoluble Residue wt %	Dolomite wt %	Calcite wt %	$\delta^{13}C$ °/₀₀ (PDB)	$\delta^{18}O$ °/₀₀ (PDB)	Temperature* °C		
						N&C	F&S	M&K
ASH1A	11	85	4	10.66	-0.66	47	34	31
ASH1B	11	85	4					
ASH1C	11	82	7					
ASH1D	12	85	3					
ASH1E	14	81	5	7.95	-1.86	53	40	37
ASH1F	20	73	7					
ASH1G	30	67	3					
ASH1H	43	48	9					
ASH1H1	29	67	4					
ASH1I	44	48	8					
ASH1I1	36	61	3					
ASH1J	45	47	8					
ASH1J1	36	60	4					
ASH1K	65	25	10					
ASH1L	70	9	21					
ASH2A	12	81	7					
ASH2B	14	82	4					
ASH2C	17	79	4					
ASH2D	17	78	5					
ASH2E	25	69	6					
ASH2F	28	64	8					
ASH2G	37	55	8					
ASH2H	32	61	7					
ASH2I	44	48	8	4.55	-4.20	67	54	49
ASH2J	51	33	16					
ASV1A	57	22	21					
ASV1B1	35	58	7	5.86	-2.76	58	45	42
ASV1B2	25	70	5					
ASV1C1	20	77	3	8.10	-2.09	54	41	38
ASV1C2	15	83	2					
ASV1D	12	85	3	7.87	-2.54	57	44	41
ASV1E	9	89	2	9.83	0.04	43	30	28
ASV1F	10	87	3					
ASV1G	9	86	5	10.65	0.30	42	29	27
ASV1H	13	84	3					
ASV1I	13	85	2	8.44	-1.96	54	41	38
ASV1J	14	84	2	6.91	-2.57	57	44	41
ASV1K	19	77	4	4.98	-3.71	64	51	47
ASV1L1	24	71	5					
ASV1L2	30	62	8					
ASV1M1	37	55	8	3.76	-4.47	69	56	51
ASV1M2	39	51	10					
ASV1N	74	3	23					
ASV2A1	38	51	11					
ASV2A2	28	65	7					
ASV2B1	20	73	7					
ASV2B2	16	82	2					
ASV2C	13	84	3	7.82	-2.61	57	44	41
ASV2D	11	86	3					
ASV2E	11	87	2	9.37	-0.48	46	33	31
ASV2F	15	82	3					
ASV2G	14	84	2	7.42	-2.84	59	46	42
ASV2H	19	78	3					
ASV2I	26	69	5	4.16	-4.10	66	53	49
ASV2J	32	61	7					
ASV2K1	40	49	11					
ASV2K2	44	48	8					
ASV30	43	52	5					
ASV3A	34	63	3	4.30	-4.20	67	54	49
ASV3B	25	72	3	5.77	-4.11	66	53	49
ASV3C	17	81	2					
ASV3D	15	82	3					
ASV3E	16	81	3	7.24	-2.47	57	44	40
ASV3F	15	82	3	6.14	-3.98	65	53	48
ASV3G	23	72	5					

Continued

Table 1, Continued

Sample	Insoluble Residue wt %	Dolomite wt %	Calcite wt %	$\delta^{13}C$ °/₀₀ (PDB)	$\delta^{18}O$ °/₀₀ (PDB)	Temperature* °C N&C	F&S	M&K
ASV3H	21	76	3					
ASV3I	25	71	4	4.00	-4.25	67	54	50
ASV4A1	47	45	8					
ASV4A2	33	59	8					
ASV4B1	27	69	4					
ASV4B2	22	74	4					
ASV4C	20	78	2					
ASV4D	19	79	2					
ASV4E	19	79	2					
ASV4F	21	75	4	5.10	-4.23	67	54	49
ASV4G	47	36	17					
ASV4H1	41	43	16					
ASV4H2	44	44	12	2.09	-5.10	73	60	54
ASV4H3	46	43	11					
ASV4H4	52	34	14					
ASV4I	61	25	14					
ASV4J	42	43	15					
ASV4K	51	36	13					
ASV4L	59	30	11					
ASV4M	65	16	19					
ASV5A	51	37	12					
ASV5B	43	49	8					
ASV5C	36	55	9	3.11	-5.01	72	59	54
ASV5D	29	67	4					
ASV5E	25	73	2					
ASV5F	22	73	5					
ASV5G	22	74	4					
ASV5H	23	69	8					
ASV5I	24	72	4					

* Calculated from experimentally-determined $\alpha_{dolomite-H_2O}$ by N&C = Northrop & Clayton (1966) and M&K = Matthews & Katz (1977), and $\alpha_{protodolomite-H_2O}$ by F&S = Fritz & Smith (1970) and using constant $\alpha_{dolomite-calcite}$ of 3.1°/₀₀ given by Friedman and O'Neil (1977). $\alpha_{CO_2-H_2O}$ at 25°C of 1.0412 (Friedman & O'Neil, 1977) was used for all the equations.

** Locations of samples are shown in Figure 2.

center outward in a peneconcentric pattern: dolomite accounts for up to 89 weight % in the center, decreasing to 16% at the periphery; insoluble residue increases simultaneously from 9 to 65%, and calcite, although generally less than 10 weight %, increases from 2 to 19%, as shown in Figures 3, 4, 5, and 6.

Uncrushed foraminifers are more abundant in the center than in the periphery of the concretion. Most are completely, but some only partially, replaced by dolomite. Pyrite, primarily framboidal, also increases in abundance toward the periphery. Micro-stylolites, enriched in clay minerals but mainly in Fe-oxyhydroxides, are present close to the periphery of the concretion, where the amount of dolomite is below 70 weight %. Early diagenetic fracturing is common throughout the concretion, as postulated by Roehl (1981) and Redwine (1981). Most of these fractures are filled with dolomite. No textural relationships between the fractures and stylolites were observed in the thin sections studied.

The dolomite is rich in Ca and Fe. Mole % $CaCO_3$ in the groundmass dolomite does not vary greatly; it roughly increases outward from 55 to 58 mole %, as shown in Table 3a. The most important systematic chemical changes are in the Mg/Fe ratios of the dolomite. $MgCO_3$ decreases outward from 39 to 27 mole %, $FeCO_3$ increases outward from 5 to 15 mole % (Table 3a).

Dolomite crystals replacing foraminifers and filling fractures and vugs are also rich in Ca and Fe. Electron microprobe analyses show that the Fe content of most dolomitized foraminifers varies between 10 and 15 mole % $FeCO_3$ (Table 3b), and of most dolomite crystals filling fractures and vugs between 14 and 18 mole % $FeCO_3$ (Table 3c).

On the basis of the average Fe content of the dolomitized foraminifers, which is equal to that of the dolomite in the groundmass at the periphery of the concretion, it is suggested that dolomitization of foraminifers occurred late, close to the cessation of dolomite formation in the concretion.

The average Fe content of the dolomite in the fractures and vugs is higher than that of the dolomite in the groundmass. This dolomite thus records the final stage of dolomite formation in this concretion.

The high Fe content of dolomite is responsible for the weathered yellow-brown color seen in outcrops.

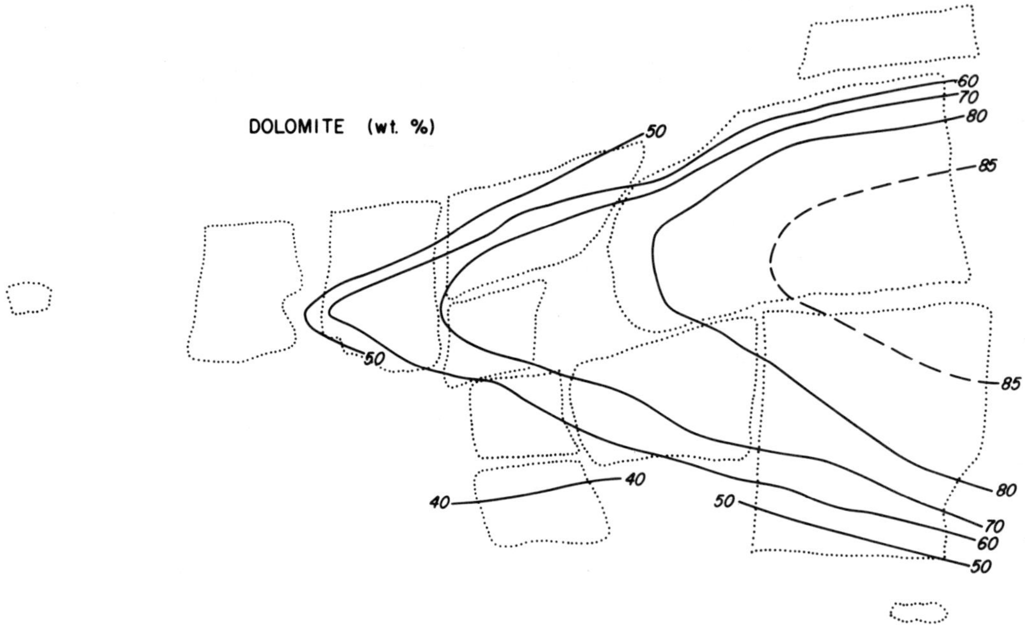

Fig. 3. Spatial distribution of weight percent dolomite throughout the eastern half of the dolomite concretion at Arroyo Seco.

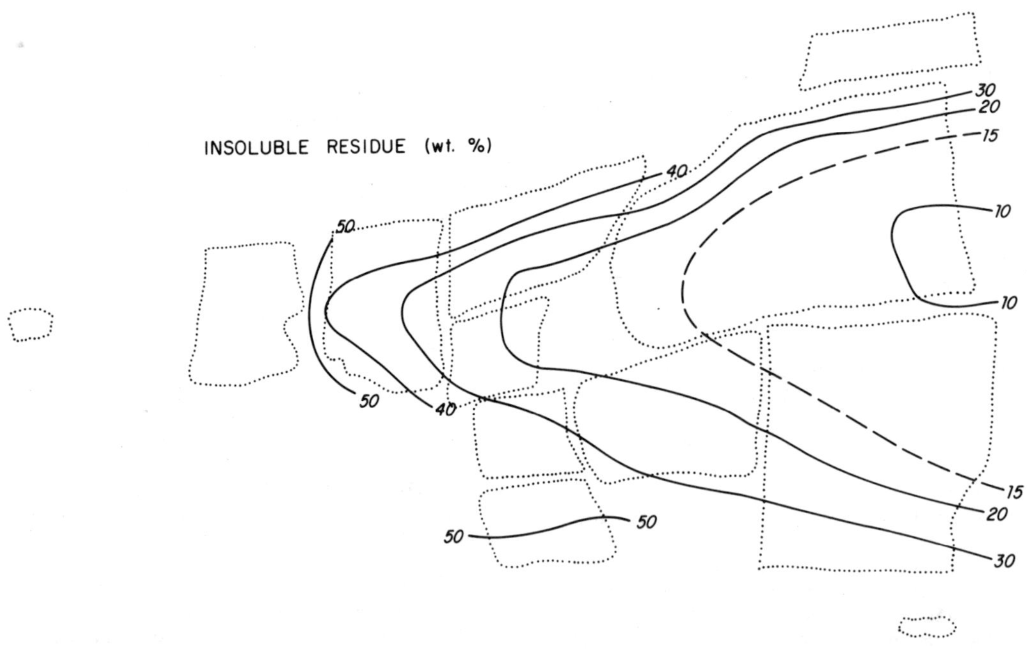

Fig. 4. Spatial distribution of weight percent insoluble residue throughout the eastern half of the dolomite concretion at Arroyo Seco.

Dolomitic Layer

The dolomitic layer is composed mainly of dolomite, averaging about 40 weight % opal-CT and quartz. Some opal-A, feldspars, and traces of pyrite and mixed-layer smectite/illite are also present. Weight percent insoluble residue and quartz are summarized in Table 2 and shown in Figure 7. Much of the quartz, in particular in the bottom 10 to 15 cm of the layer (see Fig. 7), the feldspars, and the mixed-layer smectite/illite are detrital. Dolomite, opal-CT, some of the quartz (on the basis of cathode-luminescence), and pyrite are authigenic. The euhedral dolomite rhombs are disseminated within a matrix composed of diagenetic silica phases, detrital quartz, and feldspars. No calcite was detected.

The mineralogy of the layer and the chemistry of the dolomite do not exhibit any obvious trend. The amount of dolomite varies irregularly between 55 and

Table 2

Mineralogy, Carbon and Oxygen Isotopes of Road Cut 166 Dolomite Layer. Location of samples shown in Figure 7.

cm	Sample	Insoluble Residue wt %	Quartz wt % in Insoluble Residue*	Opal-CT $d_{(101)}$ Å	Dolomite °2θ(104)**	$\delta^{13}C$ °/₀₀ (PDB)	$\delta^{18}O$ °/₀₀ (PDB)	Temperature*** °C N&C	F&S	M&K
0	RC2000	72	26	4.06	30.90					
12	RC1410	47	21	4.06	30.83	13.62	-1.83	53	40	37
13	RC1421	53	51	4.04	30.84					
14	RC1422	55	62	4.03	30.91					
15	RC1431	52	62	4.03	30.81 }	14.78	-0.94	48	35	33
16	RC1432	49	50	4.03	30.84 }					
17	RC1440	47	56	4.05	30.90					
19	RC1450	43	30	4.06	30.87	13.41	-2.13	55	42	39
21	RC1460	45	35	4.06	30.85					
22	RC1510	45	65	4.03	- }					
23	RC1520	44	54	4.05	- }	13.93	-1.60	52	39	36
25	RC1530	44	40	4.04	30.85 }					
27	RC1610	46	48	4.03	30.86	14.36	-1.50	51	38	35
29	RC1620	48	67	4.05	30.89					
30	RC1630	48	60	4.04	-					
31	RC1640	47	55	4.05	30.91					
33	RC1710	46	18	4.04	30.87	13.71	-1.47	51	37	35
36	RC1720	50	33	4.05	30.91					
37	RC1721	58	57	4.04	30.60					
38	RC1731	60	61	4.03	30.86 }					
39	RC1732	58	57	4.03	30.88 }	14.77	-1.37	51	37	35
40	RC1733	55	53	4.03	30.89 }					
41	RC1810	77	43	4.03	30.86					
43	RC1820	86	49	4.03	30.91					
45	RC1830	86	44	4.03	30.86					
47	RC1910	89	69	4.04	-					
48	RC1911	92	47	4.05	30.85					
49	RC1920	93	43	4.04	30.88					
50	RC1921	93	41	4.04	30.93					

* Determined by separation of clays and opal phases (Pisciotto, 1978).

** CuKα radiation

*** Calculated from experimentally-determined $\alpha_{dolomite-H_2O}$ by N&C = Northrop & Clayton (1966) and M&K = Matthews & Katz (1977), and $\alpha_{protodolomite-H_2O}$ by F&S = Fritz & Smith (1970) and using constant $\alpha_{dolomite-calcite}$ of 3.1°/₀₀ given by Friedman and O'Neil (1977). $\alpha_{CO_2-H_2O}$ at 25°C of 1.0412 (Friedman & O'Neill, 1977) was used for all the equations.

5 weight %. An approximate inverse relationship between $d_{(101)}$ spacing of the opal-CT and weight percent quartz is seen between 12 and 35 cm in Figure 7.

Only one dolomitized uncrushed foraminifer was observed in the twelve thin sections studied. Similarly, some diatoms are well preserved; many are dolomitized, suggesting dolomitization before extensive compaction. Framboidal pyrite is present irregularly in small quantities throughout the layer.

The composition of the dolomite crystals stays nearly constant at about 56 mole % $CaCO_3$ and a maximum of 3 mole % $FeCO_3$.

STABLE-ISOTOPE RESULTS

Dolomite Concretion

Values of $\delta^{13}C$ and $\delta^{18}O$ of the dolomite in the concretion decrease systematically from the center outward, in a peneconcentric pattern. The $\delta^{13}C$ values range from 10.66°/₀₀ in the center to 2.09°/₀₀ (PDB) at the periphery; the $\delta^{18}O$ values vary from 0.30°/₀₀ at the center to -5.10°/₀₀ (PDB) at the periphery. These results are presented in Table 1 and in Figures 5 and 6, which show a vertical and a horizontal section, respectively, across the center of the concretion; these sections are designated by arrows in Figure 2.

Carbon isotopes of diagenetic carbonates, which reflect the $\delta^{13}C$ of the pore fluid bicarbonate, measured as ΣCO_2, are sensitive indicators of organic matter degradation reactions. The most important depth-related diagenetic zones and the sources and carbon isotopic composition of the diagenetic CO_2 in each are:

1. Light $\delta^{13}C$ CO_2 (about -28°/₀₀) is produced by (a) aerobic respiration $CH_2O + O_2 \rightarrow CO_2 + H_2O$, and

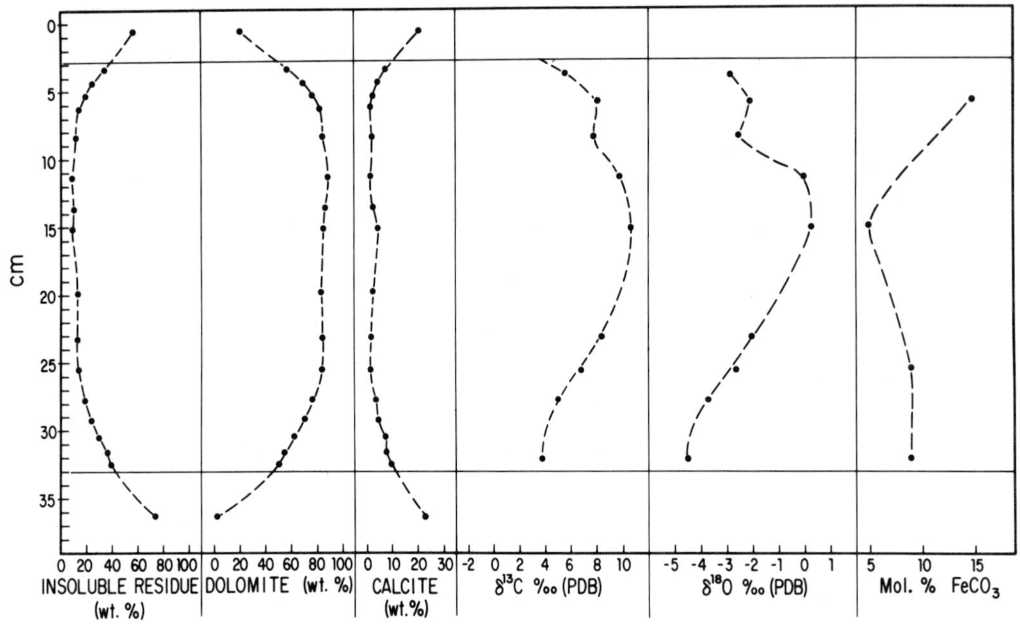

Fig. 5. Mineralogical, chemical, and isotopic changes in a vertical section through the dolomite concretion. The two horizontal lines represent boundaries of the concretion. See Figure 2 for position of section.

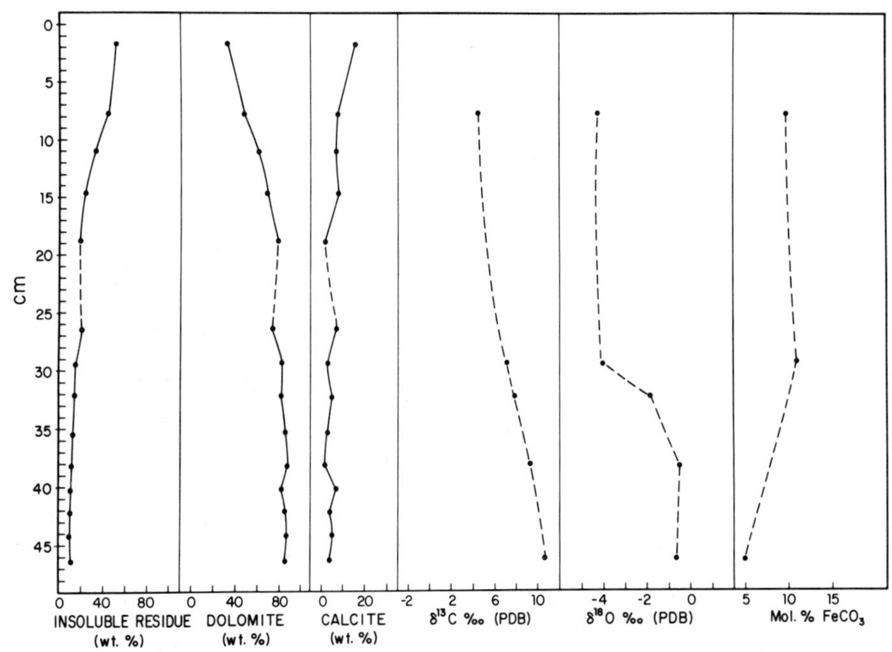

Fig. 6. Mineralogical, chemical, and isotopic changes in a horizontal section through the dolomite concretion. See Figure 2 for position of section.

(b) microbial sulfate reduction $2(CH_2O) + SO_4^{-2} \rightarrow S^{-2} + 2CO_2 + 2H_2O$.

2. Heavy $\delta^{13}C$ CO_2 (up to +15°/₀₀) is produced by anaerobic carbonate reduction (methanogenesis) $2(CH_2O) \rightarrow CH_4 + CO_2$ (for example, Rosenfeld and Silverman, 1959; Presley and Kaplan, 1968; Bottinga, 1969; Claypool and Kaplan, 1974; Nissenbaum et al., 1972; Suess, 1979.) Other important reactions that also produce light $\delta^{13}C$ CO_2 are:

(a) anaerobic methane oxidation (Barnes and Goldberg, 1967; Reeburgh and Heggie, 1974; Reeburgh, 1983, and references therein), and

(b) thermocatalytic decarboxylation ($\delta^{13}C$ CO_2 up to -20°/₀₀).

The first (most probably a biological reaction) at the depth interval between the sulfate and carbonate reduction zones, is directly or indirectly

Table 3a

Electron Microprobe Analyses of Arroyo Seco Dolomite Concretion: Dolomite in Groundmass. Location of samples shown in Figure 2.

Sample	$CaCo_3$ mole %	$MgCO_3$ mole %	$FeCO_3$ mole %
ASH1A	55.3	42.0	2.7
	54.9	39.0	6.1
	55.1	38.9	5.9
	54.7	40.7	4.6
	56.5	37.4	6.1
	54.2	40.4	5.4
	56.4	37.1	6.5
	54.7	40.1	5.3
	57.4	36.3	6.3
Average	56	39	5
ASH2I	53.2	38.8	8.1
	54.6	36.4	9.0
	56.1	34.4	9.5
	57.1	27.0	15.9
	58.1	30.3	11.6
	58.1	33.9	8.0
	57.3	33.7	9.0
	55.6	37.1	7.3
Average	56	34	10
ASV1C	58.2	26.1	15.6
	59.1	27.1	13.8
	58.2	25.2	16.7
	58.3	26.4	15.4
	56.6	29.7	13.7
Average	58	27	15
ASV1J	54.1	36.5	9.4
	59.4	25.9	14.7
	53.7	37.7	8.6
	54.4	36.7	8.9
	49.8	46.9	3.3
	54.8	38.0	7.2
	58.5	27.1	14.5
	53.6	38.0	8.4
	55.7	35.1	9.2
Average	55	36	9
ASV1M	58.8	32.5	8.7
	57.5	31.9	10.6
	56.8	35.0	8.2
Average	58	33	9
ASV3G	58.4	31.7	9.9
	54.8	36.7	8.5
	50.6	35.4	14.0
	55.5	30.9	13.6
Average	55	34	11

Table 3b

Electron Microprobe Analyses of Arroyo Seco Dolomite Concretion: Dolomitized Foraminifers. Location of samples shown in Figure 2

Sample	$CaCO_3$ mole %	$MgCO_3$ mole %	$FeCO_3$ mole %
ASH1A	58.9	29.2	11.9
	59.6	30.1	10.2
ASH2I	58.0	26.4	15.5
ASV1C	58.7	25.6	15.7
ASV1J	54.7	34.9	10.3
	58.0	34.1	7.9
	50.8	38.9	10.2
	52.0	46.1	1.9
	58.4	32.4	9.3
ASV1M	59.5	29.9	10.6
	59.1	26.1	14.8
ASV3G	57.1	29.9	13.0

Table 3c

Electron Microprobe Analyses of Arroyo Seco Dolomite Concretion: Dolomite in Fractures and Vugs. Location of samples shown in Figure 2.

Sample	$CaCO_3$ mole %	$MgCO_3$ mole %	$FeCO_3$ mole %
ASH1A	58.8	26.9	14.4
	57.4	24.2	18.4
	58.5	27.1	14.5
	58.8	26.6	14.6
	58.7	26.7	14.5
ASH2I	57.7	24.8	17.5
	56.8	25.9	17.3
ASV1C	56.8	40.8	2.4
	59.4	26.6	14.0
	53.2	39.2	7.6
ASV1J	55.0	27.6	17.5
	53.7	27.9	18.4
ASV1M	56.7	41.2	2.1
	58.1	32.5	9.4
ASV3G	56.9	26.7	16.4
	59.0	23.6	17.4
	53.6	42.3	4.2

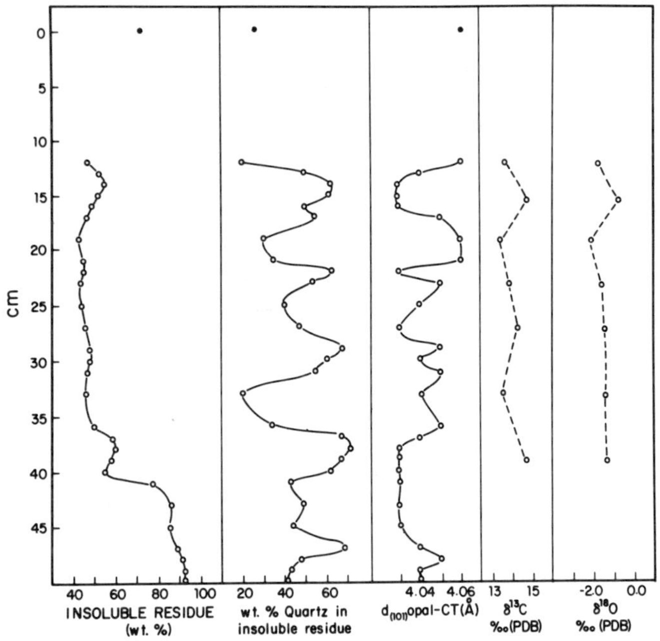

Fig. 7. Mineralogical and isotopic changes in a vertical section through the dolomitic layer at Road Cut 166, easternmost proximal side of Santa Maria Basin. The solid circles, Sample RC2000 in Table 2, are in the host rock above the layer. Top of bed starts at about 8 to 10 cm.

coupled to sulfate reduction as follows:

$$CH_4 + SO_4^{-2} \longrightarrow HS^- + HCO_3^- + H_2O$$

Thermocatalytic decarboxylation, a non-biological reaction, occurs below the carbonate reduction zone.

Accordingly, the heavy $\delta^{13}C$ of the concretion dolomite indicates formation, below the sulfate reduction zone, in the anaerobic carbonate reduction zone. The significantly lighter values toward the edges reflect either dolomite formation in a closed system, resulting in progressive in situ depletion of ^{13}C in the ΣCO_2 reservoir, or diffusion and/or advection of isotopically lighter carbonate ions from the zone of thermocatalytic decarboxylation.

The oxygen isotopic composition of diagenetic carbonates, which formed in systems of high fluid to rock ratios, reflects the $\delta^{18}O$ value of the fluid phase and the temperature of formation. As it is assumed that the concretion dolomites equilibrated isotopically with a fluid phase having $\delta^{18}O = 0.0°/_{\circ\circ}$ (SMOW), temperatures of dolomite formation can be calculated (they are summarized in Table 1). The dolomite of this concretion formed at temperatures between 27° and 54°C according to the fractionation factor by Matthews and Katz (1977), between 29° and 60°C according to Fritz and Smith (1970), or between 42° and 73°C according to Northrop and Clayton (1966), indicating continuous dolomite formation with increasing burial and temperature.

Dolomitic Layer

The $\delta^{13}C$ and $\delta^{18}O$ values of the dolomite vary slightly but simultaneously, and do not exhibit any obvious trend, as shown in Figure 7. The isotopic results are summarized in Table 2. The $\delta^{13}C$ values range from 14.78 to 13.41°/₀₀ (PDB), showing that the dolomite in this layer also formed in the anaerobic carbonate reduction zone. The $\delta^{18}O$ values of -0.94 to -2.13°/₀₀ (PDB) show that dolomitization took place in a very narrow temperature range of less than 10°C, between 33° and 39°C according to Matthews and Katz (1977), 35° to 42°C according to Fritz and Smith (1970), and 48° to 55°C according to Northrop and Clayton (1966).

The isotopic composition of the quartz separated from the insoluble fraction is 24.4°/₀₀ (SMOW) for Sample RC1440 and 23.5°/₀₀ (SMOW) for Sample RC1610. On the basis of thin section analyses and cathodeluminescence properties, these two quartz separates are composed of a mixture of about 92% diagenetic and 8% detrital quartz. The quartz of Sample RC1810 is primarily detrital. Its $\delta^{18}O$ value of 16.2°/₀₀ (SMOW) was used to calculate the $\delta^{18}O$ values of the diagenetic quartz in Samples RC1440 and RC1610. The values are 25.2°/₀₀ and 24.2°/₀₀, respectively. These values indicate maximum temperatures of quartz formation of 73° to 79°C, according to the experimentally-determined fractionation expression by Clayton et al. (1972), and of 58° to 64°C according to the fractionation expression by Knauth and Epstein (1975). Calculated temperatures of quartz formation in isotopic equilibrium with pore fluids depleted in oxygen isotopes relative to SMOW will be lower. (In this temperature range, pore water $\delta^{18}O$ depletion by 1°/₀₀ lowers the calculated temperature of formation by ∿ 6°C.) These formation temperatures are higher by 25° to 35°C than the calculated temperatures of formation of the associated dolomite. Hence, dolomite formed earlier, at shallower depth and at lower temperatures than the diagenetic quartz.

DIAGENETIC EVOLUTION OF THE DOLOMITE CONCRETION AND DOLOMITIC LAYER

The dolomite concretion records temperatures of formation between about 25°C in the center and 60°C in the periphery. According to Mertz and Garrison (1983), sedimentation rates for the Sandholdt Member of the Monterey Formation, at Arroyo Seco, were approximately 400 m/10^6 yrs. Assuming bottom temperatures of 7° to 8°C and geothermal gradients of 60° to 70°C/km (Garrison, personal communication), the dolomite concretion started to form at about 250 m below the sediment/water interface and completed its growth at about 900 m burial depth. Dolomite formation started about 0.6 x 10^6 years after deposition, and the concretion continued to grow for almost 1.7 x 10^6 years at a slow average accretion rate of 0.2 to 0.3 mm/10^3 yrs. These accretion rates are very similar (on the low side) to the rates calculated by Kelts and McKenzie (1982) of 0.2 to 0.7 mm/10^3 yrs at DSDP Site 479 in the Guaymas Basin, Gulf of California. There, where the geothermal gradients are 90°-95°C/km, dolomite started to form at a shallower depth, at greater than 80 meters burial depth. Pisciotto and Mahoney (1981) reported the occurrence of similar diagenetic dolomites offshore Southern and Baja California, between 87 and 658-meter depths.

Despite distinct differences between the dolomite concretion and dolomitic layer, (1) in the vertical distributions of mineralogical, chemical, and isotopic changes, and (2) in the range of dolomite formation temperatures, plots of $\delta^{13}C$ versus $\delta^{18}O$ for the two dolomites (in Fig. 8) show a positive linear trend, and they seem to have an almost identical slope (of about 1.5).

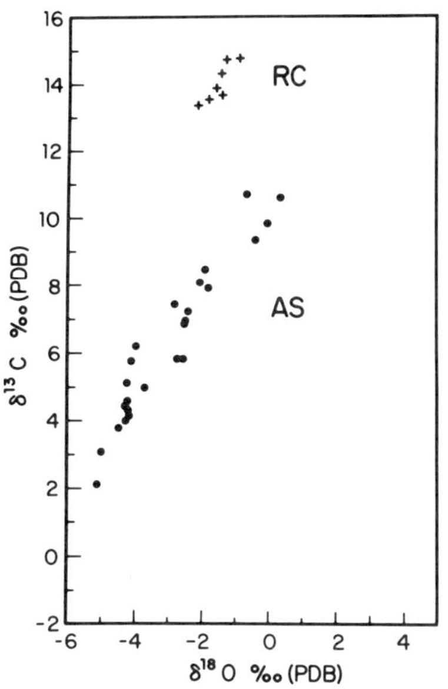

Fig. 8. $\delta^{13}C$ versus $\delta^{18}O$. Solid circles represent dolomite in concretion at Arroyo Seco (AS), and crosses represent dolomite in layer at Road Cut 166 (RC).

In organic matter-rich sediments, slopes of carbon and oxygen isotopic trends of diagenetic carbonates record diagenetic histories. These trends are primarily influenced by sedimentation rates, ratios of terrigenous to biogenic components, and geothermal gradients. For example, a positive but less steep slope would reflect a depositional environment of higher geothermal gradients and/or slower sedimentation rates; a steeper slope would imply higher sedimentation rates and/or lower geothermal gradients; a negative slope may be anticipated for dolomitization in the sulfate reduction zone, where $\delta^{13}C$ values are lighter than -5 to -8°/₀₀ (PDB). Similar and even lighter $\delta^{13}C$ values but a positive slope would imply dolomitization in the thermocatalytic decarboxylation zone.

Hence, it is assumed that sedimentation rates and geothermal gradients were similar in both the Arroyo Seco area and the proximal side of the Santa Maria Basin. The small range in the isotopic composition of the dolomite layer, but the rather clear trend in the $\delta^{13}C$ versus $\delta^{18}O$ plot (Fig. 8) suggest that this layer is composed of coalescing mini-dolomitic layers in which dolomitization started at about the same time.

At Arroyo Seco, the ratio of terrigenous to biogenic material in the host rock is higher than at Road Cut 166. In addition, the terrigenous material at Arroyo Seco contains a higher amount of clay minerals. These sedimentological differences are responsible for the observed differences in the chemistry of these dolomites; in the concretion, the dolomite is Fe-rich, and in the layer it is Fe-poor. In both areas framboidal pyrite formed throughout, in the sulfate reduction zone before the growth of the concretion and the dolomitization of the layer. At Arroyo Seco, the amount of Fe^{+3} available for reduction to Fe^{+2} and remobilization in the anaerobic zones exceeded the amount precipitated as sulfides in the sulfate reduction zone; the excess iron was available during dolomite formation. At Road Cut 166, however, most of the available iron must have been consumed by the precipitation of sulfides.

The gradual increase in iron content in the dolomite of the concretion groundmass, from the center outward, most probably reflects a slow increase in the concentration of dissolved iron in the pore fluids, due either to progressive diagenesis of the terrigenous material or to a gradual decrease in the amount and/or rate of dolomite formation.

The clear chemical trends in the dolomite of the concretion and the isotopic trends observed in both the concretion and layer (Figs. 6, 7, 8) could not have been retained or produced by subsequent chemical or isotopic exchange.

Field occurrence and fossil preservation indicate that the dolomite started to form before compaction in both the concretion and layer. At the periphery of the concretion, which became dolomitized later, at a significantly greater depth and at higher temperatures than the center of the concretion, the effect of compaction is clearly reflected in the distribution of weight % dolomite and insoluble residue, seen in the contours of Figures 3 and 4, and in the increase in the amount of pyrite close to the concretion margins. Stronger compaction is observed in the upper than in the lower half of the concretion.

The two sources of carbonate ions for dolomite formation in organic matter-rich sediments are: (1) solid $CaCO_3$ mostly as foraminifers and coccoliths, and (2) decomposition of organic matter. The carbon isotopic values indicate that much of the dolomite studied formed from organic matter carbonate. The amount of dolomite which replaced $CaCO_3$ is higher in the concretion, (estimated up to 25% ± 10% of the dolomite present; calculated on the basis of an estimation of ∿ 35% biogenic $CaCO_3$ in host sediment, and 60% porosity; the biogenic $CaCO_3$ is replaced by dolomite; about 90% of the 60% porosity is occupied by void-filling dolomite; the concretion thus contains an average of 70% of replacement plus void-filling dolomite) than in the layer (∿ 10 to 15%, calculated on the basis of ∿ 10 to 15% biogenic $CaCO_3$ in host sediment, and 60% porosity; the biogenic $CaCO_3$ is replaced by dolomite; about 75% of 60% porosity is occupied by void-filling dolomite; the layer thus contains an average of 50% of replacement plus void-filling dolomite). This difference was probably controlled by the compositions of the host sediments; at Arroyo Seco the sediment is significantly more calcareous than at Road Cut 166. The total amount of dolomite is also much greater in the concretion than in the layer. It is also suggested that sites of dolomite formation, in the form of concretions, layers, or even disseminated dolomite crystals may be guided by the availability of precursor solid carbonate.

The source of most magnesium for dolomite formation is sea water. Detailed mass balance calculations for Mg^{+2} are beyond the scope of this paper. The following are, however, a few general considerations. At high sedimentation rates, diffusion of Mg^{+2} from overlying sea water is insignificant, because communication with sea water ceases rapidly. The amount of Mg^{+2} available in the original pore fluid plus that from ion exchange sites, at porosities of 70% to 80%, could dolomitize approximately 2% of these or similar sedimentary sections. Fluid advection and/or convection must be invoked for higher volumes of dolomitization. At slower sedimentation rates, however, Mg^{+2} diffusion from overlying sea water may be important.

ACKNOWLEDGMENTS

We are grateful to Dr. R. E. Garrison, University of California, Santa Cruz, for organizing an important field trip, during which we collected the rocks described in this paper. We also benefitted from discussions with Dr. Garrison on specific issues related to the paper, and from his and Dr. D. H. Zenger's reviews of the manuscript. We thank Ms. G. Anderson for her proficient assistance in the isotopic analyses. This research was supported by ONR Contract No. USN N00014-80-C-0440 (M.K.) and by funds from Exxon Production Research Co., Houston.

REFERENCES

Baker, P. A., and Kastner, M., 1981, Constraints on the formation of sedimentary dolomite: Science, v. 213, p. 214-216.

Barnes, R. O., and Goldberg, E. D., 1967, Methane production and consumption in anoxic marine sediments: Geology, v. 4, p. 297-300.

Berner, R. A., 1968, Calcium carbonate concretions formed by the decomposition of organic matter: Science, v. 159, p. 195-197.

Bottinga, Y., 1969, Calculated fractionation factors for carbon and hydrogen isotope exchange in the system calcite-CO_2-graphite-methane-hydrogen and water vapor: Geochim. Cosmochim. Acta, v. 33, p. 49-64.

Bramlette, M. N., 1946, The Monterey Formation of California and the origin of its siliceous rocks: U. S. Geol. Survey Prof. Paper 212, 57 p.

Claypool, G. E., and Kaplan, I. R., 1974, The origin and distribution of methane in marine sediments, in Kaplan, I. R., ed., Natural Gases in Marine Sediments: Plenum Press, p. 99-139.

Clayton, R. N., and Mayeda, T. K., 1963, The use of bromine pentafluoride in the extraction of oxygen from oxides and silicates for isotopic analysis: Geochim. Cosmochim. Acta, v. 27, p. 43-52.

Clayton, R. N., O'Neil, J., and Mayeda, T. K., 1972, Oxygen isotope exchange between quartz and water: J. Geophys. Res., v. 77, p. 3057-3067.

Coleman, M. L., and Raiswell, R., 1981, Carbon, oxygen and sulphur isotope variations in concretions from the upper Lias of N.E. England: Geochim. Cosmochim. Acta, v. 45, p. 329-340.

Craig, H., 1957, Isotopic standards for carbon and oxygen and correction factors for mass spectrometric analysis of carbon dioxide: Geochim. Cosmochim. Acta, v. 12, p. 133-149.

_____ 1961, Standard for reporting concentrations of deuterium and oxygen-18 in natural waters: Science, v. 133, p. 1833.

Curtis, C. D., Petrowski, C., and Oertel, G., 1972, Stable carbon isotope ratios within carbonate concretions: a clue to place and time of formation: Nature, v. 235, p. 98-100.

Degens, E. T., Pierce, W. D., and Chilingar, G. V., 1962, Origin of petroleum-bearing fresh water concretions of Miocene age: Bull. Amer. Assoc. Petrol. Geol., v. 46, p. 1522-1525.

Deuser, W. G., 1970, Extreme $^{13}C/^{12}C$ variations in Quaternary dolomites from the continental shelf: Earth Planet. Sci. Lett., v. 8, p. 118-124.

Durham, D. L., 1974, Geology of the southern Salinas Valley: U. S. Geol. Survey Prof. Paper 819, 111 p.

Friedman, G. M., 1959, Identification of carbonate minerals by staining methods: J. Sed. Petrol., v. 29, p. 87-97

Friedman, I., and Murata, K. J., 1979, Origin of dolomite in Miocene Monterey shale and related formations in the Temblor Range, California: Geochim. Cosmochim. Acta, v. 43, p. 1357-1365.

Friedman, I., and O'Neil, J. R., 1977, Compilation of stable isotope fractionation factors of geochemical interest: U. S. Geol. Survey Prof. Paper 440-KK, 12 p.

Fritz, P., and Smith, D. G. W., 1970, The isotopic composition of secondary dolomites: Geochim. Cosmochim. Acta, v. 34, p. 1161-1173.

Goldsmith, J. R., and Graf, D. L., 1958, Relation between lattice constants and composition of the Ca-Mg carbonates: Amer. Mineral., v. 43, p. 84-101.

Graf, D. L., and Goldsmith, J. R., 1956, Some hydrothermal syntheses of dolomite and protodolomite: J. Geol., v. 64, p. 173-186.

Graham, S. A., 1976, Tertiary sedimentary tectonics of the central Salinian Block of California: Ph.D. Thesis, Stanford Univ., Stanford, CA, 510 p.

Hathaway, J. C., and Degens, E. T., 1969, Methane-derived marine carbonates of Pleistocene age: Science, v. 165, p. 690-692.

Hein, J. R., O'Neil, J. R., and Jones, M. G., 1979, Origin of authigenic carbonates in sediments from the deep Bering Sea: Sedimentol., v. 26, p. 681-705.

Hodgson, W. A., 1966, Carbon and oxygen isotope ratios in diagenetic carbonates from marine sediments: Geochim. Cosmochim. Acta, v. 30, p. 1223-1233.

Hoefs, J., 1970, Kohlenstoff und Sauerstoff - Isotopenuntersuchungen an Karbonatkonkretionen und Umgebendem Gestein: Contrib. Mineral. Petrol., v. 27, p. 66-79.

Hudson, J. D., 1978, Concretions, isotopes, and the diagenetic history of the Oxford Clay (Jurassic) of central England: Sedimentology, v. 25, p. 339-370.

Irwin, H., Curtis, C., and Coleman, M., 1977, Isotopic evidence for source of diagenetic carbonates formed during burial of organic-rich sediments: Nature, v. 269, p. 209-213.

Jarosewich, E., and MacIntyre, I. G., 1983, Carbonate reference samples for electron microprobe and scanning electron microscope analyses: J. Sediment. Petrol., v. 53, p. 677-678.

Kastner, M., and Baker, P. A., 1982, A new insight into the origin of sedimentary dolomite, in 1982/1983 Yearbook of Science and Technology: McGraw Hill, p. 406-408.

Kelts, K., and McKenzie, J. A., 1982, Diagenetic dolomite formation in Quaternary anoxic diatomaceous muds of Deep Sea Drilling Project Leg 64, Gulf of California, in Curray, J. R., Moore, D. G., et al., eds., Initial Reports of the Deep Sea Drilling Project, Vol. 64, Part 2, p. 553-569.

Knauth, L. P., and Epstein, S., 1975, Hydrogen and oxygen isotope ratios in silica from JOIDES Deep Sea Drilling Project: Earth Planet. Sci. Lett., v. 25, p. 1-10.

Kulm, L. D., Schrader, H., Resig, J. M., Thornburg, T. M., Masias, A., and Johnson, L., 1981, Late Cenozoic carbonate on the Peru continental margin: lithostratigraphy, biostratigraphy, and tectonic history: Geol. Soc. Amer. Mem. 154, p. 469-508.

Matthews, A., and Katz, A., 1977, Oxygen isotope fractionation during the dolomitization of calcium carbonate: Geochim. Cosmochim. Acta, v. 41, p. 1431-1438.

McCrea, J. M., 1950, On the isotopic chemistry of carbonates and a paleotemperature scale: J. Chem. Phys., v. 18, p. 849-857.

Mertz, K. A., and Garrison, R. E., 1983, Notes on the Sandholdt Member of the Monterey Formation in the Arroyo Seco area, Monterey County, California, in Fieldguide to the Monterey Formation, Northern Santa Lucia Mountains and Monterey Area, Monterey County, California: Soc. Econ. Paleontol. Mineral. Pac. Sec., p. 89-96.

Murata, K. J., Friedman, I., and Madsen, B. M., 1969, Isotopic composition of diagenetic carbonates in the marine Miocene formations of California and Oregon: Geol. Surv. Prof. Paper 614B, 24 p.

Nissenbaum, A., Presley, B. J., and Kaplan, I. R., 1972, Early diagenesis in a reducing fjord, Saanich Inlet, British Columbia I. Chemical and isotopic changes in major components of interstitial water: Geochim. Cosmochim. Acta, v. 36, p. 1007-1027.

Northrop, D. A., and Clayton, R. N., 1966, Oxygen-isotope fractionations in systems containing dolomite: J. Geol., v. 74, p. 174-196.

Pisciotto, K. A., 1978, Basinal sedimentary facies and diagenetic aspects of the Monterey Shale, California: Ph.D. Thesis, University of California, Santa Cruz, 450 p.

_____ 1981, Review of secondary carbonates in the Monterey Formation, California, in Garrison, R. E., and Douglas, R. G., eds., The Monterey Formation and Related Siliceous Rocks of California: Soc. Econ. Paleont. Mineral. Spec. Publ., p. 273-284.

Pisciotto, K. A., and Garrison, R. E., 1981, Lithofacies and depositional environments of the Monterey Formation, California, in Garrison, R. E., Douglas, R. G., Pisciotto, K. A., Isaacs, C. M., and Ingle, J. C., eds., The Monterey Formation and Related Siliceous Rocks of California: Soc. Econ. Paleont. Mineral., Pac. Sec. Spec. Publ., p. 97-122.

Pisciotto, K. A., and Mahoney, J. J., 1981, Isotopic survey of diagenetic carbonates, Deep Sea Drilling Project Leg 63, in Yeats, R. S. et al., eds., Initial Reports of the Deep Sea Drilling Project, Vol. 63, p. 595-609.

Presley, B. J., and Kaplan, I. R., 1968, Changes in dissolved sulfate, calcium and carbonate from interstitial water of near shore sediments: Geochim. Cosmochim. Acta, v. 32, p. 1037-1048.

Redwine, L. E., 1981, Hypothesis combining dilation, natural hydraulic fracturing, and dolomitization to explain petroleum reservoirs in Monterey shale, Santa Maria area, California, in Garrison, R. E., Douglas, R. G., Pisciotto, K. A., Isaacs, C. M., and Ingle, J. C., eds., The Monterey Formation and Related Siliceous Rocks of California: Soc. Econ. Paleont. Mineral. Pac. Sec. Spec. Pub., p. 221-248.

Reeburgh, W. S., 1983, Rates of biogeochemical processes in anoxic sediments: Ann. Rev. Earth Planet. Sci., v. 11, p. 269-298.

Reeburgh, W. S., and Heggie, D. T., 1974, Microbial methane consumption reactions and their effect on methane distributions in freshwater and marine environments: Limnol. Oceanogr., v. 22, p. 1-9.

Roehl, P. O., 1981, Dilation brecciation - a proposed mechanism of fracturing, petroleum expulsion and dolomitization in the Monterey Formation, California, in Garrison, R. E., Douglas, R. G., Pisciotto, K. A., Isaacs, C. M., and Ingle, J. C., eds., The Monterey Formation and Related Siliceous Rocks of California: Soc. Econ. Paleont. Mineral. Pac. Sec. Spec. Pub., p. 285-315.

Rosenfeld, W. D., and Silverman, S. R., 1959, Carbon isotope fractionation in bacterial production of methane: Science, v. 130, p. 1658-1659.

Royse, C. F., Wadell, J. S., and Peterson, L. E., 1971, X-ray determination of calcite-dolomite: an evaluation: J. Sediment. Petrol., v. 41, p. 483-488.

Russell, K. L., Deffeyes, K. S., and Fowler, G. A., 1967, Marine dolomite of unusual isotopic composition: Science, v. 155, p. 189-191.

Sass, E., and Kolodny, Y., 1972, Stable isotopes, chemistry and petrology of carbonate concretions (Mishash Formation, Israel): Chem. Geol., v. 10, p. 261-286.

Sippel, R. F., 1968, Sandstone petrology, evidence from luminescence petrography: J. Sed. Petrol., v. 38, p. 530-554.

Spotts, J. H., and Silverman, S. R., 1966, Organic dolomite from Point Fermin, California: Amer. Mineral., v. 51, p. 1144-1155.

Suess, E., 1979, Mineral phases formed in anoxic sediments by microbia decomposition of organic matter: Geochim. Cosmochim. Acta, v. 43, p. 339-352.

Supko, P. R., Stoffers, P., and Coplen, T. B., 1974, Petrography and geochemistry of Red Sea dolomite, in Whitmarsh, R. B. et al., eds., Initial Reports of the Deep Sea Drilling Project, Vol. 23, p. 867-878.

Syers, J. K., Chapman, S. L., Jackson, M. L., Rex, R. W., and Clayton, R. N., 1968, Quartz isolation from rocks, sediments and soils for determination of oxygen isotope composition: Geochim. Cosmochim. Acta, v. 32, p. 1022-1025.

Weber, J. N., Williams, E. G., and Keith, M. L., 1964, Paleoenvironmental significance of siderite nodules in some shales of Pennsylvanian age: J. Sed. Petrol., v. 34, p. 814-818.

Woodring, W. P., and Bramlette, M. N., 1950, Geology and paleontology of Santa Maria district, California: U. S. Geol. Survey Prof. Paper 222, 185 p.

Zinkernagel, U., 1978, Cathodoluminescence of quartz and its application to sandstone petrology, in Füchtbauer, H., Lisitzyn, A. P., Milliman, J. D., and Seibold, E., eds., Contrib. Sedimentol., v. 8: Schweizerbart'sche Verlagsbuchhandlung, 69 p.

RECOVERY OF SILICEOUS MICROFOSSILS BY THE DISAGGREGATION OF DOLOMITE

C. W. Lagle
Department of Geological Sciences
University of California
Santa Barbara, California 93106

Current Address: Neuston Rand-West
P.O. Box 1514
Goleta, California 93117

ABSTRACT

This article describes a new technique to extract siliceous microfossils from well indurated marine (and non-marine) dolomites. Fossil extraction is achieved through a two phase process, consisting of five steps. This recovery technique works well with a variety of dolomites in the Monterey Formation, including dolomites which occur in the opal-CT and quartz diagenetic zones. This process will not degrade lightly silicified frustules or morphological features required for taxonomic accuracy. Disaggregation by this method has provided enhanced biostratigraphic resolution of a Monterey lithofacies not previously considered usable. Observation and utilization comments relative to this procedure, a processing flow chart, a table of methods universally employed to clean material containing siliceous microfossils, and two plates containing photomicrographs of selected diatoms from studies employing this process are also included.

INTRODUCTION

Much of the Monterey Formation of California consists of diagenetically altered siliceous or siliceous-calcareous rocks in which biostratigraphically useful microfossils (principally diatoms and foraminifera) have been largely destroyed. Burial diagenesis has promoted dissolution of the opal-A diatom frustules and reprecipitation of the silica as either opal-CT or microquartz, converting the sediments into hard, dense porcelanites or cherts. Concurrent with these silica phase changes, calcitic foraminiferal tests are commonly replaced by microquartz, or remaining unreplaced, their calcite tests are firmly cemented in a siliceous matrix of opal-CT or microquartz so that the fossils cannot be extracted from porcelanites and cherts for micropaleontological analysis.

As a result of the factors noted above, accurate biostratigraphic dating and paleoenvironmental analysis by means of diatoms and foraminifera have, in the past, not been possible in large parts of the Monterey Formation. However, utilizing the procedures described below it is now possible to recover very well preserved diatoms from dolomite concretions and beds in the Monterey Formation. The diatoms illustrated on Plates 1 and 2 were recovered using this process from the opal-CT and quartz diagenetic zones during a recent study of the Monterey Formation in the western Transverse Ranges (Hornafius, Grivetti and Lagle, 1982; Hornafius and Lagle, 1984). These dolomites apparently formed early during burial as cements, which in some instances encased the diatom frustules, protecting them from dissolution in heated pore waters during subsequent burial (cf. Plate 11 in Bramlette, 1946). Because dolomites are very widespread in the Monterey Formation, the methods outlined below should allow much more precise biostratigraphic age dating of Monterey rocks than has heretofore been possible. Furthermore, they should also provide a suite of insoluble residue minerals amenable to study by sedimentologists, mineralogists and geochemists.

BACKGROUND

The taxonomy of siliceous microfossils, particularly diatoms, is based largely upon observations of the fine morphological features of the organism's siliceous skeleton. Such detail is visible only after the fossil has been thoroughly cleaned of rock, chemical and organic debris.

There are many published accounts that describe methods used to clean siliceous skeletons (see Table 1). The simplest is described by Patrick and Reimer (1966) who used heat to induce oxidation of organic materials, leaving a bare, charred siliceous skeleton for permanent mounting and study. Reimann (1960) used the enzyme "pancreatin" to digest the organic material that interferes with microscopy. Schrader (1974) improvised upon the general use of hydrogen peroxide (30% H_2O_2) by the addition of heat to disaggregate loosely consolidated Tertiary sediments. The other processes outlined on Table 1 are also designed to disaggregate loosely consolidated samples. However, these techniques often subject the worker to strong acids, harmful vapor and other dangerous reagents. Frequently these chemicals even degrade the siliceous materials sought. The two phase procedure described on the processing flow chart (see Table 2) is specifically designed to disaggregate well indurated dolomite and limestone, and of course may be used also on uncemented diatomites. Basically, an acid digestion of the rock sample precedes the oxidation of any residual organic materials contained in the rock. This method is advantageous over previously described methods in that it is not dangerous.

This process worked remarkably well on Miocene dolomites collected from the Monterey Formation of California (Hornafius, Grivette and Lagle, 1982; Lagle, 1982; Hornafius and Lagle, 1984), and also on Eocene dolomites of the Moler ("Cementstein") Formation near Silstrup, Denmark (Lagle, unpublished data). When subjected to this disaggregation process dolomites from both units yielded a clean residue of well preserved diatoms, silicoflagellates, radiolarians,

in Garrison, Robert E.; Kastner, Miriam and Zenger, Donald H., eds., 1984, **Dolomites of The Monterey Formation and Other Organic-Rich Units:** Pacific Section S.E.P.M., Vol. 41, p. 185-194.

TABLE 1

ACTIVE MECHANISM	EQUIPMENT/ PROCEDURE	REAGENTS	REFERENCES
Heat Treatment	Hot Plate.	Heat.	Patrick & Reimer, 1966.
Enzyme Digestion		Pancreatin	Reimann, 1960.
Chemical Digestion	Acid Treatment.	H_2SO_4+ heat.	Kolbe, 1916.
		50% HNO_3+ heat.	Crawford, 1972.
		$K_2S_2O_2$.	Ma & Jefferey, 1978.
	Acid & Oxidation Treatment.	H_2SO_4, NaOH, HNO_3.	Mandra et al, 1973.
		HCl, H_2O_2+ heat.	
		H_2SO_4, $KMnO_4$, $(COOH)_2$	Swatman, 1941, Hasle & Fryxell, 1970.
		HCl, H_2SO_4, $NaOH_3$.	Hendey, 1964.
		H_2O_2, $(COOH)_2$, $K_2Cr_2O_7$.	Van der Werf, 1953.
		H_2O_2, $(COOH)_2$ + heat.	Schrader, 1974.
		H_2SO_4, $K_2Cr_2O_7$.	Patrick & Reimer, 1966.
		$KMnO_4$, HCl.	Fryxell, 1975.
		HNO_3, NaOH or H_2O_2.	Hargraves & Guillard, 1974.
	Acid etching in constant temperature H_2O bath;	0.5% HCl + heat,	Lagle, this paper.
	Acid digestion;	11.5% HCl,	
	Oxidation digestion;	15% H_2O_2,	
	Differential density solution.	$(Na_4P_2O_7 \cdot 10H_2O)$ 0.5%.	

TABLE 1: Summary of cleaning methods commonly used on materials bearing siliceous microfossil skeletons.

and additionally, sponge spicules, siliceous spores and cysts from the Moler Formation. Lightly silicified representatives of each group were also present without degradation features, allowing detailed taxonomic study and also stratigraphic age interpretation of a lithofacies not previously considered dateable. This widely occurring and important rock type can now be studied in a new light.

MATERIALS AND METHODS

Described below in detail and outlined on the processing flow chart (Table 2) is the step by step method for processing dolomite samples, from the initial disaggregation to the final recovery of siliceous fossil material. Adherence to the outlined procedure will produce a cleaned residue suitable for making permanent mounts for electron and light microscopy. Only those acids and other reagents that do not attack opal-A and biogenic silica are used.

Step A Sample Preparation

Procedure A-1: Roll up the sample in its cloth collection bag. Use a hammer or vice to break up and crush the sample, obtaining small pieces, 15 mm. or smaller. Transfer the small fraction to an appropriately identified 500 ml. beaker. The remaining (large) fraction should be retained for future study by placing it into a properly identified bottle or envelope. Procedural Note: Experience has shown that well indurated dolomite will shatter and sliver much like glass when broken and crushed by hammer or vice. After each sampling discard the sample bag and use a wire brush to thoroughly clean the hammer and/or vice jaws to prevent sample contamination.

Procedure A-2: Transfer the small fraction from its beaker into a jar or similar vessel of a small rotary mill. With the grinder and sample in the sealed vessel, mill the sample until the chips and pieces become pellets, no larger than 2 mm. Weigh out and separate 5 gms. of the milled sample, (the more dolomite dust the better). Transfer the weighed, small fraction back into its beaker. The remaining unused small fraction is placed in its storage enveloped. Cover the beaker containing the sample with a pyrex type beaker cover and proceed to Step B.

Step B Pre-Acid Digestion Etching

Procedure B-1: Remove the 500 ml. sample beaker cover and add a quantity of 0.5% HCl solution (by volume), heated to 40^oC., sufficient to cover the dolomite pellets and dust to one centimeter in depth. The covered beaker is transferred to the fume hood and held at 40^oC. until all signs of any reaction have ceased.

Procedure B-2: When the effervesence of Procedure B-1 is no longer observed, wash the acid solution from the remaining dolomite pellets and residue using the following techniques;

Technique B-2a: Carefully swirl the contents of the 500 ml. beaker, placing the dolomite pellets and residue into suspension. After removing the beaker cover, quickly, but carefully decant the suspended contents into an appropriately identified thick wall 50 ml. pyrex centrifuge tube. Place the filled tube into its carrier and spin the sample at 1200 revolutions per minute for 2 minutes. At the end of the two minute cycle, remove the sample bearing tube and examine it for the following criteria:

Criterion A: The wash water and/or acid solution is clear.

Criterion B: The dolomite pellets and residue are firmly in place in the bottom of the centrifuge tube.

If either of Criteria A or B are not met, respin the sample for an additional 2 minutes again at 1200 revolutions per minute.

Technique B-2b: If after respining, the sample still does not meet Criteria A and B, return the sample to its beaker with fresh 0.5% HCl solution. This will allow the etching process of Procedure B-1 to go to full completion.

Technique B-2c: When Criteria A and B have been met, proceed by decanting and discarding the supernatant contained in the 50 ml. centrifuge tube. Then resuspend the remaining sample residue and pellets by forcefully directing a stream of distilled H_2O from a washbottle into the bottom of the centrifuge tube, nearly filling it. Again centrifuge the sample as in Technique B-2a. Continue washing the sample until a neutral litmus is obtained from the supernatant. This will usually require 5 or 6 washings.

Procedure B-3: Decant and discard the final supernatant. Invert the sample bearing centrifuge tube over the sample beaker. With the washbottle, rinse the residue and pellets back into the beaker in preparation for Step C. Use 5 ml. of distilled H_2O or less.

Step C Complete Acid Digestion and Evaluation

Procedure C-1: Return the 500 ml. beaker with cover in place to the fume hood. Lower the fume hood door approximately two-thirds closed for additional protection during Procedure C-2.

Procedure C-2: Using a graduated acid transfer pipette, measure out 40 ml. of 11.5% HCl solution and aspirate it into a clean 250 ml. beaker. Using rubber gloves for protection, with one hand lift the cover from the 500 ml. sample beaker, while with the other hand VERY SLOWLY decanting all 40 ml. of the HCl solution on to the etched sample residue and pellets. Immediately replace the pyrex beaker cover. At room temperature reaction C-2 may last upwards of 5 hours.

Procedure C-3: Monitor the ongoing C-2 reaction for scum formation as the digestion proceeds. Use a 9 inch fine-bore disposable pipette to aspirate any scum formation through the beaker pore spout of the 500 ml. sample beaker, with the cover in place. Discard the scum and the pipette. When reaction C-2 comes to completion continue on to Procedure C-4.

Procedure C-4: Post Acid Digestion Evaluation. Examine the sample for the following criterion:

Criterion C: Is there sufficient residue in the bottom of the 500 ml. sample beaker to have a working fraction for Step D?

Procedural Note: Criterion C requires 1 part of C-3 residue to 4 parts of undigested dolomite pellets in the 500 ml. sample beaker, as a minimum. This ratio of material is sufficient to produce a working sample for oxidation in Step D.

If the answer to Criterion C is NO additional dolomite must be dissolved to obtain more residue. In this case proceed by washing and centrifuging the

Table 2, PROCESSING FLOW CHART

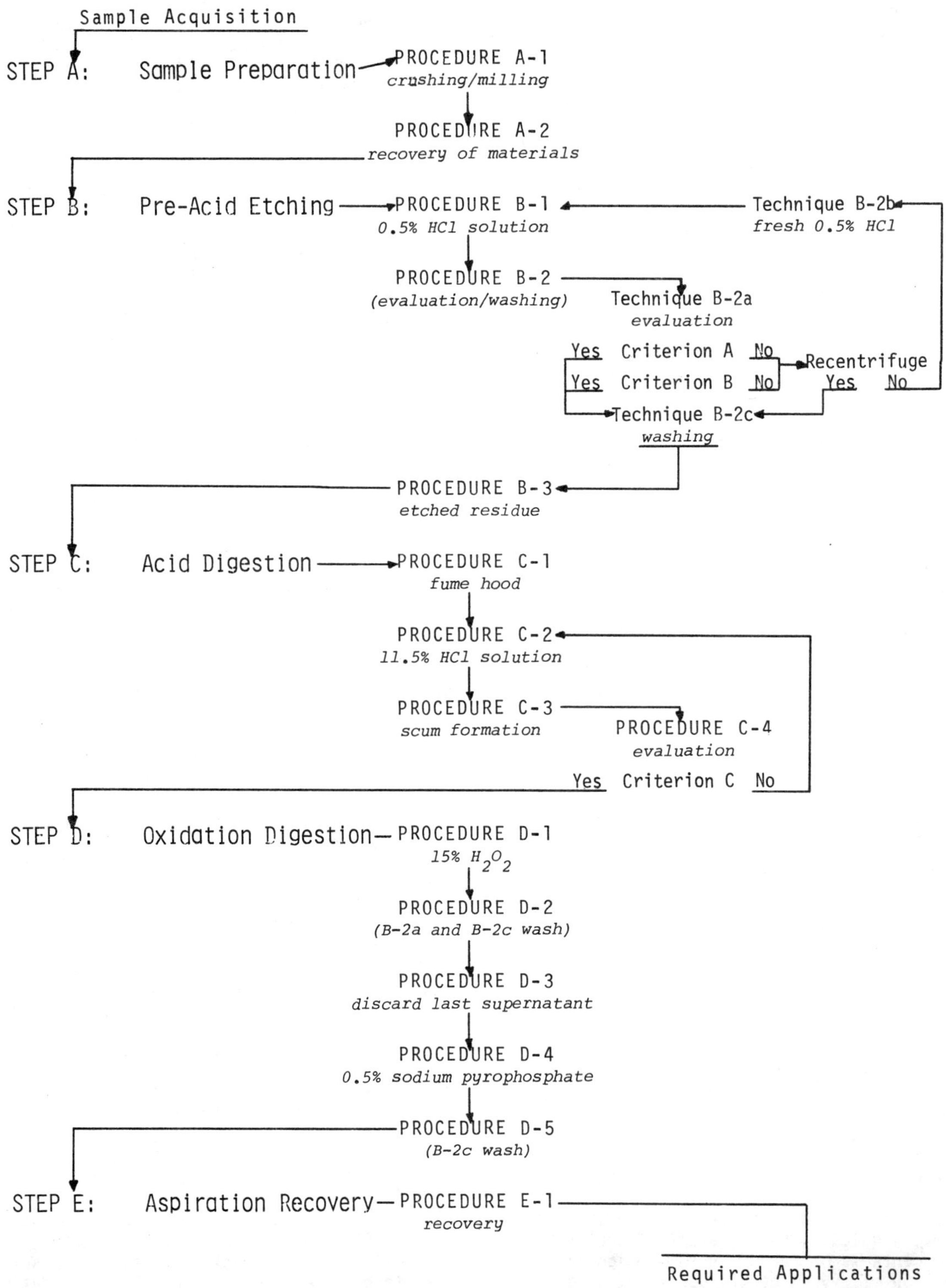

sample as in Technique B-2a and B-2c. Then repeat the acid digestion as outlined in Procedures C-1, C-2 and C-3. Following a complete second acid digestion, again wash the sample using Technique B-2a and B-2c. Return the acid-free washed residue and pellets (if any remain) into a clean, appropriately identified 250 ml. beaker. Return the beaker with cover to the fume hood in preparation for Step D.

If the answer to Criterion C is YES, then wash the sample as in Techniques B-2a and B-2c. Return the acid-free washed residue and pellets (if any remain) into a clean, appropriately identified 250 ml. beaker with cover to the fume hood in preparation for Step D.

Step D Oxidation Digestion of Residual Organic Substances

Procedure D-1: Prepare a solution of 15% H_2O_2 using distilled H_2O and 30% hydrogen peroxide (technical grade or better). Carefully add to the 250 ml. sample beaker, 50 ml. of the 15% H_2O_2 solution. Cover the sample beaker and place it in the fume hood until reaction D-1 is finished. Reaction D-1 may last up to fourteen days.

Procedure D-2: At the completion of reaction D-1, wash and centrifuge the sample residue and pellets following Technique B-2a and B-2c. However, in this instance swirl, suspend and decant ONLY the residue, leaving all pellets and pellet fragments (if any remain) in the oxidation beaker.

Procedure D-3: After the final washing, decant and discard the supernatant, leaving the residue in the bottom of the centrifuge tube.

Procedure D-4: Resuspend the sample residue by directing a stream of 0.5% sodium pyrophosphate from a wash bottle. Allow the sample to soak in the sodium pyrophosphate solution undisturbed for 2 minutes before centrifuging as in Procedure B-2.

Procedure D-5: At the end of the centrifuge cycle decant and discard the used sodium pyrophosphate solution. Wash and centrifuge the remaining sample residue 5 times, using Technique B-2c. This will remove all traces of the sodium pyrophosphate from the oxidized sample. However, in this instance, do not decant the 5th or final wash water. Proceed to Step E. Procedural Note: Once the oxidation of residual organic substances of Step D is complete, it is very important that the final residue containing the cleaned microfossil skeletons be thoroughly washed of any remaining clays or mineral grains. Schrader (1974) reports that a solution of 0.5% sodium phyrophoshpate will suspend clay size material. This will result in clean sample residues. A 0.5% solution of sodium pyrophosphate will not degrade siliceous microfossil frustules or important ultrastructures.

Step E Aspiration Recovery of Fossil Bearing Residue

Procedure E-1: Using a clean 9 inch, fine-bore disposable pipette, exercise the greatest care and aspirate only that distinct and generally thin veneer (1 mm. thick or less) of fossil bearing material off the top of the remaining residue. Empty the pipette of the fossiliferous residue into a glass vial of convenient size. Add 2 drops of buffered, 37% formaldehyde to the vial. Then fill the vial with enough distilled H_2O to top off and cap tightly. A cleaned residue is now ready for slide making or for placing on grids for TEM or SEM applications. The insoluble residue left in the centrifuge tube may also be retained for possible mineralogical, geochemical or sedimentological applications by resuspension in distilled H_2O and placed in an appropriately identified container.

DISCUSSION OF PROCEDURES

This disaggregation process is merely a complex extension of a basic field test for dolomite, where the powdered sample is submitted to dilute hydrochloric acid. However, unlike the field test, the disaggregation environment is rigidly controlled. The following comments are intended to amplify the procedures outlined above and help the processor achieve successful recovery.

When preparing dolomite sample materials for Step A, retain as much fine sample chips and dust as possible. These very small fragments are minute amounts of $CaMg(CO_3)_2$ and matrix attached to the microfossil skeleton. SEM observations of the crushed sample material reveal small fissures in the dolomite matrix and selective matrix fracturing around the siliceous fossil skeleton. Weak acid etching in Procedure B-1 enlarges the matrix fractures, forming interstices around the microfossil skeleton and into the dolomite matrix. This weakens the general matrix strength of the sample pellet and establishes increased sample surface area for additional sites for acid dissolution, in Step C. Due largely to the initial acid etching of Step B, sample pellets and dust often undergo complete matrix dissolution during the reaction of Procedure C-2. This will yield an abundant amount of a dark, jelly-like organic-rich fossil residue in the digestion beaker.

Dolomite sample color and smell provide reliable indications of what the processor may expect during the acid digestion of Step C. Vitrinite reflectance studies (Lagle, unpublished data) indicate dark colored dolomites are organic-rich. During the acid digestion reaction of Procedure C-2 kerogen is released from the disaggregated rock. The reaction heat forces kerogen maturation. These freshly released organic compounds are less dense than the acid solution and rock materials in the digestion beaker. The matured kerogen becomes cohesive and rises to the acid solution-air interface, forming scum during the reaction. In the covered beaker, scum formation impedes the dissipation of digestion reaction heat. Retention of reaction heat increases the reaction rate. Reaction acceleration will cause additional scum formation. This phenomena will perpetuate the loss of sample material needed for Step D by boilover. It will also promote frustule pitting and partial dissolution of the siliceous skeleton. Occasionally, samples collected from the Monterey Formation contain abundant sulfides. These sample materials are also dark and produce an odor similar to spend gunpowder when crushed during Procedure A-1. The C-2 reaction rates generated by these dolomites can become highly accelerated. This will usually occur within 10 minutes of beginning the C-2 procedure. The addition of several milliliters of 0.5% ETOH solution (in distilled H_2O), directed through the beaker pour spout from a washbottle will effectively reduce reaction temperature and rate.

OBSERVATIONS AND SUMMARY

1) The most remarkable aspect of this process is the discovery that many dolomites in the Monterey and Moler Formations contain well preserved siliceous microfossils (Plates 1 and 2).

2) X-ray diffraction studies showed that many of the Monterey dolomites studied occur within the opal-CT and quartz diagenetic zones. These dolomites invariably provided well preserved and abundant opal-A diatoms and silicoflagellates. The dolomites occur as concretions and lenticular beds; they may have formed very early in diagenesis, thus encasing and protecting the opal-A microfossil skeletons from dissolution in pore waters during burial.

3) In a recent study Hornafius, Grivetti and Lagle (1982) demonstrated enhanced stratigraphic control for utilizing dolomites in the Monterey Formation of the western Transverse Ranges. A multidisciplinary approach incorporating diatom biostratigraphy and magnetic stratigraphy provided increased age resolution and greater accuracy in calculating sedimentation rates, and it also helped clarify regional correlation studies.

4) As in all biostratigraphic studies recognizing reworked fossils, especially microfossils, can be a problem. The criteria established for recognition of reworked fossil materials are no different for fossils recovered from dolomite, except that the phenomenon is more easily identified. Reworked siliceous microfossils from the dolomite samples processed for these studies, when encourtered, showed very severe effects of dissolution and frustule pitting, making them almost unrecognizable.

5) Massive, concretionary and laminated dolomites have been analyzed with these procedures. All of these types of dolomite have well preserved remanent magnetization and/or well preserved siliceous microfossils. It should be noted that samples taken from a single bed may vary markedly in fossil abundance or magnetic signal strength. While the reasons for this random or selective preservation remain unclear, repetitive lateral sampling of the same bed will usually insure recovery of both good magnetic signals and well preserved siliceous microfossils.

6) The procedures outlined in this paper should be applicable to the study of insoluble residues (other than siliceous microfossils), a type of investigation commonly applied to carbonate rocks but only rarely utilized up till now by sedimentologists and geochemists in studies of Monterey dolomites and limestones.

ACKNOWLEDGEMENTS

The author wishes to thank David Pierce of the Preston Cloud Laboratory, Department of Geological Sciences, University of California, Santa Barbara, for his stimulating and thought provoking comments on this subject. Robert Garrison, Donald Zenger and Miriam Kastner reviewed the initial manuscript and offered numerous helpful suggestions. I would also like to recognize the contributions and encouragement offered by John Barron, Scott Hornafius and Wade Hampton. Not to be forgotten is the unforgettable impetus offered by Professor Donald W. Weaver during the initial investigation which led to this paper.

EXPLANATION OF PLATE 1

Diatoms from dolomite samples of the Monterey Formation collected at Point Pedernales, California. The dolomites were processed utilizing the techniques described in this paper. For sample locations see Hornafius, Grivetti and Lagle (1982), Hornafius (1984).

Figure 1. Actinocyclus ellipticus Grunow. Sample PP191A. Length 42 μm.

Figure 2. Actinocyclus ellipticus var. javanicus Reinhold. Sample PP81. Length 37 μm.

Figure 3. Thalassiosira oestrupii (Ostenfeld). Sample PP191A. Diameter 28 μm.

Figure 4. Thalassionema hirosakiensis (Kanaya) Schrader. Sample PP81. Length 31 μm.

Figure 5, 6. Thalassiosira antiqua (Grunow) Cleve. Sample PP81. Diameter 39 μm, 28 μm.

Figure 7. Nitzschia fossilis (Frenguelli). Kanaya and Schrader. Sample PP96. Length 42 μm.

Figure 8. Nitzschia miocenica Burckle. Sample PP96. Length 23.5 μm.

Figure 9, 10. Nitzschia praereinholdii Schrader. Sample PP96. Length 47 μm, 53 μm.

Figure 11. Lithodesimum minusculm Grunow. Sample PP81. Length of base 29.5 μm.

REFERENCES CITED

Bramlette, M. N., 1946, The Monterey Formation of California and the origin of its siliceous rocks: U.S. Geological Survey Prof. Paper 212, 57 p.

Crawford, R., 1972, The structure and formation of the siliceous wall of the diatom Meolsira nummuloides (Dillw.); Second Symposium on Recent and Fossil Marine Diatoms, Sept. 4-9, London, Proceedings, p. 131-146.

Fryxell, G. A., 1975, Morphology, taxonomy and distribution of selected diatom species of Thalassiosira (Cleve), in the Gulf of Mexico and the Antartic: Ph.D. dissertation, Texas A&M University, College Station, 189 p.

Hargraves, P. E., and Guillard, R. R. L., 1974, Structure and physiological observations on small marine diatoms: Phycologia, v. 13, p. 163.

Hasle, G. R., and Fryxell, G. A., 1970, Diatoms, cleaning and mounting for light and electron microscopy: Trans. Am. Microscopal Soc., v. 89, no. 4, p. 469-474.

PLATE 1

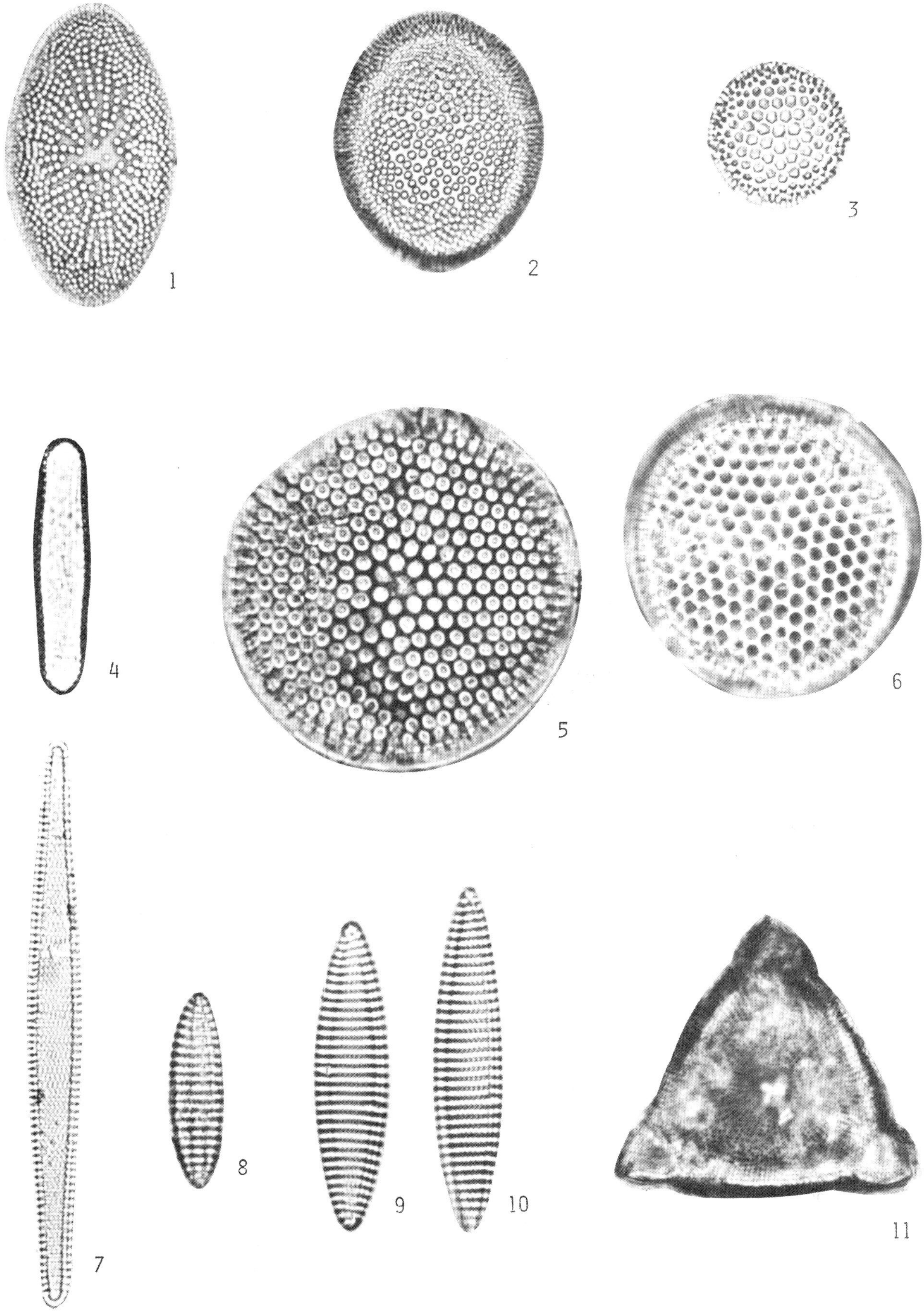

Hendey, N. E., 1964, An introductory account of the smaller algae of the British coastal waters. Part V. Bacillariophyceae (Diatoms): Fishery Investigations, ser. 4, Her Majesty's Stationary Office, London, 317 p.

Hornafius, J. S., 1984, Paleomagnetism of the Monterey Formation in the Western Transverse Ranges, California: Ph.D. dissertation, University of California, Santa Barbara.

Hornafius, J. S., Grivette, M. S., and Lagle, C. W., 1982, Structure and stratigraphy of the Monterey Formation near Point Pedernales, Santa Barbara County, California, in Fife, D. L., and Minch, J. A. (eds.), Geology and Mineral Wealth of the California Transverse Ranges: South Coast Geological Soc., Santa Ana, California, 800 p.

Hornafius, J. S., and Lagle, C. W., 1984, Magnetic stratigraphy and diatom biostratigraphy of the Monterey Formation at selected localities in the western Transverse Ranges, California (abs), in Garrison, R. E., Kastner, M., and Zenger, D. H. (eds.), Dolomites of the Monterey Formation and Other Organic-Rich Units: Pacific Section SEPM (this volume).

Lagle, C. W., 1982, Field correlation and paleontological evaluation of selected rock materials found at the Santa Barbara Presidio archaeological excavation, Santa Barbara, California, in The Archaeology of the Royal Presidio of Santa Barbara, Phase VIII Archaeological Excavation of the Chapel Site: Santa Barbara, Santa Barbara Trust for Historical Preservation, p. 123-129.

Ma, J. W. C., and Jeffrey, L. M., 1978, Description and comparison of a new cleaning method of diatom frustules for light and electron microscope studies: Jour. Microscopy, v. 112, no. 2, p. 235-238.

Mandra, Y. T., Brigger, A. L., and Mandra, H., 1973, Chemical extraction techniques to free fossil silicoflagellates from marine sedimentary rocks: Proc. California Acad. Sci., 4th ser., v. 39, no. 15, p. 273-284.

Patrick, R., and Reimer, C. W., 1966, The diatoms of the United States, Volume 1: Monographs of the Academy of Natural Sciences of Philadelphia, 688 p.

Reimann, V. B., 1960, Bildung, Bau und Zuzammenhang der Bacillariophyceenschalen: Nova Hedwigia, v. 2, p. 349-355.

Schrader, J.-H., 1974, Proposal for cleaning diatom sediments and land exposed marine sediments: Nova Hedwigia, v. 45, p. 403-409.

Swatman, C. C., 1941, Simple method for cleaning fresh diatoms: Jour. Quekett Microscopial Club, ser. 4, no. 1, p. 191.

Van der Werf, A., 1955, A new method of concentrating diatoms and other organisms: Applied Limnonology, v. 12, p. 276-279.

EXPLANATION OF PLATE 2

Diatoms from dolomite samples of the Monterey Formation collected at Hondo Canyon, Santa Barbara Co., California. For sample locations see Hornafius and Lagle (1984), Hornafius (1984).

Figure 1. *Actinocyclus ingens* Rattray. Sample WB78. Diameter 79 µm.

Figure 2. *Actinocyclus ingens* var. *nodus* Bauldauf and Barron. Sample WB78. Diameter 61 µm.

Figure 3. *Coscinodiscus lewisianus* Greville. Sample WB167. Length 33 µm.

Figure 4. *Denticulopsis hyalina* (Schrader) Simonsen. Sample WB129. Length 32 µm.

Figure 5, 6, 7. *Denticulopsis lauta* (Bailey) Simonsen. Samples WB167, WB78, WB89A. Length 16 µm, 14 µm, 12.5 µm.

Figure 8. *Rhaphoneis miocenica* Schrader. Sample WB129. Length 34 µm.

Figure 9, 10, 11. *Annellus californicus* Hanna. Sample WB78. Diameter valve view 44 µm, 42.5 µm; height valve margin view 41 µm.

Figure 12, 13. Unidentified cysts. Sample WB129A. Diameters 9 µm, 12 µm.

PLATE 2

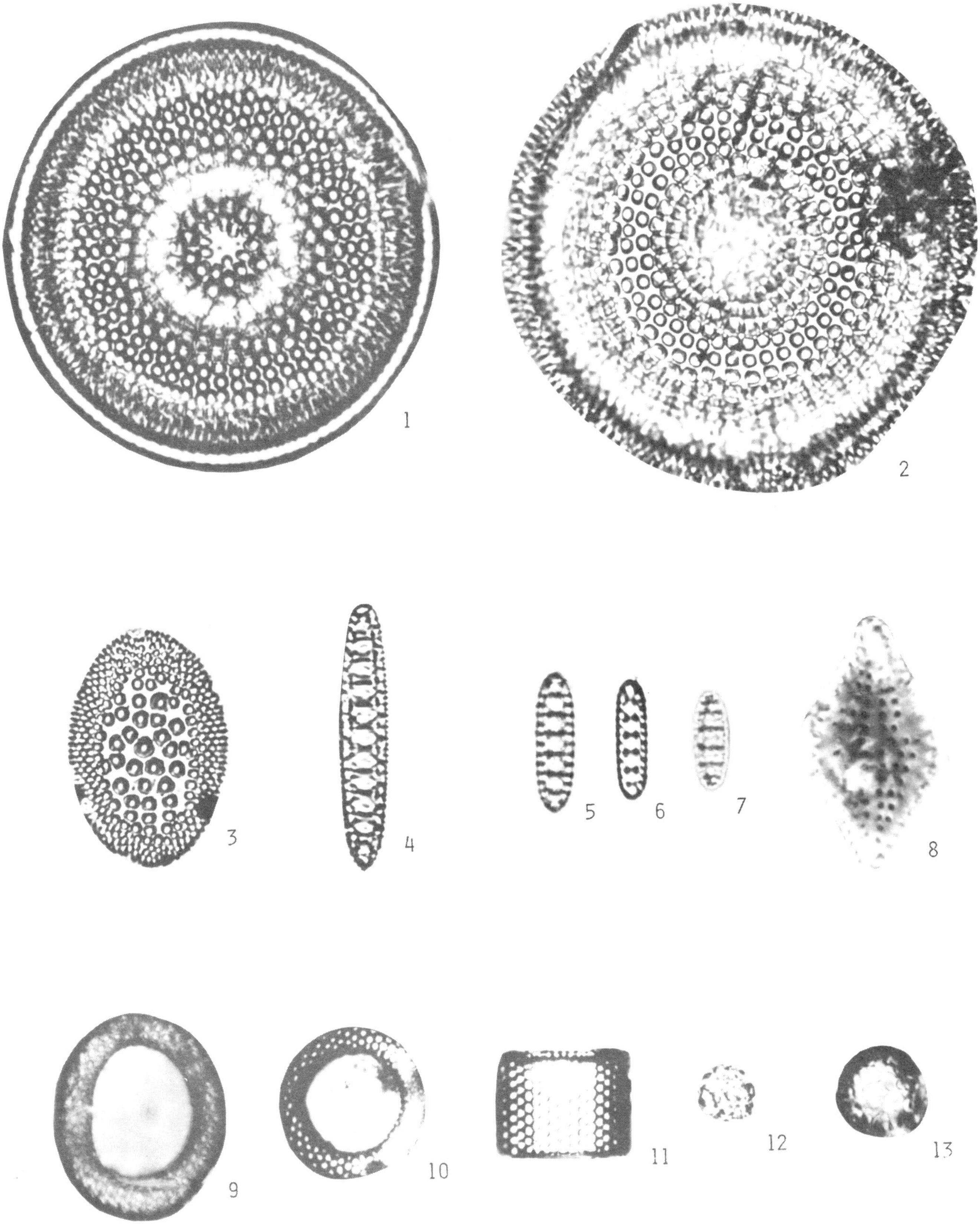

ORIGIN OF REMANENT MAGNETIZATION IN DOLOMITE FROM THE MONTEREY FORMATION

J. Scott Hornafius
Department of Geological Sciences
University of California
Santa Barbara, California 93106

ABSTRACT

Dolomites and limestones in the Monterey Formation generally possess either a stable primary magnetic remanence acquired prior to Plio-Pleistocene folding, or a stable secondary magnetic remanence acquired subsequent to Plio-Pleistocene folding. Chemical analyses, thermal demagnetization, and IRM saturation experiments suggest that the primary remanence is carried by detrital magnetite, whereas the secondary remanence is carried by authigenic magnetite ± hematite.

The detrital origin of the primary magnetite is suggested by a positive correlation between the IRM saturation intensity of the dolomite (proportional to the magnetite content) and the aluminum content of the dolomite (proportional to the abundance of aluminosilicate minerals in the dolomite). The ratio of IRM saturation intensity to aluminum content in Monterey dolomites implies that the volume ratio of detrital magnetite to aluminosilicate minerals is approximately 1 part in 25,000. The hysteresis properties of five dolomites suggests that the magnetite occurs as pseudo-single domain grains (grain size = 0.1 to 10 microns). A primary magnetic remanence carried by authigenic pyrrhotite (or maghemite) was only suspected in one dolomite bed.

Two soft-sediment slump folds that were subsequently impregnated with carbonate cement failed the fold test, which suggests that the primary remanence is locked in more than a few tens of centimeters below the seafloor. However, correlation of the magnetic polarity sequence and the diatom biostratigraphy within a 50-m-thick stratigraphic section suggests that the primary remanence is locked in less than a few meters below the sediment-water interface. The primary remanence is probably acquired during initial compaction and dewatering of the diatomaceous sediment. Impregnation of the sediment with carbonate cement at shallow burial depths then protects the primary remanence from alteration during subsequent burial diagenesis, tectonic deformation, and weathering.

Although most dolomite beds were either entirely normal or reversed in polarity over a distance of kilometers, a few dolomite beds possessed a random distribution of normal and reversed polarities, with few stable intermediate directions. The mixed polarities recorded by these beds cannot be ascribed to mineralogic differences between the normally and reversely magnetized samples, and are probably due to reversals of the earth's magnetic field during the acquisition of remanent magnetization in these beds. The failure of these dolomite beds to record transient magnetic field directions during a magnetic field reversal suggests that the time required for the dolomite to acquire a stable magnetic remanence is longer than the duration of a field reversal, i.e., longer than approximately 4000 years. On the other hand, the ability of the dolomite to record polarity events as short as 100,000 years suggests that the magnetization process does not greatly exceed 100,000 years in duration.

Remagnetization of the dolomite often results in an increase in magnetic intensity. Acquisition of a strong secondary remanence is attributed to partial oxidation of pyrite in the rock to magnetite ± hematite. Remagnetization is probably caused by post-orogenic introduction of meteoritic waters to the formation via aquifers during the Pleistocene, as well as weathering at the outcrop during the Recent.

INTRODUCTION

The Monterey Formation is a middle and upper Miocene siliceous hemipelagic deposit which is widespread in coastal California. The formation is comprised of rhythmically bedded siliceous mudstones, calcareous mudstones, diatomites, porcelanities, cherts, limestones, and dolomites (Bramlette, 1946). A reconnaissance paleomagnetic study of these rock types revealed that only the dolomites and limestones in the Monterey Formation possess a stable primary magnetic remanence (Hornafius et al., 1981). Other rock types are either remagnetized in the present-day field direction, unstably magnetized, or too weakly magnetized to be measured ($J_{NRM} < 10^{-8}$ emu/cc).

Dolomites and limestones in the Monterey Formation occur either as continuous 0.1-1.0 meter thick beds or as nodules along discrete bedding planes. Numerous lines of evidence suggest that the dolomite and limestone forms by carbonate impregnation of the unconsolidated diatomaceous sediment in the early stages of burial diagenesis. Differences in the thickness of layers which pass through dolomite nodules indicate that the sediment outside the nodules has been compacted three to four times more than the sediment within the dolomite (Bramlette, 1946). Carbon isotope studies suggest that the dolomite forms at shallow burial depths and that it does not experience solution-reprecipitation reactions during later burial diagenesis (Friedman and Murata, 1979). Oxygen isotope ratios in the dolomite suggest that the dolomite precipitates at temperatures ranging from 17-86°C, which is equivalent to burial depths ranging from ~80 to ~660 meters below the seafloor (Pisciotto, 1981).

A paleomagnetic investigation was recently conducted on the carbonates in the Monterey Formation, in order to determine the timing and extent of tectonic rotation in the western Transverse Ranges (Hornafius, 1984). As a part of this research effort, field and laboratory work was conducted in order to determine the origin of the remanent magnetization in the Monterey dolomites and limestones. This paper is a summary of that work. Although both dolomites and limestones were examined in this study,

in Garrison, Robert E.; Kastner, Miriam and Zenger, Donald H., eds., 1984, **Dolomites of The Monterey Formation and Other Organic-Rich Units:** Pacific Section S.E.P.M., Vol. 41, p. 195-212.

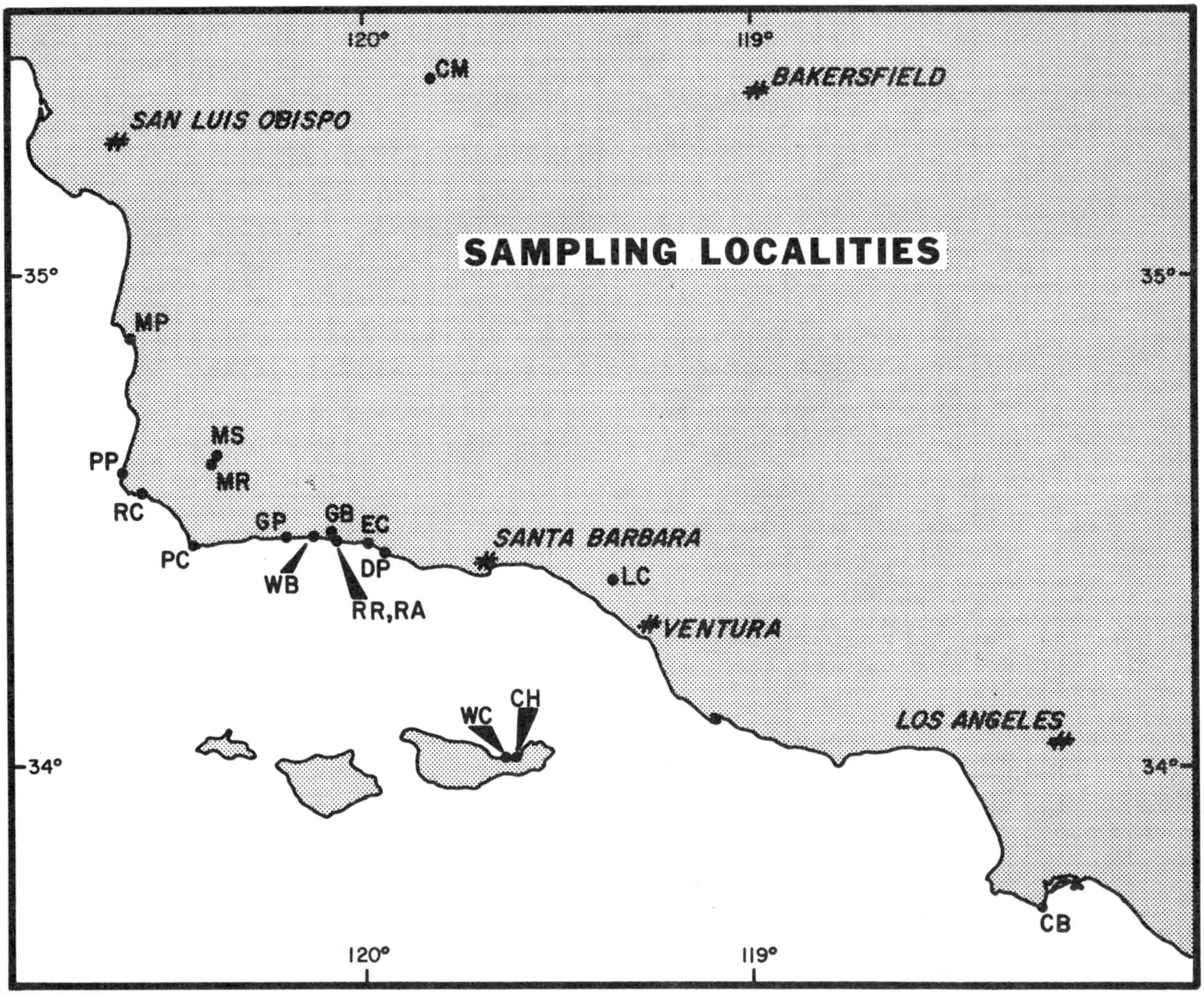

Fig. 1. Sites where Monterey dolomite samples examined in this study were collected. Rincon dolomite samples were collected at sites GB and LC. Detailed descriptions of each sampling locality are given in Hornafius (1984).

no difference between the paleomagnetic or rock magnetic properties of the limestones and dolomites could be discerned. Because dolomite is far more abundant than limestone, the carbonates within the Monterey Formation will be referred to simply as dolomite for the remainder of the paper. The sampling localities used in this study are illustrated in Fig. 1.

In southern California, the Monterey Formation locally overlies the lower Miocene Rincon Shale, a clay shale which also contains abundant dolomite nodules (Dibblee, 1966). The initiation of Monterey sedimentation corresponds to an increase in the rate of upwelling along the continental margin of California approximately 16 m.y. ago (Ingle, 1981). This resulted in an increase in diatom production, which caused the deposition of mudstones with a large siliceous biogenic component. Although Rincon dolomites contain a stable magnetic remanence, the paleomagnetic directions obtained from these dolomites were much more dispersed than the directions obtained from the Monterey dolomites (see Hornafius, 1984). The rock magnetic properties of three dolomite samples from the Rincon Shale were also investigated, for purposes of comparison with Monterey dolomites.

PALEOMAGNETIC PROPERTIES

Field Stability Tests

The Monterey Formation in southern California was folded during the Plio-Pleistocene. This recent deformational event enables a fold test to be applied to the dolomite beds in the Monterey Formation in order to distinguish magnetic directions acquired in the Miocene (primary directions) from directions resulting from remagnetization in the Pleistocene to Recent (secondary directions). Primary directions generally were greater than 20° from the present-day dipole field direction for the region (I = 54°, D = 0°; and I = -54°, D = 180°) prior to structural correction, and passed fold tests and reversal tests after structural correction (Fig. 2). Secondary directions were generally within 20° of the dipole field direction (both normal and reversed) and failed fold tests and reversal tests after structural correction (Fig. 3).

Demagnetization Behavior

Dolomite beds in both the Monterey Formation and the Rincon Shale possess similar paleomagnetic pro-

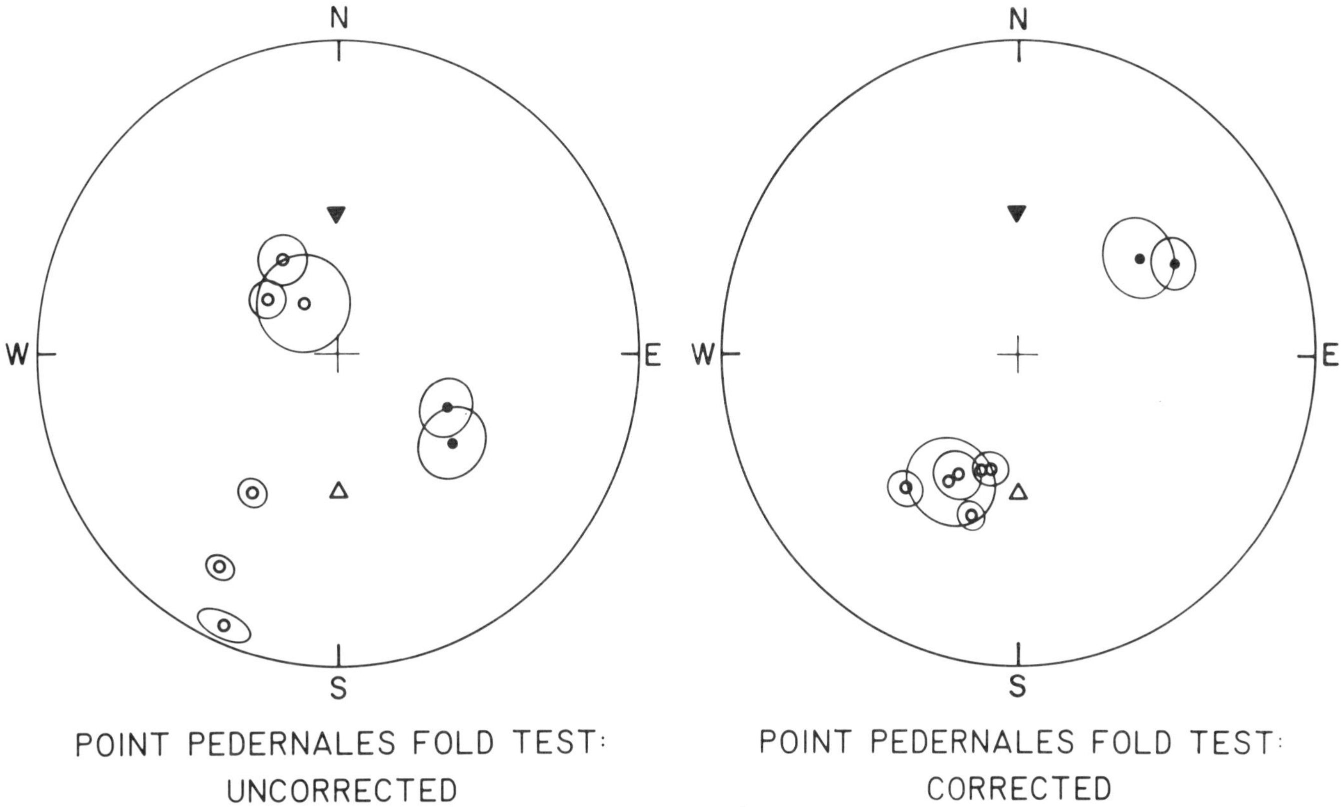

Fig. 2. Equal-area stereographic projection of the mean paleomagnetic directions from 5-8 m.y. old dolomite beds sampled at Point Pedernales (site PP), before and after structural correction. Reversed magnetic directions (hollow circles; upper hemisphere), which are dispersed about a N60°W trending fold axis prior to applying a bedding tilt correction (left), become tightly clustered after structural correction (right), and therefore pass the fold test. Normal magnetic directions (solid circles; lower hemisphere) are nearly antiparallel to the reversed directions after structural correction, and therefore pass a reversal test. This successful fold test and reversal test indicates that the magnetization was acquired prior to Plio-Pleistocene folding. Ellipses around the circles indicate the uncertainty in the mean directions at the 95% confidence level. Triangles indicate the present-day dipole field directions for the region.

perties, although the stable primary directions from Monterey dolomites are much more tightly clustered than the directions from Rincon dolomites. NRM intensities in both formations range from 10^{-7} emu/cc to 10^{-6} emu/cc. All samples possess a large viscous component pointing in the present-day dipole field direction. This component is removed by AF demagnetization to 100-200 Oe or by thermal demagnetization to 200-250°C. A stable component of magnetization remains after this viscous component is removed. This higher coercivity and higher blocking temperature component decays rapidly upon further AF demagnetization, but sometimes resists further thermal demagnetization until 400-500°C (Fig. 4). This stable component is usually either entirely primary or entirely secondary in origin.

MAGNETIC MINERALOGY

Thermal Demagnetization Curves

Thermal demagnetization experiments demonstrate that the blocking temperature spectrum of the magnetic mineral carrying both the primary and secondary remanence in the majority of Monterey dolomites ranges from 300 to 550°C (Fig. 4). This range of blocking temperatures suggests that magnetite (Curie temperature = 580°C) is the dominant mineral carrying the magnetic remanence in Monterey dolomites. The possibility that pyrrhotite or maghemite is the magnetic carrier can be ruled out because there is generally no appreciable drop in intensity at or immediately below the Curie temperature for pyrrhotite (325°C) or the breakdown temperature of maghemite (350°C).

Out of approximately twenty dolomite beds for which detailed thermal demagnetization experiments were conducted, only one bed displayed paleomagnetic behavior which suggested that a component of the primary remanence is carried by pyrrhotite or maghemite. Samples from a single bed in the Lion's Head section (site MP; Fig. 1) experienced an increase in magnetic intensity between 200 and 300°C (Fig. 4). Vector differences indicated that the component which was removed between 200 and 300°C is antiparallel to the dominant normal-polarity primary component in the samples (Fig. 5). This observation suggests that the antiparallel reversed-polarity component is also primary in origin, and is not due to a secondary overprint acquired in the Pleistocene to Recent. IRM saturation of samples from this bed occurred by 3 kOe (e.g., MP38A and MP40; see Fig. 6), which suggests that hematite is not responsible for this reversed component. This primary reversed component is probably due to the presence of early-formed pyrrhotite in this dolomite bed.

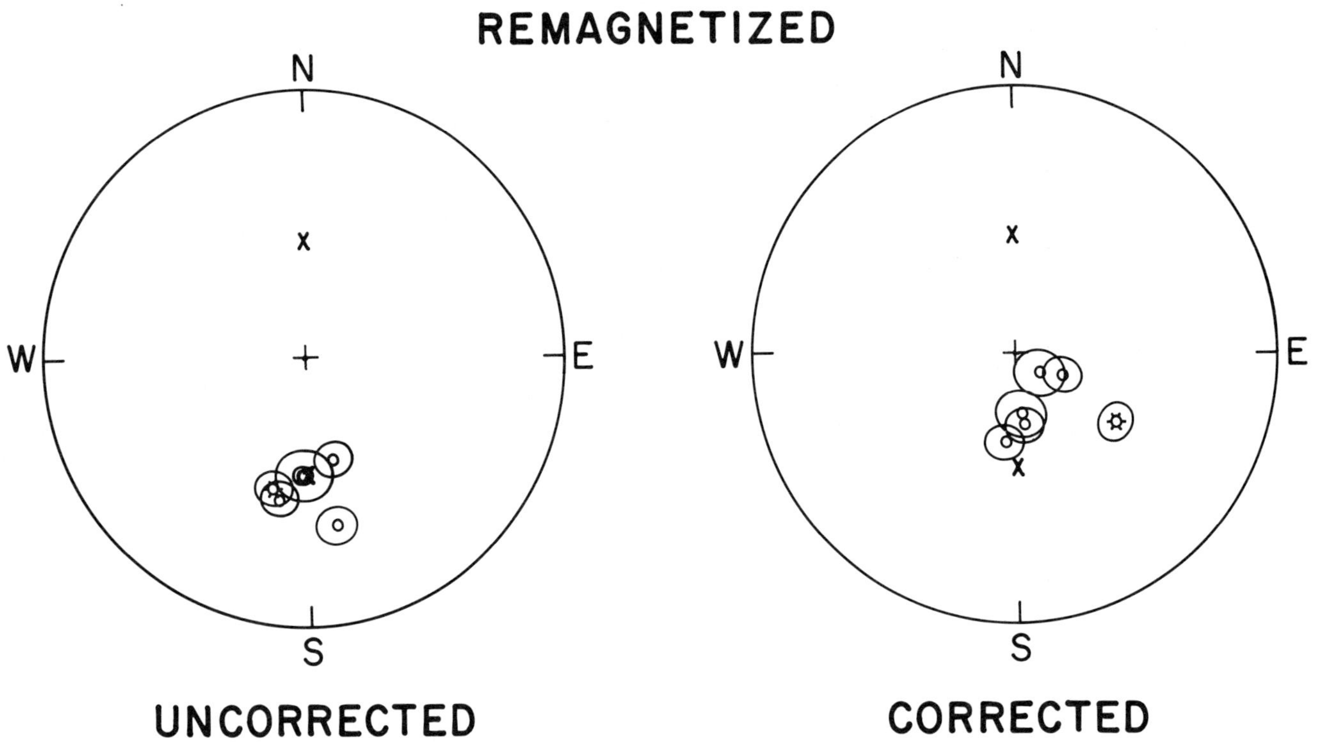

Fig. 3. Equal-area stereographic projection of the mean paleomagnetic directions from reversely remagnetized dolomite beds at sites WB (circles) and EC (star). The magnetic directions in these beds were within 20° of the reversed dipole field direction for these sites (x) prior to structural correction (left). The dispersion in the magnetic directions increased after structural correction (right), which suggests that the magnetization was acquired subsequent to Plio-Pleistocene folding. Ellipses indicate the uncertainty in the mean magnetic directions at the 95% confidence level.

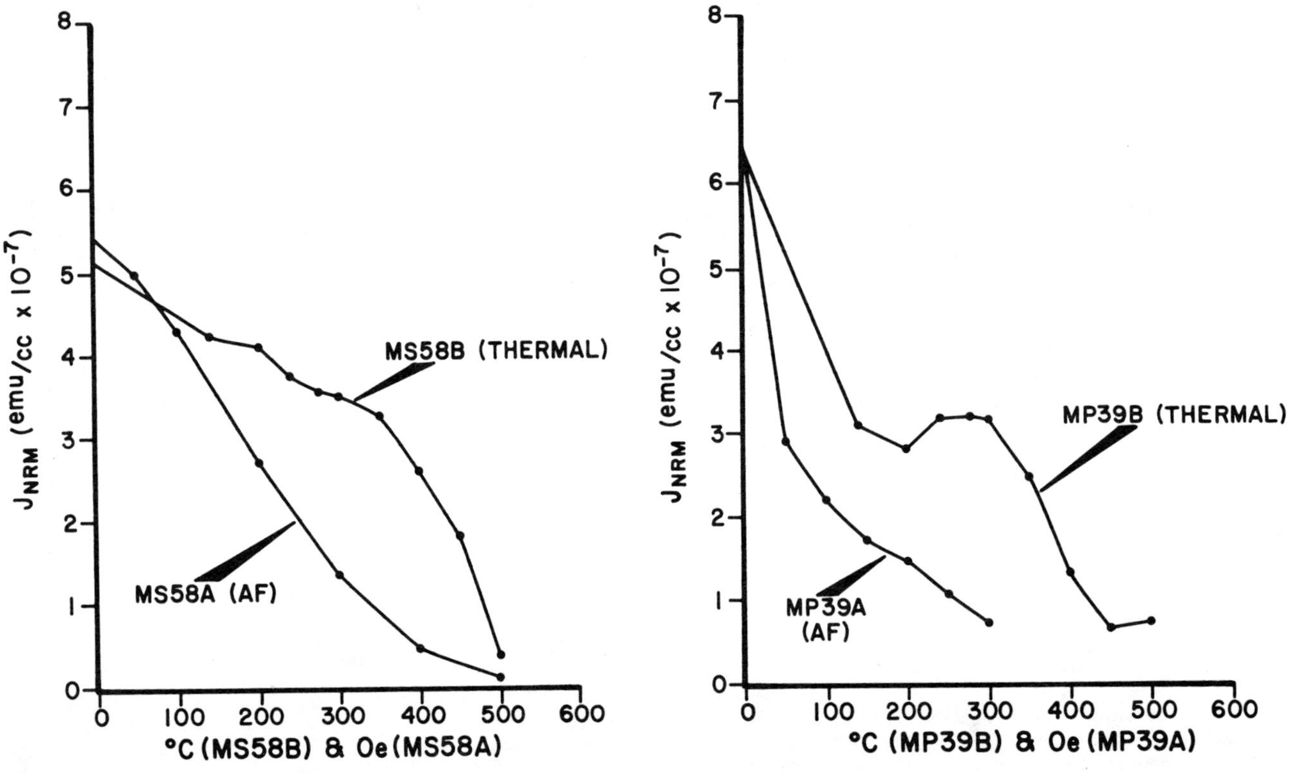

Fig. 4. Demagnetization curves of the natural remanent magnetization (NRM) in sister samples from two Monterey dolomite cores during thermal and alternating field (AF) demagnetization.

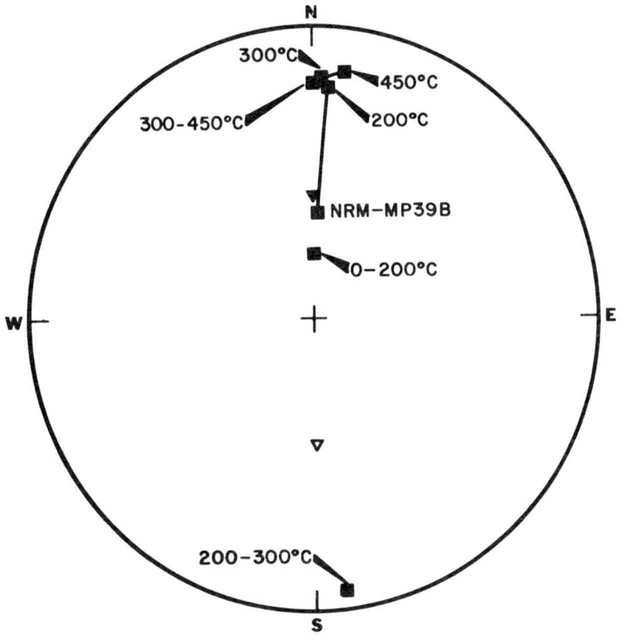

Fig. 5. Equal-area stereographic projection of the demagnetization behavior of sample MP39B, which displayed an anomalous increase in magnetic remanence between 200 and 300°C (Fig. 4).

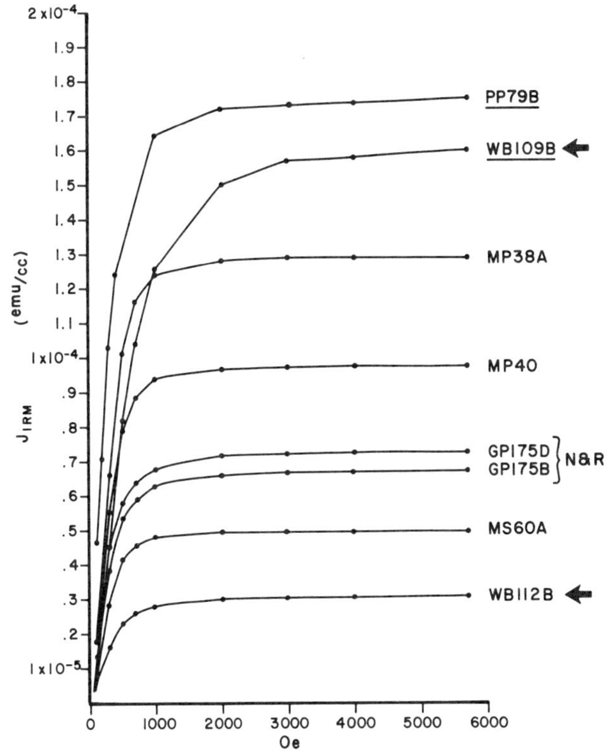

Fig. 6. Isothermal remanent magnetization (IRM) acquisition curves for Monterey dolomites. All samples with primary remanent directions become magnetically saturated by 3 kOe, whereas some samples with secondary remanent directions require higher fields (e.g., PP79B and WB109B). Samples WB112B and WB109B (indicated by arrows), which were sampled from the same bed, contain primary and secondary remanent directions, respectively.

IRM Saturation (J_{RS})

The IRM saturation intensities of dolomites in the Monterey Formation range from 10^{-5} to 10^{-3} emu/cc. Dolomites from the Monterey Formation and the Rincon Shale that have a primary remanence usually become 99% magnetically saturated between 2 and 3 kOe, which again suggests that magnetite is the dominant magnetic mineral carrying the remanence in these rocks (Fig. 6 and Table 1), in agreement with the results of the thermal demagnetization experiments. Some dolomites containing secondary directions saturate completely by 3 kOe (Table 1), while others require higher fields (Fig. 6). This observation suggests that hematite or goethite, in addition to magnetite, are magnetic carriers in some remagnetized rocks.

Remagnetized dolomites generally have higher NRM and IRM intensities than dolomites carrying a primary remanence (Fig. 7). In beds containing both primary and secondary directions, the remagnetized samples have IRM saturation intensities as much as five times greater than the samples containing a primary remanence (Fig. 6). The increase in both NRM and IRM intensity upon remagnetization requires that new magnetic grains are formed or are growing through their critical blocking volumes during the remagnetization process (see below).

Remanent Coercivity (H_{RC})

The remanent coercivity of dolomite beds possessing primary directions in the Monterey Formation generally ranges from 220 to 360 Oe (Fig. 7 and Table 1). Tuff beds from the Monterey Formation generally have higher remanent coercivities than the dolomites with which they are interbedded (Fig. 7), which suggests that primary volcanogenic magnetite is not the principal source of the magnetite in the dolomites. Three samples of Rincon dolomite had remanent coercivities in the same range as the remanent coercivities observed in Monterey dolomites (Fig. 7).

Monterey dolomites possessing secondary directions have remanent coercivities that are equal to or greater than the remanent coercivities displayed by the majority of dolomites containing primary directions (Fig. 7). Remagnetized dolomites which become magnetically saturated by 3 kOe have remanent coercivities indistinguishable from dolomites with primary directions. On the other hand, remagnetized dolomites which do not become magnetically saturated by 3 kOe have higher coercivities (H_{RC} > 400 Oe). Since hematite and goethite have higher coercivities than magnetite, this observation again suggests that hematite and/or goethite are present in some remagnetized dolomites.

Chemical Composition vs. Magnetic Intensity

Primary magnetization

The major oxide abundances in 24 samples of Monterey dolomite and 3 samples of Rincon dolomite were analyzed in order to determine the relationship

Table 1. Chemical composition[1] and magnetic properties of carbonates from the Monterey Formation and Rincon Shale

Sample[2] Number	Al_2O_3 (wt.%)	SiO_2 (wt.%)	CaO (wt.%)	MgO (wt.%)	Fe (wt.%)	Na_2O (wt.%)	K_2O (wt.%)	CO_2 (wt.%)	P_2O_5 (wt.%)	J^3_{RS} (emu/g) x 10^{-5}	J_{RS} (3 kOe) / J_{RS} (5.7 kOe)	H_{RC} (Oe)
					Monterey - Primary Magnetization							
DP58A	.42	5.5	32.06	15.77	.11	.12	.06	43.3	.46	2.88	-	259
DP67A	.55	4.0	35.56	11.45	.17	.20	.10	42.5	.51	1.69	.989	339
GP36A	1.45	8.5	26.46	16.60	.55	.39	.28	42.5	.12	9.10	.996	323
GP42A	.53	3.4	47.74	1.10	.22	.15	.13	41.0	.14	1.98	.976	325
GP175B	.45	4.2	29.82	18.09	.14	.18	.07	44.7	.07	2.15	.993	228
GP175D	.57	2.2	29.68	17.76	.19	.22	.11	43.3	.37	2.85	.994	229
MP40*	.25	2.4	30.10	18.76	.06	.09	.04	41.8	.04	3.03	.996	250
MS60A*	.53	2.6	29.40	18.59	.15	.18	.08	44.0	.07	1.66	-	238
RA14A	.79	2.9	40.46	7.14	.54	.19	.16	41.0	.62	3.03	-	230
RR03A	1.30	3.5	35.70	11.45	.95	.22	.24	44.0	.09	2.63	-	270
WB04A	.57	2.7	30.80	16.77	.40	.20	.12	47.0	.44	1.85	.993	242
WB74A	1.17	4.6	39.48	7.30	1.20	.26	.23	44.7	.36	4.31	.973	356
WB81A	.94	17.3	33.18	6.64	1.00	.26	.20	37.3	.58	2.41	-	313
WB112B*	.21	3.5	30.52	17.76	.06	.05	.02	44.0	.03	.98	.987	225
WB128B	.23	4.0	31.78	16.27	.07	.09	.04	42.5	.02	.75	.992	238
WB163	.30	5.6	30.94	16.77	.09	.09	.05	44.7	.04	.85	.985	272
					Monterey - Secondary Magnetization							
CM06A*	.64	2.5	26.74	18.59	2.50	.20	.10	43.3	.12	33.7	1.000	313
DP01A	2.08	12.0	27.44	14.28	.64	.40	.41	40.3	.04	17.0	.959	602
EC04A	.53	2.7	32.48	15.11	.95	.14	.10	44.7	.05	2.86	.950	410
GP10A	.74	14.5	27.30	14.61	.42	.22	.14	38.0	.05	.87	.971	498
GP175C	.60	9.9	28.42	16.10	.23	.18	.12	41.0	.03	1.15	.984	324
PP79B*	.68	2.7	29.96	17.10	.26	.18	.11	41.8	.41	5.63	.983	221
WB30A*	.60	2.9	30.66	17.10	.42	.19	.12	44.7	.29	6.87	1.000	260
WB109B	.26	1.9	32.76	16.77	.41	.08	.04	45.9	.04	5.17	.981	458
					Rincon - Primary Magnetization							
GB02A	1.40	8.2	25.20	13.94	4.60	.20	.24	41.8	.19	1.92	-	317
LC05A	1.64	5.3	30.10	14.94	2.30	.20	.29	38.8	.16	2.10	.975	362
LC07A	1.64	6.0	27.44	15.94	2.00	.27	.25	40.3	.20	2.50	.990	274

[1]Chemical analyses performed by Skyline Labs, Inc., 12090 W. 50th Place, Wheat Ridge, Colorado 80033.
[2]Asterisks indicate samples on which hysteresis measurements were made (Table 2 and Fig. 9).
[3]Saturation remanent magnetization intensity after treatment in a 5.7 kOe field (see Fig. 6).

between the magnetic intensity and the chemical composition of the dolomite (Table 1). If the magnetite in the dolomite is detrital in origin, then there should be a positive correlation between the saturation remanent magnetization intensity (J_{RS}) of the dolomite with primary magnetization and the abundance of oxides which are found exclusively in the detrital minerals in the dolomite. Isaacs (1980) has demonstrated that the oxide abundances in the detrital component in Monterey rocks are relatively constant within the area where the majority of samples used in this study were collected. She showed that Al_2O_3 constituted approximately 24% of the detrital component by weight. The abundance of Al_2O_3 in Monterey dolomites is therefore assumed to be proportional to the detrital component in the dolomite.

In Fig. 8, the wt. % Al_2O_3 in 16 samples of Monterey dolomite that possess a primary magnetic remanence is plotted against the saturation remanent magnetization intensity of the dolomite. The positive correlation between these two parameters (J_{RS}/wt. % Al_2O_3 = 4.14 x 10^{-5} emu/gram wt. %; correlation coefficient = 0.78; N = 15) suggests that the magnetite in the dolomite is detrital in origin. The sample with the highest J_{RS}/wt. % Al_2O_3 ratio (MP40A; not included in the linear regression) is from the dolomite bed which is believed to contain authigenic pyrrhotite (Figs. 4 and 5). Rincon dolomites, on the other hand, have lower J_{RS}/wt. % Al_2O_3 ratios than are observed in Monterey dolomites with primary magnetization.

The variation in the J_{RS} vs. wt. % Al_2O_3 ratio observed in Monterey dolomites with primary magnetization could have a number of causes, including:

(1) A variation in the abundance of magnetite in the detrital component: Higher J_{RS}/wt. % Al_2O_3 ratios would occur in dolomites that have a higher percentage of magnetite in their detrital fractions. This would occur if there were differences in the provenance of the detrital component of the sediment within the study area (e.g., mafic vs. granitic source rocks) or a

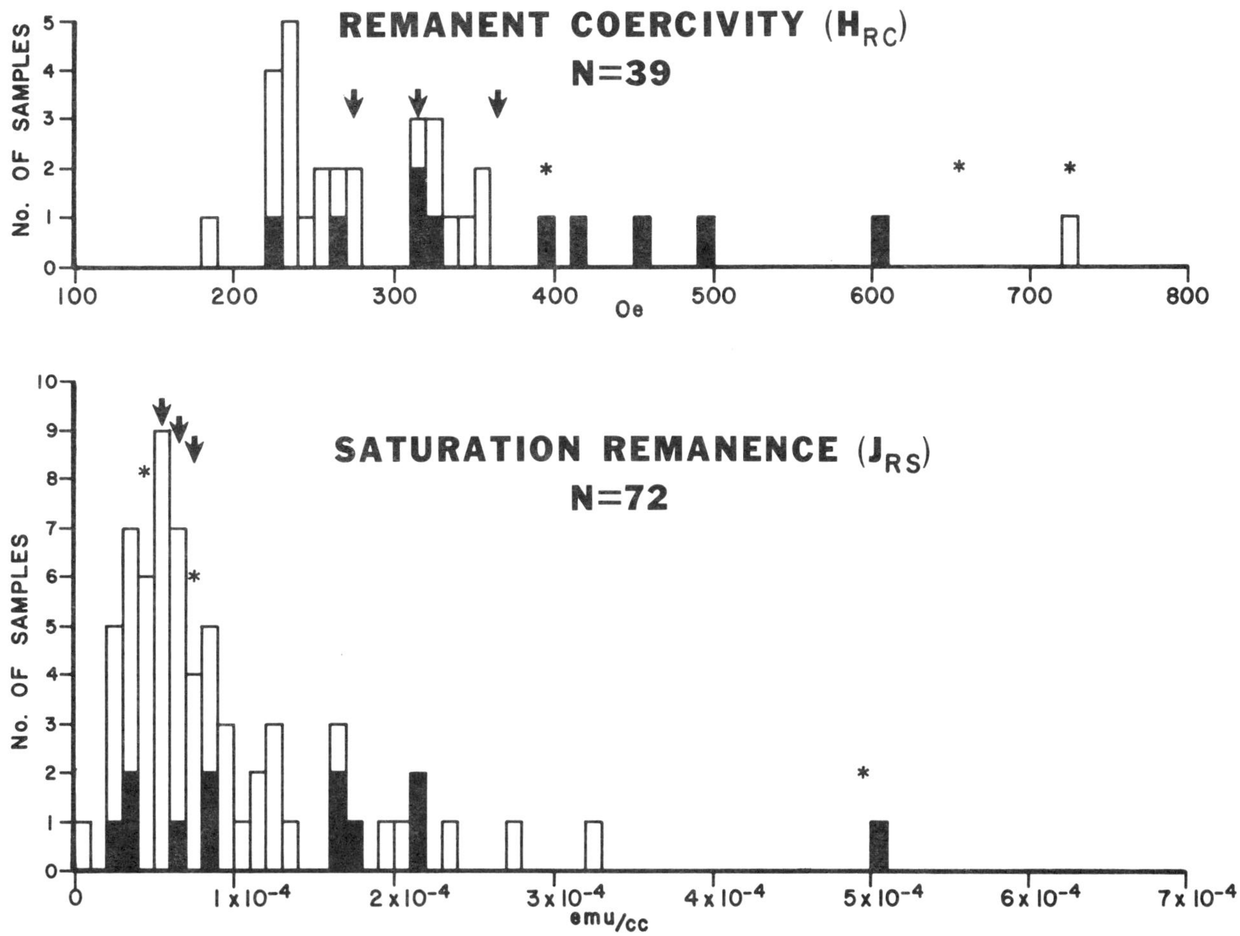

Fig. 7. Histograms of remanent coercivity (top) and saturation remanent intensity (bottom) of Monterey dolomites with primary remanence (open bars) and secondary remanence (solid bars). Arrows represent Rincon dolomites. Asterisks indicate Monterey tuff beds.

significant difference in the distance of transport from the source terrain to the sites of deposition (due to the difference in hydrolic equivalence between magnetite and the aluminosilicate minerals which comprise the bulk of the detrital fraction).

(2) The presence of undetected authigenic magnetic phases: Authigenic pyrrhotite, magnetite, maghemite, hematite, or jacobsite would increase the J_{RS}/wt. % Al_2O_3 ratio. The high value observed in sample MP40 (denoted by a cross in Fig. 8) is probably due to authigenic pyrrhotite (see Fig. 5).

(3) Variations in the grain size and composition of the detrital magnetite: A decrease in grain size or an increase in titanium content (x) would cause an increase in the J_{RS} value of the magnetite. For example, the J_{RS} of titanomagnetite (x = 0.4) decreases by a factor of four between grains that are 1 and 10 microns in diameter (Day, 1977). Similarly, J_{RS} can increase by a factor of four between pure magnetite (x = 0) and titanomagnetite with x = 0.4, since J_{RS}/J_S increases by a factor of two (Day, 1977) and J_S decreases by a factor of two (Nagata, 1961) between these compositional extremes.

The significantly lower J_{RS}/wt. % Al_2O_3 ratios in the Rincon dolomites compared to the Monterey dolomites (Fig. 8) could be due to (1) or (3) above.

Secondary magnetization

The variation in the J_{RS}/wt. % Al_2O_3 ratio observed in Monterey dolomites with secondary remanent magnetization is much greater than the variation observed in dolomites with primary magnetization (Fig. 8). Samples collected from two dolomite beds with both primary and secondary remanent directions showed a 3.8-fold and a 5.3-fold increase in J_{RS} intensity upon remagnetization, although the Al_2O_3 wt. % remained the same (samples WB112B-WB109B and WB04A-WB30A; Table 1). The increase in J_{RS} upon remagnetization of these beds probably resulted from partial oxidation of the pyrite in the dolomite to magnetic phases. Thermal demagnetization of sample WB30B suggests that magnetite was the dominant magnetic phase formed during the remagnetization process in this sample, although a small amount of maghemite or pyrrhotite may also be present (Fig. 11). Although hematite (or goethite) was not formed during remagnetization of sample WB30 (which became completely magnetically saturated by 2000 Oe), it was formed during remagnetization of sample WB109 (Fig. 6 and Table 1).

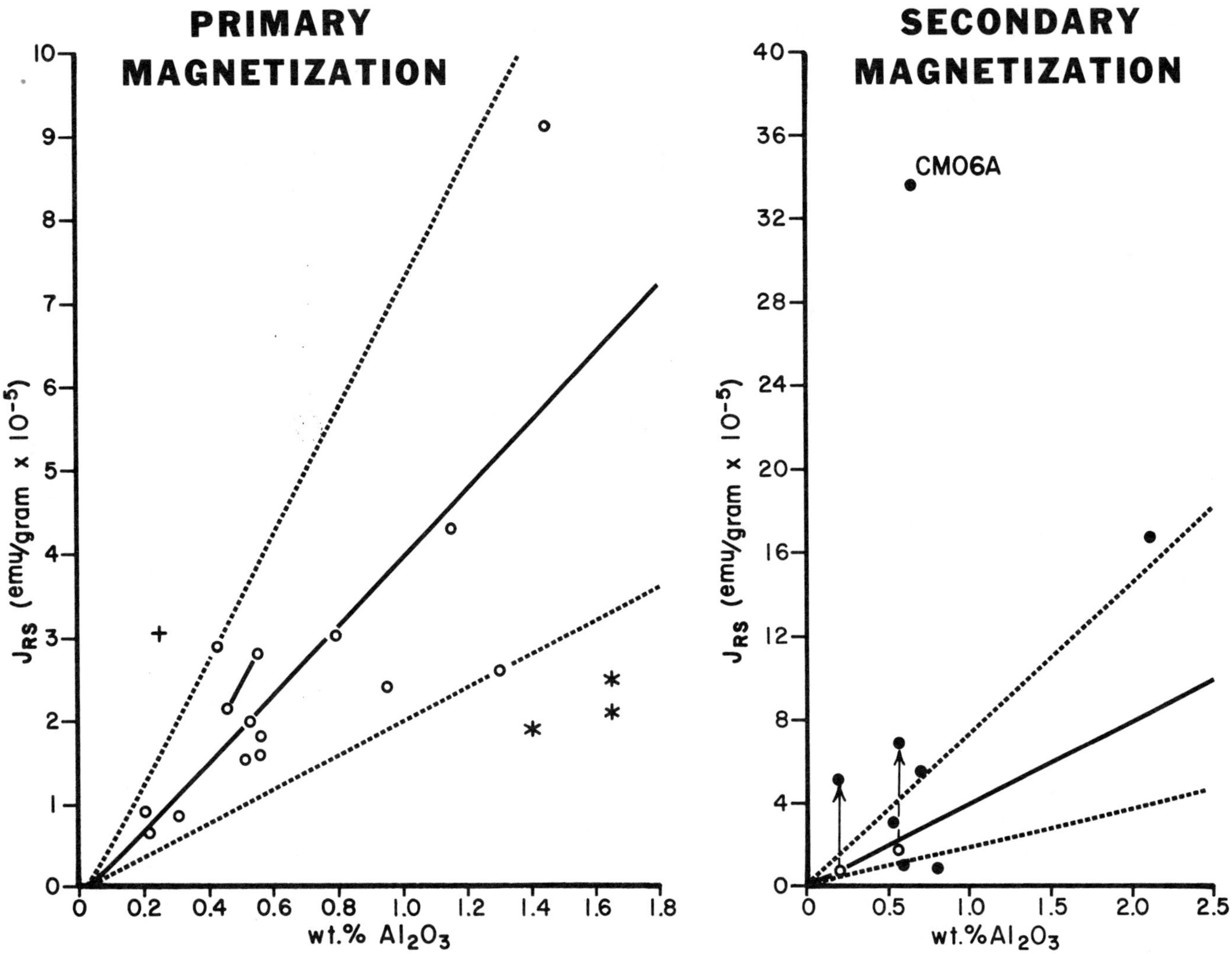

Fig. 8. Saturation remanent magnetization intensity (J_{RS}) vs. aluminum content in Monterey and Rincon dolomites (data from Table 1). Left: Monterey dolomites with primary magnetization (open circles). Heavy solid line is the least-squares linear regression to the open circles. Dashed line indicates the extremes in the J_{RS}/wt. % Al_2O_3 ratio. Rincon dolomites (asterisks) and sample MP40 (cross) from the bed suspected to contain authigenic pyrrhotite (Figs. 4 and 5) were not included in the linear regression. Short solid line connects samples with normal and reversed polarities from the same dolomite bed (GP175B and GP175D, respectively). Right: Monterey dolomites with secondary magnetization (solid circles). Solid and dotted lines are the same as on the left (note change in scales). Arrows connect samples with primary and secondary magnetization from the same bed (WB04A-WB30A and WB112B-WB109B). Sample CM06A is the most strongly magnetized sample discovered in the Monterey Formation.

Domain State

The hysteresis properties of six samples of Monterey dolomite were measured on the horizontally vibrating sample magnetometer at UCSB (Table 2 and Fig. 9). H_{RC}/H_C and J_{RS}/J_S ratios indicate that the magnetite in these rocks falls in the pseudo-single domain field of Day et al. (1977). Three of the six samples were remagnetized dolomites which became magnetically saturated by 3 kOe. Two of these remagnetized samples (WB30A and PP79B) were indistinguishable from the dolomites with primary directions on the basis of their hysteresis parameters (Fig. 9). The other remagnetized sample (CM06A), whose J_{RS} intensity was an order of magnitude greater than any other sample (Table 1), did not show any hysteresis. The lack of hysteresis in this magnetically strong sample may be due to a large super-paramagnetic population of magnetite grains in this sample.

The domain state of the magnetite in two of the six samples for which hysteresis measurements were made was investigated using the test of Johnson, Lowrie, and Kent (1975). An I_SRM was first induced in each sample and stepwise AF demagnetized. An ARM was then induced in the same sample, in an alternating field of 1000 Oe with a biasing field of 0.5 Oe, and stepwise demagnetized. In both samples, the ARM was more resistant to AF demagnetization than the I_SRM (Fig. 10), again suggesting that the magnetite is either single domain or pseudo-single domain. Comparison of the hysteresis parameters of the dolomites with the hysteresis parameters of synthetic magnetite of known size indicates that the magnetite in the dolomite has a mean grain size of 0.1 to 10 microns (Day et al., 1977).

Table 2. Hysteresis properties of Monterey dolomites

Sample Number	H_{RC}/H_C	J_{RS}/J_S	J_{RS}/J_{NRM}
	Primary magnetization		
MS60A	1.45	0.129	48.1
MP40	1.66	0.138	232.7
WB112B	1.88	0.125	155.0
	Secondary magnetization		
WB30A	1.65	0.169	143.5
CM06A	-	-	195.0
PP79B	1.35	0.108	691.7

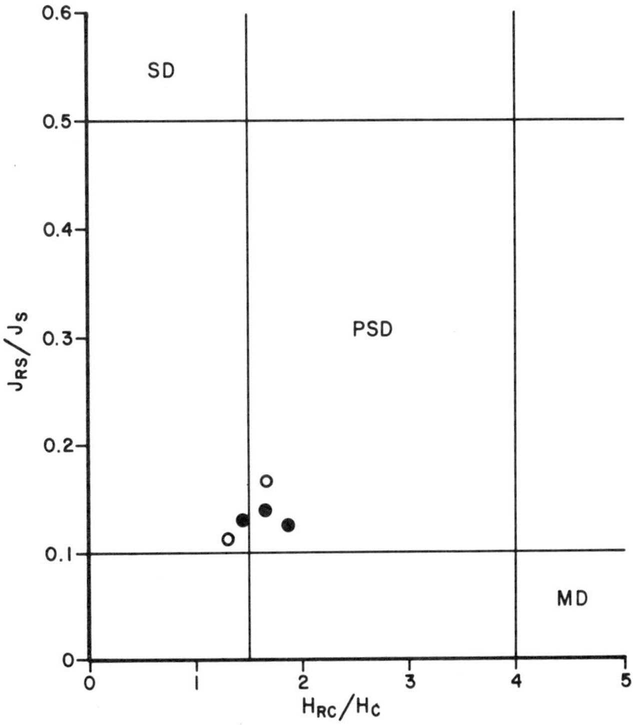

Fig. 9. Hysteresis properties of Monterey dolomites with primary magnetization (solid circles) and secondary magnetization (open circles); data from Table 2. Samples fall in the pseudo-single domain (PSD) field of Day et al. (1977), rather than the single domain (SD) or multidomain (MD) field.

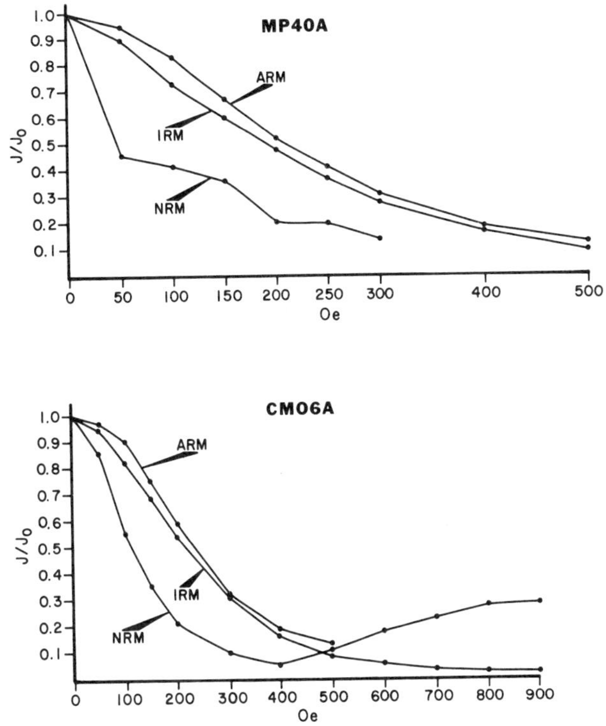

Fig. 10. Normalized AF demagnetization curves of Monterey dolomites with primary magnetization (MP40A) and secondary magnetization (CM06A). Anhysteretic remanent magnetization (ARM) is more resistant to AF demagnetization than isothermal remanent magnetization (IRM), which suggests that the remanence is carried by SD or PSD grains (Johnson et al., 1975). Sample MP40 is from the same bed as MP39 (Figs. 4 and 5). The demagnetization path of CM06A is illustrated in Fig. 22.

Abundance of Magnetite in the Dolomite

The amount of magnetite present in a particular sample of dolomite can be determined from the saturation remanent magnetization intensity (J_{RS}) and the hysteresis properties of the dolomite (J_{RS}/J_S), since the saturation magnetization of magnetite (J_S = 92 emu/gram; Nagata, 1961) is independent of grain size.

The ratio of dolomite to magnetite (by weight) in the samples for which the J_{RS}/J_S ratio was determined (Table 1) ranges from 1.2×10^6 for a sample with primary magnetization (WB112B) to 7.4×10^4 for a sample that is remagnetized (WB30A). Since the positive (ferrimagnetic) susceptibility of magnetite is approximately a million times greater than the negative (diamagnetic) susceptibility of pure dolomite, Monterey dolomites would be expected to possess a very weak negative susceptibility or a small positive susceptibility, which is what is observed (see below).

The amount of detrital magnetite in the detrital component in the dolomite can be determined from the Al_2O_3 analyses, J_{RS}, and J_{RS}/J_S values, by assuming that Al_2O_3 comprises 24% of the detrital fraction by weight (Isaacs, 1980). In the two samples with primary magnetization in Table 2 which do not contain authigenic phases (WB112B and MS60), the magnetite to detrital ratio is 1/10,000 and 1/15,000 by weight, respectively. Since the density of aluminosilicate minerals is approximately half as great as the density of magnetite, these values imply that the volume ratio of magnetite to aluminosilicate minerals in the dolomite is approximately one part in 25,000.

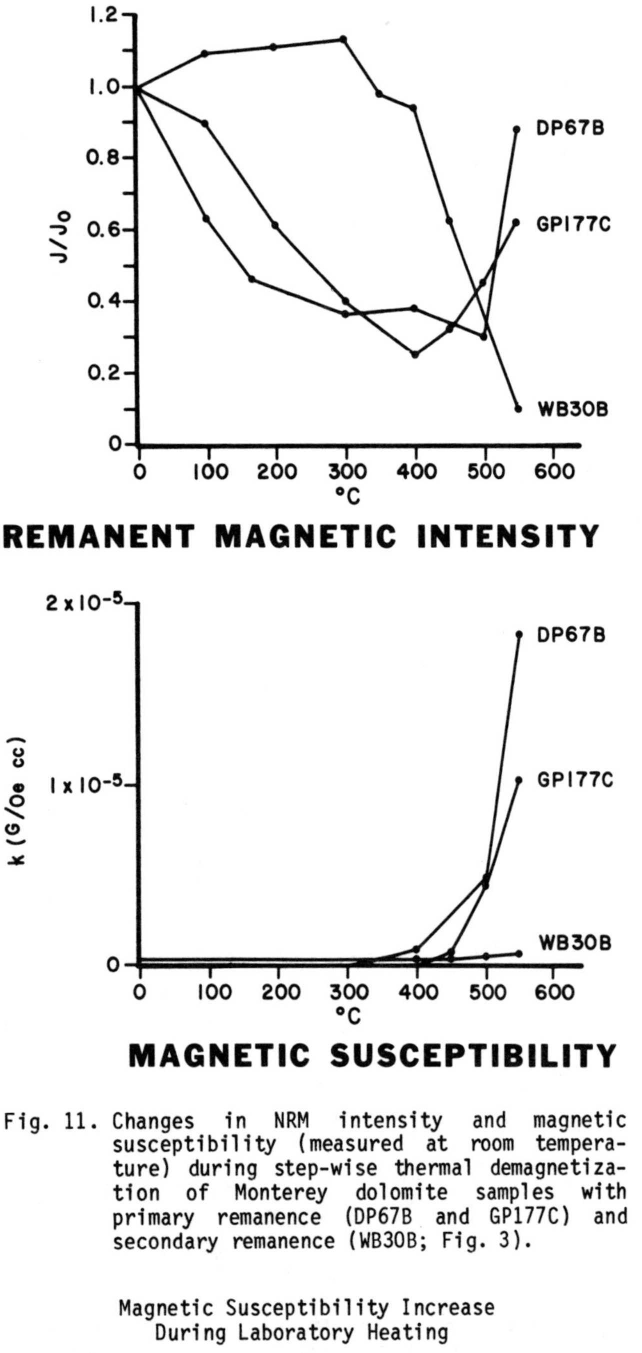

Fig. 11. Changes in NRM intensity and magnetic susceptibility (measured at room temperature) during step-wise thermal demagnetization of Monterey dolomite samples with primary remanence (DP67B and GP177C) and secondary remanence (WB30B; Fig. 3).

Magnetic Susceptibility Increase During Laboratory Heating

Most dolomite samples from the Monterey Formation experienced erratic changes in direction and intensity after heating above 300°C (Fig. 11). In order to determine the cause of this behavior, the magnetic susceptibility of a few samples was measured after each thermal demagnetization step. Monterey dolomites with a primary magnetic remanence display either a very weak negative susceptibility or a weak positive susceptibility ($k < 10^{-6}$ G/Oe cc). However, after heating above 300°C, the dolomite begins to acquire a strong positive susceptibility (Fig. 11), which suggests that a ferrimagnetic mineral is being created in the heating process. The new magnetic phase is probably magnetite, which forms from the breakdown of pyrite at elevated temperatures (Lowrie et al., 1982). The creation of new magnetic phases at temperatures above 300°C rendered directions obtained with thermal demagnetization above this temperature suspect. Thermal demagnetization was therefore only carried to 300°C during routine thermal demagnetization. Remagnetized dolomites did not show a dramatic increase in susceptibility after heating at high temperatures (Fig. 11), which suggests that the pyrite in these dolomites had already broken down during the remagnetization process (see below).

DEPTH OF ACQUISITION OF PRIMARY MAGNETIZATION

A number of field tests were conducted in order to determine the depth beneath the seafloor at which the primary magnetization in Monterey dolomites is acquired. These tests included paleomagnetic fold tests, correlations of the magnetic stratigraphy in the Monterey Formation with diatom datum events, and an examination of the records of field reversals in Monterey dolomite beds.

Synsedimentary Slump-Fold Tests

Two soft-sediment slump folds were sampled to determine if the primary remanent magnetization in the Monterey dolomite beds is acquired at the seafloor (prior to slumping) or during burial of the sediment (after slumping). Both slumps occurred in well-laminated, unconsolidated sediment, which was subsequently impregnated with carbonate cement. Slumping produced sinusoidal folds (amplitude = 10 cm) with planed-off tops in one case (Rodeo Canyon), and a fold with upright limbs (amplitude = 30 cm) in the other (Point Pedernales). Both synsedimentary slumps failed the fold test, indicating that the majority of the magnetization was locked in subsequent to slumping (Figs. 12 and 13). The slump fold at Rodeo Canyon (Fig. 12) suggests that the magnetization is acquired at depths of more than 10 cm below the seafloor, whereas the slump fold at Point Pedernales (Fig. 13) implies that magnetization of the sediment occurred at burial depths of more than 30 cm.

Lateral Continuity of Remanent Magnetization

A number of experiments were conducted in order to determine the lateral continuity of the primary remanent magnetization recorded in Monterey dolomite beds. At Hondo Beach and Tajiguas Beach (sites WB and RA; Fig. 1), Monterey dolomite beds are continuously exposed over a distance of approximately one hundred and two hundred meters, respectively. Although the beds in these stratigraphic sections were alternately normal and reversed in polarity (Fig. 14), the remanent magnetization within individual beds was constant in polarity over the outcrop area, and the dispersion in magnetic directions within individual beds was small (kappa = 70 to 1165). Correlation of a distinctive set of tuff beds in these two sections enabled the dolomite beds at 23, 28.5, 33, and 37 meters in the Hondo Beach section (Fig. 14) to be matched up with correlative dolomite beds in the Tajiguas Beach section, where they displayed the same polarity sequence (see Hornafius, 1984). The agreement between the polarity of the dolomite beds in the Tajiguas Beach and Hondo Beach sections implies that polarity events that occur over a stratigraphic interval of only a few meters are laterally continuous for a distance of at least 4 kilometers. The tremendous lateral continuity of the polarity and direction of remanence in the dolomite suggests that the acquisition of remanence in the Monterey Formation occurs within a narrow lock-in zone at a constant depth beneath the sediment-water interface.

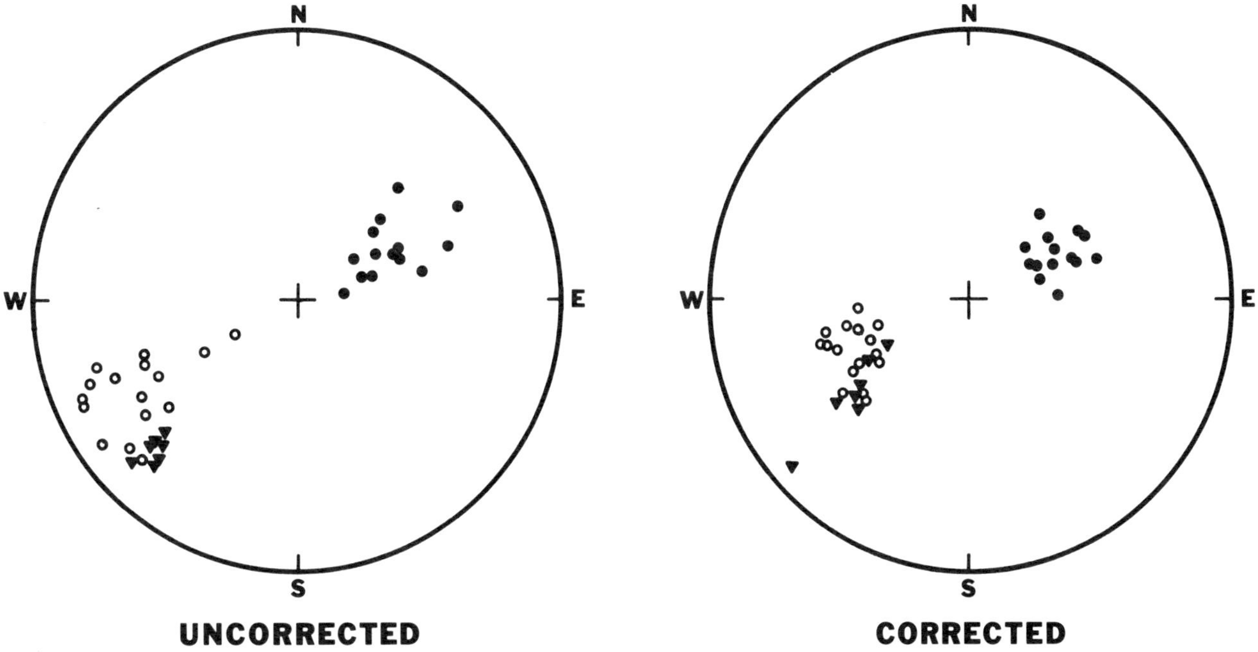

Fig. 12. Equal-area stereographic projection of the stable paleomagnetic directions of individual samples from three Monterey dolomite beds at Rodeo Canyon (site RC). Two dolomite beds from tectonic folds (open circles = upper hemisphere; closed circles = lower hemisphere) pass the fold test, whereas a dolomite-cemented soft-sediment slump fold (solid triangles = upper hemisphere) fails the fold test.

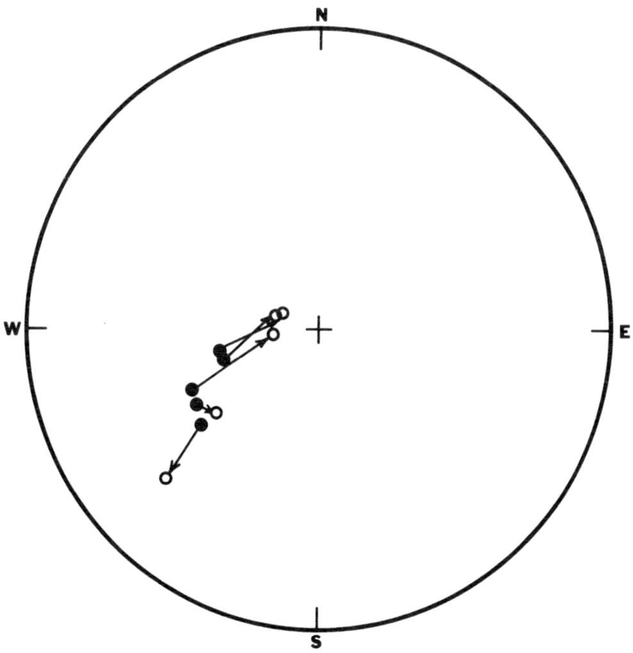

Fig. 13. Equal-area stereographic projection of a soft-sediment slump fold test on a dolomite bed at Point Pedernales (site PP). Solid circles are the structurally uncorrected magnetic directions of individual samples. Open circles are the same directions after structural tilt corrections were applied. All symbols plot on the upper hemisphere. The increase in dispersion after structural correction suggests that the magnetization was acquired subsequent to slumping.

Magnetic Stratigraphy

A 50-m-thick stratigraphic section of the Monterey Formation at Hondo Beach was found to possess a sufficient amount of dolomite to correlate the magnetic reversal sequence in this section with the magnetic timescale (Fig. 14). Diatoms were extracted from the dolomites in this section by Charles W. Lagle, using an acid dissolution technique described by Lagle (this volume). The first occurrence of <u>Denticulopsis hyalina</u> within this section (see plate 2, Fig. 4 in Lagle, this volume) was found to agree with its first occurrence relative to the magnetic stratigraphy observed in deep-sea cores (Hornafius and Lagle, in preparation). This observation suggests that the time elapsed between deposition and magnetization of the dolomites in the Monterey Formation is not appreciably greater than the time lag between deposition and magnetization of deep-sea sediments.

Records of Field Reversals

Out of approximately one hundred dolomite beds in the Monterey Formation which were found to possess stable primary magnetic directions, all but six were either entirely normal or entirely reversed in polarity. However, three dolomite beds at sites GP and WC and three dolomite nodules at sites PP, MS, and CH possessed both normal and reversed polarities, as well as some unstable or intermediate directions. The distribution of the normal and reversed polarities in a pair of dolomite beds at Gaviota Pier (site GP) and in a dolomite nodule at Sweeney Road (site MS) was documented in detail.

At Gaviota Pier, a pair of 10-cm-thick dolomite beds that are separated by 10 cm stratigraphically were extensively sampled for a distance of 40 meters

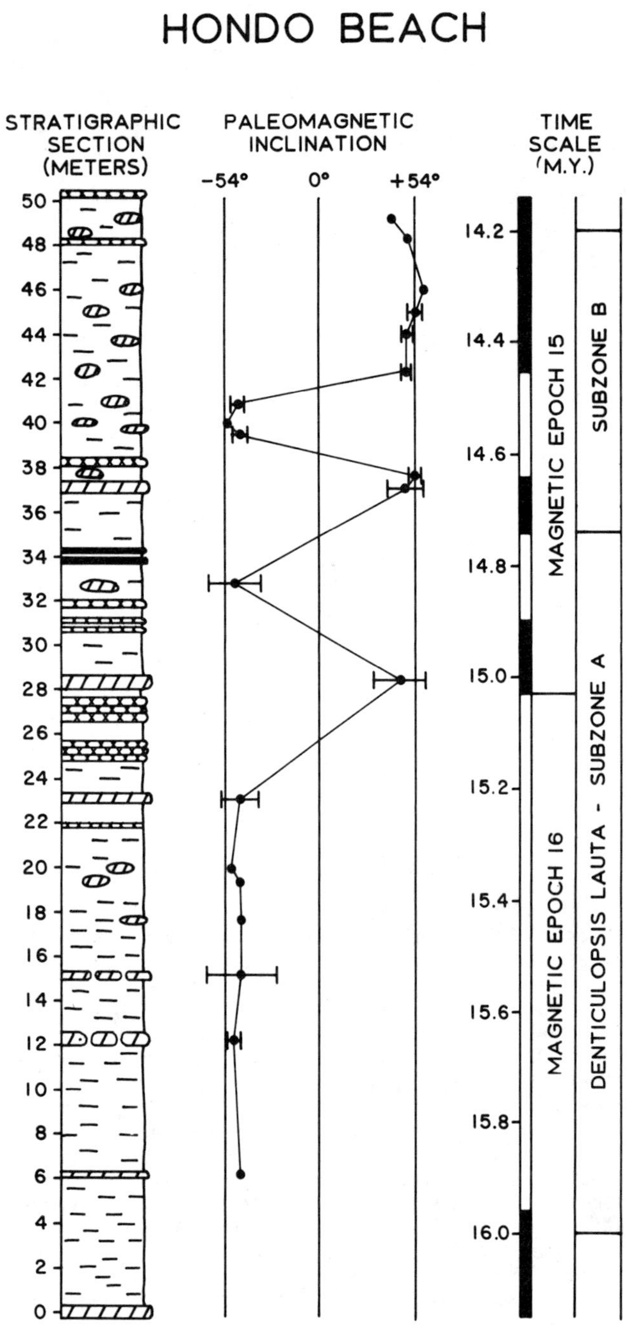

Fig. 14. Magnetic inclination vs. stratigraphic height in the Hondo Beach section (site WB). Magnetic timescale is from Ness et al. (1980). Correlation of the D. lauta diatom zone with the magnetic timescale is from Barron (1981). The first occurrence of D. hyalina at 37 meters in this stratigraphic section (C. W. Lagle, personal communication, 1983) is in agreement with the first occurrence of this diatom datum at the top of the lowest reversed interval in magnetic epoch 15 in deep-sea cores (J. Barron, personal communication, 1983). Diagonal lines = dolomite. Cross-hatching = porcelanite. Dashes = calcareous siliceous shale. $\Delta I = \alpha_{95}$.

along strike (Figs. 15 and 16). Out of 42 dolomite samples collected from these beds, 17 were normally magnetized, 18 were reversely magnetized, six had intermediate or partially remagnetized directions, and one sample was unstably magnetized (Figs. 17 and 18). Surprisingly, the distribution of normal and reversed polarities did not display stratigraphic order, and samples collected from the same sediment laminae were often opposite in polarity (Figs. 15 and 16). In some cases, sister samples from the same core were opposite in polarity (Figs. 15 and 17). Sediment laminations are completely undisturbed throughout both beds, which rules out bioturbation as the reason for this stratigraphic inconsistency. A pair of sister samples with opposite polarities (samples GP175B and GP175D) had identical remanent coercivities (Table 1), saturation magnetization curves (Fig. 6) and J_{RS}/wt. % Al_2O_3 ratios (Fig. 8), which suggests that both the normal and reversed polarities are carried by magnetite with a similar grain size distribution.

The pair of dolomite beds illustrated in Figs. 15 and 16 are again exposed 0.8 km along strike, where they also possess both normal and reversed polarities. At both localities, stratiform dolomite nodules occur 6 meters stratigraphically above this pair of dolomite beds. Extensive sampling of these overlying dolomite nodules (17 cores) yielded only normal polarities at both localities. The mixed polarities in the pair of dolomite beds in Fig. 15 therefore appear to be restricted to a narrow stratigraphic horizon which extends laterally for a distance of at least 0.8 kilometers. The stratigraphically limited extent of the mixed polarity interval and the similar rock magnetic properties of the normally and reversely magnetized dolomite samples suggest that these beds acquired their magnetization during a reversal of the earth's magnetic field.

At Sweeney Road, a dolomite nodule also contains normal and reversed polarities (Fig. 19). Since the nodule is underlain by reversely magnetized beds and is overlain by normally magnetized beds, the mixed polarities in this bed were probably acquired during an R to N field reversal. The normally magnetized samples all occur in one corner of the nodule, although some unstable intermediate directions are found in the center of the nodule (Fig. 20). The magnetization pattern in this nodule is not concentric about the center of the nodule, and it does not show stratigraphic order. As in the Gaviota Pier case, cores that contain the same sediment laminae are opposite in polarity. However, a greater proportion of samples had intermediate or unstable directions in the Sweeney Road dolomite nodule than in the mixed polarity dolomite beds at Gaviota Pier (compare Figs. 18 and 21).

DISCUSSION

Origin of the Primary Magnetization

A number of mechanisms could be responsible for the primary remanent magnetization of the dolomite beds in the Monterey Formation. Since the magnetite carrying the remanence in these rocks is probably detrital, and not authigenic, these mechanisms include:

(1) Depositional remanent magnetization (DRM) acquired during settling of the magnetite particles onto the ocean bottom.

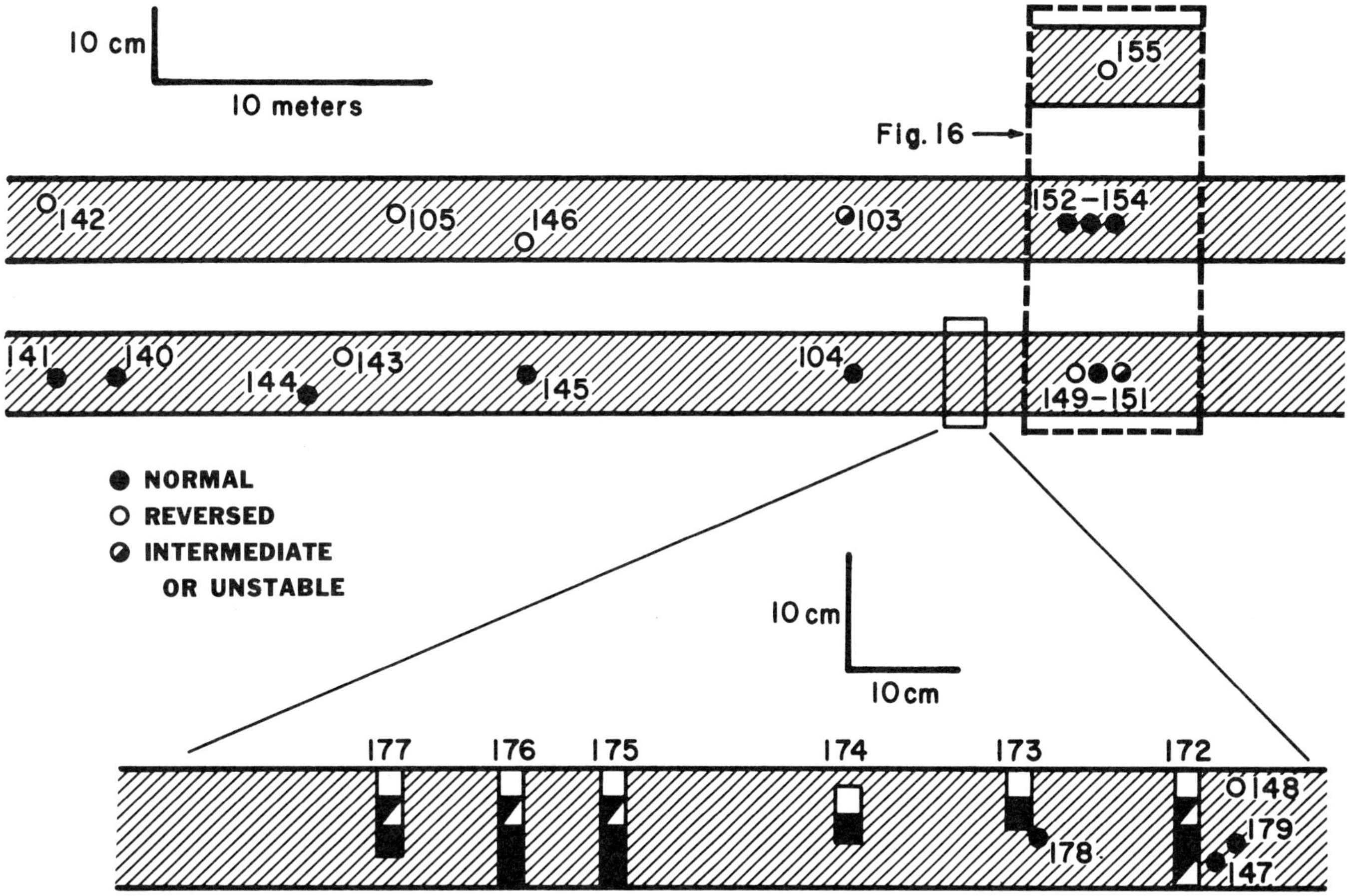

Fig. 15. Distribution of normal and reversed polarity samples from dolomite beds at Gaviota Beach (site GP).

(2) Post-depositional remanent magnetization (pDRM) acquired during compaction and dewatering of the sediment.

(3) Post-depositional remanent magnetization acquired during carbonate cementation of the host sediment.

Each of these mechanisms implies a different burial depth for the magnetization process. The mechanism of acquisition of magnetization in Monterey dolomites can therefore be evaluated with the field evidence presented in the preceding section.

The remanent magnetization in Monterey dolomites must have been acquired more than a few tens of centimeters below the seafloor, since carbonate-impregnated soft-sediment slump folds in the Monterey Formation failed the paleomagnetic fold test (Figs. 12 and 13). A pDRM mechanism is therefore suggested. However, the correlation between the diatom biostratigraphy and the magnetic polarity sequence at Hondo Beach (Fig. 14) suggests that the magnetization is acquired less than a few meters below the seafloor.

The oxygen isotopes ratios observed in Monterey dolomites suggest that carbonate impregnation of the origin diatomaceous sediment occurs at burial depths ranging from approximately 80 to 660 meters (Pisciotto, 1981). Although authigenic dolomite has been observed to form at burial depths as shallow as 15 meters in the Cariaco Basin off Venezuela (Friedman and Murata, 1979), dolomite has never been observed to form at burial depths of less than a few meters in modern depositional environments that are analogs to the inferred depositional setting of the

Fig. 16. Photograph of the mixed-polarity dolomite beds illustrated in Fig. 15. Samples GP149, GP150, and GP151 from the lower bed were (from left to right) reversed, normal, and intermediate in polarity, respectively (see Fig. 17).

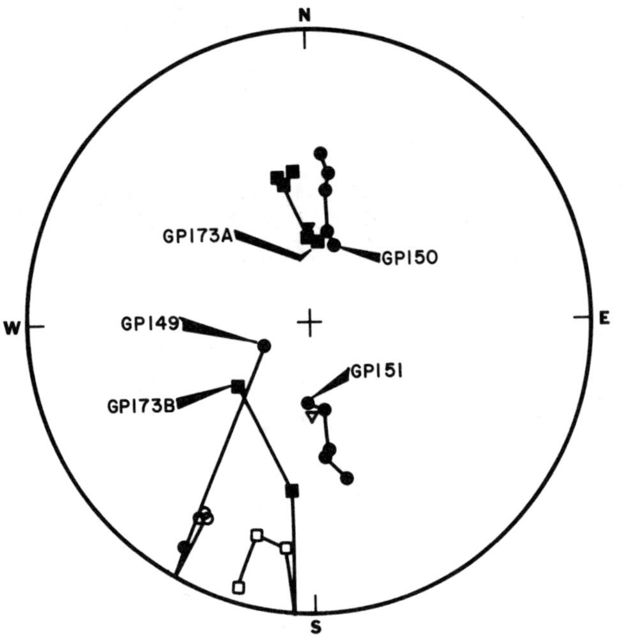

Fig. 17. Equal-area stereographic projection of the demagnetization paths of samples from the lower dolomite bed illustrated in Fig. 15. Samples GP149, GP150, and GP151 were sampled along strike from the same sediment laminae (Fig. 16). Samples GP173A and GP173B are sister samples from the same core which was drilled perpendicular to bedding (Fig. 15). Arrows point to the NRM directions of the samples. Squares indicate the NRM directions after stepwise thermal demagnetization at 100, 200, 250, and 300°C. Circles indicate the NRM directions after AF demagnetization at 50, 100, 150, and 200 Oe. Triangles indicate the present-day dipole field direction for the site. Solid symbols plot on the lower hemisphere; open symbols plot on the upper hemisphere.

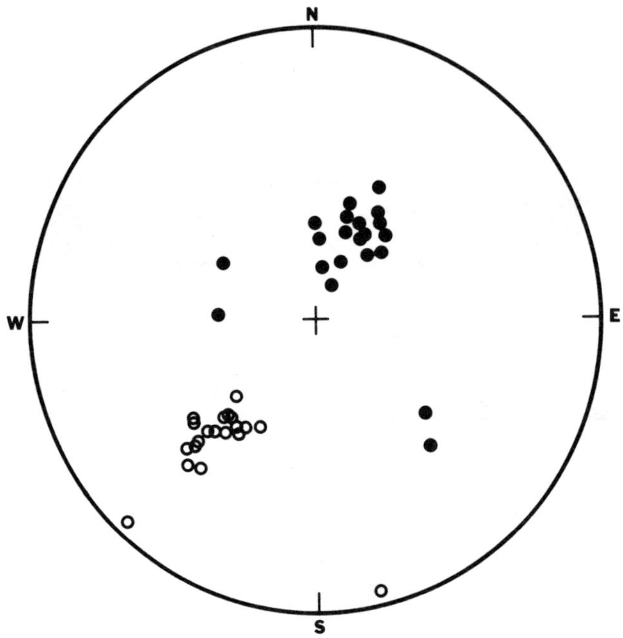

Fig. 18. Equal-area stereographic projection of the structurally-corrected magnetic directions of the samples collected from the mixed-polarity dolomite beds illustrated in Fig. 15, after demagnetization at 200 Oe or 250°C. Solid circles = lower hemisphere. Open circles = upper hemisphere.

Monterey Formation (e.g., Hein et al., 1979; Pisciotto and Mahoney, 1981; Kelts and McKenzie, 1982). The possibility that the magnetization in the Monterey dolomites is acquired during carbonate cementation of the original diatomaceous ooze therefore seems unlikely, because of the shallow burial depths at which the magnetization in the dolomite appears to have been acquired.

The mechanism of acquisition of primary magnetic remanence in Monterey dolomites that is most consistent with the field evidence presented in the preceding section is a pDRM acquired during compaction and dewatering of the sediment. This process has been proposed as the mechanism responsible for the magnetic remanence carried by deep-sea sediments in general (Verosub, 1977) and by diatomaceous sediments in particular (Kent, 1973). Both theoretical arguments and experimental observations suggest that the magnetic remanence in marine sediments is locked in due to an increase in the viscosity of the sediment during the earliest stages of sediment compaction (Lovlie, 1974, 1976; Denham and Chave, 1982). The depth and thickness of the magnetization lock-in zone in this case would be a function of the sedimentation rate, grain-size distribution, initial porosity, and resistance to mechanical compaction of the sediment.

The records of field reversals preserved in the Monterey dolomites differ markedly from the records of field reversals preserved in deep-sea sediments and lake sediments. Field reversals recorded by rapidly deposited deep-sea clays and nannofossil oozes (e.g., Fuller et al., 1979) and in dry lake sediments (Hillhouse and Cox, 1976) display an orderly stratigraphic succession of stable remanent directions acquired during rapid migration of the transient magnetic field vector during a polarity change. In contrast, field reversals recorded by Monterey dolomite beds contain few stable intermediate directions (Figs. 18 and 21) and the normal and reversed polarities in the dolomite are not distributed in a stratigraphically consistent fashion (Figs. 15 and 19).

The stratigraphically inconsistent polarity distribution in the Monterey dolomite beds illustrated in Figs. 15 and 19 becomes sensible if a long time interval is required for the original diatomaceous ooze to acquire its magnetization. Assuming that the magnetization is continuously acquired during compaction and dewatering of the sediment, between, say, 1 and 3 meters below the sediment-water interface, the length of time required to magnetize the sediment would depend on the sedimentation rate. In sections with low sedimentation rates (e.g., ~20 m/m.y. for Gaviota Pier), the sediment would "see" the ambient magnetic field for 100,000 years, which is much longer than the duration of a magnetic field reversal (~4,000; Fuller et al., 1979). Under these circumstances, sediments which were magnetized at the time of a reversal would contain both normal and reversed components of magnetization. The stratigraphically inconsistent distribution of normal and reversed polarities depicted in Figs. 15 and 19 might then be ascribed to the precarious balance between competing components of normal and reversed polarity

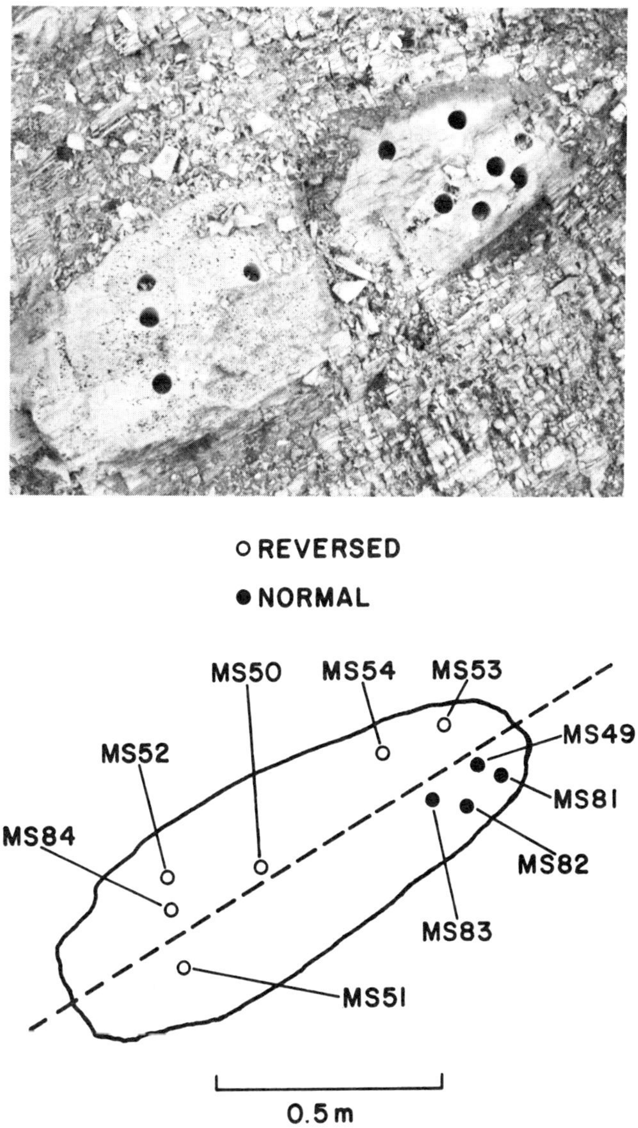

Fig. 19. Distribution of normal and reversed polarity samples from a dolomite nodule at Sweeney Road (site MS).

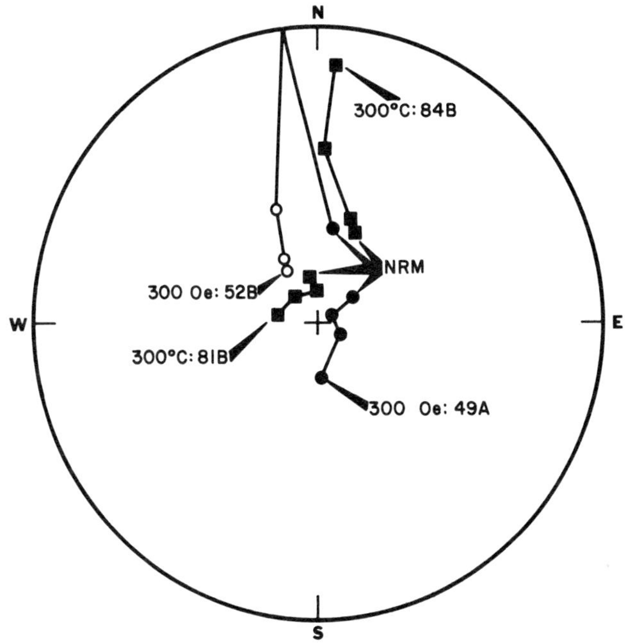

Fig. 20. Equal-area stereographic projection of the demagnetization paths of samples from the mixed-polarity dolomite nodule illustrated in Fig. 19. Symbols as in Fig. 17, except that demagnetization was conducted at 100 Oe or 100°C steps.

within each core. Which polarity predominates in this case would depend upon the inhomogeneities of the sediment compaction process. The longer inferred time interval required for compaction and dewatering of the Monterey diatomaceous oozes as compared with clay-rich deep-sea sediments may be a function of the higher porosity and resistance to mechanical compaction displayed by diatomaceous oozes as compared with clay-rich deep-sea sediments (Hamilton, 1976).

Origin of Secondary Magnetization

Both the NRM and I_SRM intensities of the dolomite often increase as a consequence of remagnetization of the dolomite in the Monterey Formation (Fig. 8). IRM saturation experiments indicate that magnetite ± hematite are magnetic carriers in remagnetized dolomite (Fig. 6). If remagnetization of the dolomite simply involved oxidation of primary magnetite to maghemite or hematite, J_{RS} would decrease, rather than increase, since the saturation magnetization intensity is greater for magnetite (480 emu/cc)

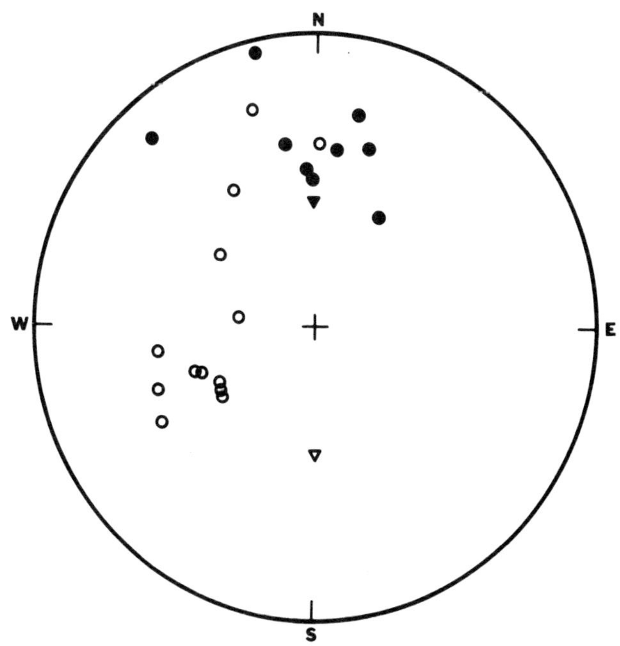

Fig. 21. Equal-area stereographic projection of the structurally corrected magnetic directions of the samples collected from the mixed-polarity dolomite nodule illustrated in Fig. 19, after demagnetization at 300 Oe or 300°C.

than for maghemite (450 emu/cc) or hematite (2.2 emu/cc). Thermal demagnetization suggests that magnetite, rather than maghemite or pyrrhotite, is the dominant newly formed phase in remagnetized dolomites which become magnetically saturated by 3 kOe (e.g., WB30B; Fig. 11). The increase in J_{RS} suggests that the iron in the newly formed magnetite is coming from pyrite.

Pyrite is a nearly ubiquitous constituent of dolomites in the Monterey Formation. During laboratory heating experiments, pyrite breaks down to magnetite above 300°C, resulting in a dramatic increase in susceptibility in dolomites with primary magnetization (Fig. 11). However, remagnetized dolomites do not show a comparable susceptibility increase during laboratory heating (Fig. 11). This observation suggests that the majority of the pyrite in the remagnetized dolomites has already been converted to magnetic phases through natural processes.

Both normally and reversely remagnetized dolomite beds were observed in the Monterey Formation in southern California. The presence of remagnetized dolomites with reversed polarities requires that the remagnetization process in these beds occurred during the Matuyama reversed epoch (0.72 - 2.47 m.y. ago; Ness, 1980), and not in the present weathering cycle. In any one stratigraphic section, remagnetized directions are generally of only one polarity, which suggests that the remagnetization process was only operative for a short time interval in each locality. However, at one site (Chico Martinez Creek, CM, along the western side of the San Joaquin Valley), some samples had both a normal and a reversed secondary direction (Fig. 22), indicating that remagnetization was occurring during both the Matuyama and Brunhes epochs.

Samples from Chico-Martinez Creek (e.g., CM06A) were the most strongly magnetized samples observed in the Monterey Formation (Fig. 8). Sample CM06A from Chico-Martinez Creek also had the highest iron content of any Monterey sample that was analyzed (Table 1). Murata et al. (1972) found that the iron content of dolomites in the Monterey Formation is strongly correlated with the oxygen isotope composition of the dolomite. They concluded that iron enrichment in Monterey dolomites is a consequence of the solution-reprecipitation of dolomite brought about by the introduction of oxygenated meteoric waters to the formation during post-orogenic uplift of the California Coast Ranges. In the Santa Ynez Range, strongly remagnetized dolomites occur preferentially in stratigraphic horizons near structurally complex zones (particularly intraformational breccias and faults), or along the axis of plunging synclines (Fig. 3), horizons which probably served as aquifers for meteoric groundwater. The occurrence of entirely normal or entirely reversed overprints in different stratigraphic sections suggests that the aquifers were active at different times in different places. In contrast, the remagnetization process in the dolomite beds sampled at Chico Martinez Creek was much more prolonged in duration (spanning both the Matuyama and Brunhes epochs) which may account for the strong iron enrichment in these beds.

In contrast to the remagnetized dolomites with a strong stable secondary remanence, many Monterey dolomites are only weakly remagnetized (e.g. GP10A and GP175C), and their J_{RS}/wt. % Al_2O_3 ratio is lower than that observed in dolomites with primary magnetization (Fig. 8). These dolomites are always normally

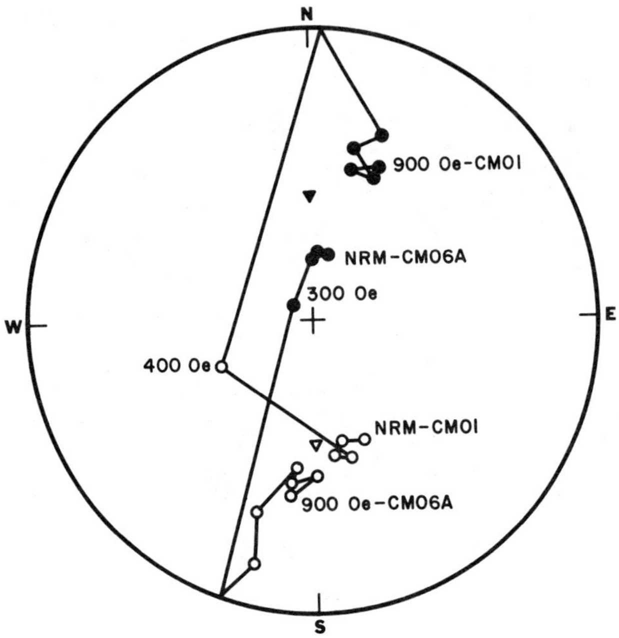

Fig. 22. Equal-area stereographic projections of the NRM demagnetization paths of remagnetized samples from Chico-Martinez Creek (site CM). Each sample contains both normal and reversed secondary directions with nonoverlapping coercivity spectra. The demagnetization curve for sample CM06A is illustrated in Fig. 10.

remagnetized, and their saturation curves and high remanent coercivities suggest that hematite and/or geothite, in addition to magnetite, are the magnetic carriers in these rocks. The weak secondary remanence in these dolomite beds was probably acquired during the Recent, due to weathering in the outcrop. The low intensity of magnetization in these samples suggests that that the iron in the magnetite and pyrite in these dolomites was converted primarily to hematite or was leached out of the rock during the remagnetization process.

SUMMARY

The primary remanent magnetization of the dolomites in the Monterey Formation appears to be a post-depositional remanent magnetization (pDRM) acquired during compaction and dewatering of the sediment. Soft-sediment fold tests and magnetic stratigraphy suggest that the magnetization is acquired greater than a few tens of centimeters, but less than a few meters, below the sediment-water interface. The failure of dolomite beds to record field reversal paths suggests that the duration of their magnetization process is greater than the duration of a field reversal (~4,000 years). On the other hand, polarity events as short as 100,000 years are recorded by the dolomite (Fig. 14), which suggests that the magnetization process did not greatly exceed 100,000 years in duration.

Carbonate cementation of the unconsolidated Monterey sediment at burial depths between ~80 and ~660 m below the sea floor (Pisciotto, 1981) permanently sealed in the magnetic direction acquired during dewatering of the original diatomaceous ooze by fixing the orientation of the already aligned magnetite grains within the sediment. Once embedded within the carbonate matrix, the magnetite grains could not become realigned during the continued porosity reduction and silica phase changes which accompanied further burial diagenesis, or by shear during tectonic deformation. The relatively impermeable dolomite matrix also helped protect the magnetite and pyrite within the rock from becoming oxidized to hematite when the formation was uplifted and exposed to weathering.

Elevation of the Monterey Formation above sea level in the Pleistocene caused the formation to be flushed with oxygenated meteoric groundwater. In zones which acted as aquifers, the dolomite was strongly remagnetized, due to partial oxidation of the pyrite in the dolomite to magnetite, or its complete oxidation to hematite. Weathering in the outcrop during the Recent also remagnetized many Monterey dolomites. The sometimes pervasive remagnetization of Monterey dolomites severely limits the feasibility of conducting a successful paleomagnetic investigation of the Monterey Formation in some areas.

Miocene formations that are lithologically equivalent to the Monterey Formation are found rimming the entire North Pacific Ocean, from Mexico to Alaska to Japan (Ingle, 1981). Since high quality paleomagnetic results have been obtained from dolomites in the Monterey Formation in southern California, the common occurrence of this formation elsewhere around the Pacific margin provides great promise for future paleomagnetic investigations.

ACKNOWLEDGEMENTS

I am grateful to Mike Fuller and Bruce Luyendyk for the use of their paleomagnetic labs, and to Bob Dunn and Mark Stein for providing technical assistance in performing some of the rock magnetic measurements. Bruce Luyendyk, Robert Coe, Bob Garrison, and Miriam Kastner provided helpful criticisms of the manuscript. This research was supported by Mobil Oil Corporation, the University of California and the National Science Foundation (grants EAR78-23426 and EAR81-16507 to Bruce Luyendyk).

REFERENCES

Barron, J. A., 1981, Late Cenozoic diatom biostratigraphy and paleoceanography of the middle-latitude eastern North Pacific, DSDP Leg 63, in Initial Reports of the Deep Sea Drilling Project: Washington, U. S. Government Printing Office, v. 63, p. 507-531.

Bramlette, M. N., 1946, The Monterey Formation of California and the origin of its siliceous rocks: U. S. Geol. Surv. Prof. Pap. 212, 57 p.

Day, R., 1977, TRM and its variation with grain size: Adv. Earth Planet. Sci., v. 1, p. 1-33.

_____, Fuller, M., and Schmidt, V. A., 1977, Hysteresis properties of titanomagnetites: grain-size and composition dependence: Phys. Earth Planet. Inter., v. 13, p. 260-267.

Denham, C. R., and Chave, A. D., 1982, Detrital remanent magnetization: viscosity theory of the lock-in zone: J. Geophys. Res., v. 87, p. 7126-7130.

Dibblee, T. W., 1966, Geology of the central Santa Ynez Mountains, Santa Barbara County, California: Calif. Div. Mines Bull., v. 186, 99 pp.

Friedman, I., and Murata, K. J., 1979, Origin of dolomite in Miocene Monterey Shale and related formations in the Temblor Range, California: Geochim. Cosmochim. Acta, v. 43, p. 1357-1365.

Fuller, M., Williams, I., and Hoffman, K. A., 1979, Paleomagnetic records of geomagnetic field reversals and the morphology of the transitional fields: Rev. Geophys. Space Phys., v. 17, p. 179-203.

Hamilton, E. L., 1976, Variations of density and porosity with depth in deep-sea sediments: J. Sed. Petrol., v. 46, p. 280-300.

Hein, J. R., O'Neil, J. R., and Jones, M. G., 1979, Origin of authigenic carbonates in sediments from the deep Berring Sea: Sedimentology, v. 26, p. 681-705.

Hillhouse, J., and Cox, A., 1976, Bruhnes-Matuyama polarity transition: Earth Planet. Sci. Lett., v. 29, p. 51-64.

Hornafius, J. S., 1984, Paleomagnetism of the Monterey Formation in the western Transverse Ranges, California: Ph.D. thesis, University of California, Santa Barbara.

_____, Luyendyk, B. P., Weaver, D. W., Fuller, M., and Wornardt, W. W., 1981, Paleomagnetic results from the Monterey Formation in the western Transverse Ranges: EOS (Trans. Am. Geophys. Un.), v. 62, p. 855.

Ingle, J. C., 1981, Origin of Neogene diatomites around the North Pacific rim, in Garrison, R. E. et al. (eds.), The Monterey Formation and related siliceous rocks of California: Soc. Econ. Paleontologists and Mineralogists Pac. Sec., p. 159-179.

Isaacs, C. M., 1980, Diagenesis in the Monterey Formation examined laterally along the coast near Santa Barbara, California: Stanford University, Ph.D. thesis, 329 p.

Johnson, H. P., Lowrie, W., and Kent, D. V., 1975, Stability of anhysteretic remanent magnetization in fine and coarse magnetite and maghemite particles: Geophys. J. R. Astron. Soc., v. 41, p. 1-10.

Kelts, K., and McKenzie, J., 1982, Diagenetic dolomite formation in Quaternary anoxic diatomaceous muds of Deep Sea Drilling Project Leg 64, Gulf of California, in Initial Reports of the Deep Sea Drilling Project: Washington, U. S. Government Printing Office, v. 64, p. 553-570.

Kent, D. V., 1973, Post-depositional remanent magnetization in a deep-sea sediment: Nature, v. 246, p. 32-34.

Lovlie, R., 1974, Post-depositional remanent magnetization in a redeposited deep-sea sediment: Earth Planet. Sci. Lett., v. 21, p. 315-320.

_____, 1976, The intensity pattern of post-depositional remanence acquired in some marine sediments deposited during a reversal of the external magnetic field: Earth Planet. Sci. Lett., v. 30, p. 209-214.

Lowrie, W., Alvarez, W., and others, 1982, Paleogene magnetic stratigraphy in Umbrian pelagic carbonate rocks: The Condessa sections, Gubbio: Geol. Soc. Amer. Bull., v. 93, p. 414-432.

Murata, K. J., Friedman, I., and Cremer, M., 1972, Geochemistry of diagenetic dolomites in Miocene marine formations of California and Oregon: U. S. Geol. Surv. Prof. Pap. 724-C, 12 p.

Nagata, T., 1961, Rock Magnetism, Maruzen, Tokyo, 350 p.

Ness, G. E., Levi, S., and Couch, R. W., 1980, Marine magnetic anomaly timescales for the

Cenozoic and late Cretaceous - a precis, critique, and synthesis: Rev. Geophys. Space Phys., v. 18, p. 753-770.

Pisciotto, K. A., 1981, Review of secondary carbonates in the Monterey Formation, California, in Garrison, R. E. et al. (eds.), The Monterey Formation and related siliceous rocks of California: Soc. Econ. Paleontologists and Mineralogists Pac. Sec., p. 273-284.

_____, and Mahoney, J. S., 1981, Isotopic survey of diagenetic carbonates, DSDP Leg 63, in Initial Reports of the Deep Sea Drilling Project: Washington, U. S. Government Printing Office, v. 63, p. 595-609.

Verosub, K. L., 1977, Depositional and post-depositional processes in the magnetization of sediments: Rev. Geophys. Space Phys., v. 15, p. 129-143.

ABSTRACTS

On the following pages are reproduced additional abstracts presented in April 1984 at the joint AAPG-SEPM Pacific Section meeting in San Diego. Included are two abstracts of papers given in the Symposium on Dolomites in the Monterey Formation and Other Organic-Rich Units (Compton and Siever; Coe, Petit, Mertz and Hollander) and two given in the Symposium on Paleomagnetism of Tertiary and Mesozoic Sequences in California, Oregon and Washington (Hornafius and Lagle; Coe, Mertz, and Petit).

MAGNESIUM MASS BALANCE CALCULATIONS FOR THE MONTEREY FORMATION, SANTA MARIA BASIN, CALIFORNIA

Compton, J. S., and Siever, R., Dept. of Geological Sciences, Harvard University, 20 Oxford St., Cambridge, MA 02138

ABSTRACT

Detailed, measured stratigraphic sections of a 1200 m thick continuous section of the Monterey Formation at Pt. Pedernales average 9 vol % dolostone, range in bulk density from 2.2 to 2.6 gcm^{-3}, and contain 40 to 94 wt % dolomite. From these data the total amount of dolomite present is calculated to be between 0.5 and 1.5 x 10^6 moles dolomite. Two possible sources of Mg are evaluated: (1) advection of pore water due to compaction, and (2) diffusion of Mg from overlying seawater. We assume little or no isochemical transfer from one phase to another within dolomite beds.

Porosity as a function of depth has been determined both for the Pt. Pedernales section and for diatomaceous marine sediments (DSDP data). The Pt. Pedernales section was decompacted to an original thickness of 1708 m. The absolute maximum seawater that may have advected through the section is 2846 m^3 (1.6 x 10^5 moles Mg); however, the maximum amount of seawater that any individual bed could have seen is only 1144 m^3 (6.3 x 10^4 moles Mg.) Therefore, advection could supply a maximum of about 15%, but more likely only 6% of the total Mg observed.

Diffusion of Mg from the overlying seawater is calculated assuming steady state porosity, constant sedimentation rates of 50 and 200 m Myr^{-1}, and taking into account upward advecting pore water. Diffusion is capable of supplying 10^6 moles Mg, within the amount observed, though certainly limited to the upper 300 m of sediment. We conclude that diffusion can account for early dolomitization without invoking large scale formation fluid flow.

ESTIMATE OF MAXIMUM BURIAL TEMPERATURE OF THE MONTEREY FORMATION AT ARROYO SECO FROM PALEOMAGNETISM OF DOLOMITE

Coe, R. S., Pettit, R. L., Mertz, K. A., Jr., Hollander, D. J., Earth Sciences, UCSC, Santa Cruz, CA 95064

ABSTRACT

We have undertaken a detailed paleomagnetic study of secondary dolomite concretions and lensoid beds, exposed in a section at Arroyo Seco, that have grown in organic-rich shales of the Sandholt Member of the Monterey Formation. Three components of remanent magnetization are clearly recognized during thermal demagnetization experiments, with blocking temperatures between (1) 20 to 100°C, (2) 100 to 250°C, and (3) 250 to 400°C. The first component, which is not consistent between samples and is easily removed at 100 degrees or less, was probably acquired during and after sampling. The intermediate component was acquired after folding in a normal geomagnetic field, probably by a viscous mechanism. The high-blocking temperature component passes fold and reversal tests, and thus must be an ancient remanence. The apparently faithful recording of reversals by this component, as well as its random orientation in dolomite clasts of identical lithology that are included in sandy grain or debris flows interbedded with the Sandholt shales, suggests that it was acquired relatively soon after deposition and prior to dolomitization. Thus we think this component is the primary remanence.

The lowest blocking temperature of the primary component that we observe during thermal demagnetization in the laboratory, 250±15 °C, is remarkably uniform in unweathered samples throughout the section. Because viscous decay is probably the mechanism that was responsible for erasing that portion of the primary remanence with blocking temperatures lower than 250 °C, we can use this cut-off temperature and Neel's theory of remanence in thermally activated assemblages of uniaxial single-domain magnetic grains to estimate the maximum temperature experienced during burial. For instance, if the remanence is carried by pure magnetite, we derive a temperature for one million years. Studies to determine which mineral actually carries the primary remanence are currently in progress.

MAGNETIC STRATIGRAPHY AND DIATOM BIOSTRATIGRAPHY OF THE MONTEREY FORMATION AT SELECTED LOCALITIES IN THE WESTERN TRANSVERSE RANGES, CALIFORNIA

Hornafius, J. Scott, and Lagle, Charles W., Department of Geological Sciences, University of California, Santa Barbara, California 93106

ABSTRACT

Dolomite beds in the Monterey Formation often possess a primary magnetic remanence. Monterey dolomite beds also contain well-preserved diatoms, even in stratigraphic sections in which the interbedded siliceous rocks have been diagenetically altered to quartz. By using the age-diagnostic diatoms extracted from the dolomites in combination with the magnetic polarity sequence recorded by the dolomite beds, extremely accurate ages can be assigned to individual beds within the Monterey Formation (± 0.1 m.y.). Immediately west of the mouth of Arroyo Hondo, 40 km west of the city of Santa Barbara, six polarity intervals in upper magnetic epoch 16 and lower epoch 15 (15.9-14.2 m.y. old) are recorded by 20 dolomite beds within a 43-meter-thick stratigraphic interval (post-compaction sedimentation rate=25 meters/m.y.). Correlation of the first occurrence of Denticulopsis hyalina in this section with the first occurrence of this diatom datum in deep-sea cores suggests that the primary magnetic remanence in Monterey dolomites is acquired within a few meters of the seafloor. The magnetic polarity sequence and diatom biostratigraphy of the upper Relizian and Luisian foraminiferal stages at Naples Beach, 25 km west of Santa Barbara, suggests that the provincial Relizian-Luisian stage boundary is equivalent in age to the magnetic epoch 15/16 boundary (15.0 m.y. old), and that the post-compaction sedimentation rate between 14.2 and 15.0 m.y.b.p. was approximately 100 meters/m.y. at this locality. At the mouth of La Honda Canyon, 3.5 km north of Point Arguello, Thalassiosira antiqua (1st occurrence 8.4 m.y.b.p.) was recovered from dolomites near the base of an 1150-meter-thick section of Monterey Formation. The magnetic polarity sequence recorded by 15 dolomite beds north of La Honda Canyon suggests that the lower 260 meters of this section was deposited during upper paleomagnetic epoch 7 (6.7-7.3 m.y.b.p.) at a post-compaction sedimentation rate of 430 meters/m.y. Age-diagnostic diatoms were recovered from dolomites in 12 other stratigraphic sections of the Monterey Formation, but the dolomite beds in these sections were too widely spaced to correlate the magnetic polarity sequence recorded by the dolomite beds with the magnetic timescale. Diatoms extraced from dolomites at the top of the exposed section of Monterey Formation in China Harbor on Santa Cruz Island are 12.4-13.3 m.y. old. Diatoms recovered from near the base and top of the upper shale member of the Modelo Formation at Piru Lake are 8.4-8.7 and 6.0-6.7 m.y. old, respectively. Diatoms recovered from a dolomite nodule 4 meters below the conglomerate beds at the top of the Monterey Formation at Gaviota Beach are 6.0-6.7 m.y. old, which suggests that initial uplift of the Santa Ynez Range occurred into the latest Miocene.

PALEOMAGNETIC RESULTS FROM DOLOMITE ROCKS OF THE MONTEREY FORMATION, ARROYO SECO AREA, CALIFORNIA

Coe, R. S., Mertz, K. A., Jr., Pettit, R. L., Earth Sciences, UCSC, Santa Cruz, CA 95064

ABSTRACT

We are beginning a detailed paleomagnetic study of the Early to Middle Miocene Sandholt Member of the Monterey Formation in the Arroyo Seco area of the Central California Coast Range. At present we have obtained results from dolomite concretions and lensoid beds sampled at 23 sites distributed over 160 meters of stratigraphic section. Both thermal and AF demagnetization reveal the presence of a large normal overprint that fails the fold test. Thermal demagnetization does a better job of removing this overprint than AF, and treatment of the sample in a small AF field (e.g., 25 oe) after each thermal step works even better. In most samples this secondary component appears to be a VRM produced during the present polarity epoch, but in some tan-colored samples goethite formed during weathering carries a strong secondary CRM.

Thermal demagnetization to 250 °C is often sufficient to isolate a characteristic component that decays unidirectionally toward the origin of its orthogonal vectore diagram until swamped by an unstable compoenent appearing around 375 to 425 °C. This characteristic component passes fold, conglomerate, and reversal tests, and hence is almost certainly primary. In samples that do not clean up completely by 250 °C, the direction of remanence usually moves along great circle paths that pass near the primary direction.

Results to date, though preliminary, are interesting. About 25 meters above the base of the section a reversal (normal below, reversed above) occurs. Nannofossils identified in these rocks place this section in the upper portion of the Helicosphaera ampliaperta Zone of Ellis (1979) and Bukry (1973). Reversed remanence is found throughout the overlying 135 meters of section examined above this point, which probably extends well into the Sphenolithus heteromorphus Zone. These data suggest that this rather long reveresed zone is equivalent to Chron 5Br of the magnetostratigraphic time scale, and that the age of the reversal boundary discovered at its base is 16.2 Ma (Harland et al., 1982). This boundary should enable precise stratigraphic correlations to be made between equivalent sections.

The mean direction for the primary component is not significantly different from that expected for this location in the Early Miocene. Thus, neither large north-south displacement nor large azimuthal rotation of these rocks with respect to stable North America appears to have occurred since that time.